Advances in Geographical and Environmental Sciences

Series Editors

Advances in Geographical and Environmental Sciences synthesizes series diagnostigation and prognostication of earth environment, incorporating challenging interactive areas within ecological envelope of geosphere, biosphere, hydrosphere, atmosphere and cryosphere. It deals with land use land cover change (LUCC), urbanization, energy flux, land-ocean fluxes, climate, food security, ecohydrology, biodiversity, natural hazards and disasters, human health and their mutual interaction and feedback mechanism in order to contribute towards sustainable future. The geosciences methods range from traditional field techniques and conventional data collection, use of remote sensing and geographical information system, computer aided technique to advance geostatistical and dynamic modeling.

The series integrate past, present and future of geospheric attributes incorporating biophysical and human dimensions in spatio-temporal perspectives. The geosciences, encompassing land-ocean-atmosphere interaction is considered as a vital component in the context of environmental issues, especially in observation and prediction of air and water pollution, global warming and urban heat islands. It is important to communicate the advances in geosciences to increase resilience of society through capacity building for mitigating the impact of natural hazards and disasters. Sustainability of human society depends strongly on the earth environment, and thus the development of geosciences is critical for a better understanding of our living environment, and its sustainable development.

Geoscience also has the responsibility to not confine itself to addressing current problems but it is also developing a framework to address future issues. In order to build a 'Future Earth Model' for understanding and predicting the functioning of the whole climatic system, collaboration of experts in the traditional earth disciplines as well as in ecology, information technology, instrumentation and complex system is essential, through initiatives from human geoscientists. Thus human geoscience is emerging as key policy science for contributing towards sustainability/survivality science together with future earth initiative.

Advances in Geographical and Environmental Sciences series publishes books that contain novel approaches in tackling issues of human geoscience in its broadest sense—books in the series should focus on true progress in a particular area or region. The series includes monographs and edited volumes without any limitations in the page numbers.

Praveen Kumar Rai · Shalini Rai
Editors

Climate, Environment and Agricultural Development

A Sustainable Approach Towards Society

Springer

Editors
Praveen Kumar Rai
Department of Geography
Khwaja Moinuddin Chishti Language University
Lucknow, Uttar Pradesh, India

Shalini Rai
Department of Microbiology
Dr Shakuntala Misra National Rehabilitation University
Lucknow, Uttar Pradesh, India

ISSN 2198-3542 ISSN 2198-3550 (electronic)
Advances in Geographical and Environmental Sciences
ISBN 978-981-97-8362-5 ISBN 978-981-97-8363-2 (eBook)
https://doi.org/10.1007/978-981-97-8363-2

This Springer imprint is published by the registered company Springer Nature Singapore Pte Ltd.
The registered company address is: 152 Beach Road, #21-01/04 Gateway East, Singapore 189721, Singapore

Preface

This book will discuss an emerging concent and technology of recent aspects not only for the climate but also about its impactr on the regional and local environment. Also various studies and their findings highlighted the sustainable finding of environmentalmonitoring & managment, agricultural vulnerability and its development and adaptation from local to global scale. Climate change greatly influence on the environment not only on the local, regional or global scale but also on the natural resurces like land, agriculture mainlly crop production, fisheries etc. As per various studies done in this book show that the changing climate is imposing the variopus challenges to human society of the world at various level. Chapter in this book clearly reveals that the agro ecosystem-management practices and technologies leads to higher productivity with degradation of various natural resources like agricultural, forest and also environmental habitat leading to soil-water-air pollution. Studies also shows that the agricultural biodiversity is now a days an extremely significant factor and results from the interaction of the external environment, various genetic resources, and the natural resource management system (land and water). Climate change also influece the loss of specific species, genes, and loss of agrobiodiversity. In the present scenerio, a major challenge for the agriculture is to feed a huge world population not only for the developed countries but also for the developing and underdeveloped countries, while at the same time reducing the environmental impact and conserving our natural resources for future generations in the sustainable ways.

The finding of chapters in this book reflect that agriculture resource can have significant impacts on the local and regional environment and its negative impacts are very serious not for the enviornbment but also for the human health. The book clealry explain the concrete methodology in a simple way so that a researcher can easily use and apply it in their work. Various individual chapters are dedicated to discipline such as monitoring of climate change and its impacts on environment, study of land use land cover, crops conditions, impacts of climate change on water resources and morphomtric analysis, soil fertility, erosion, insects, pests, weeds, microbes and diseases; greenhouse gas emission assessment etc.

Lucknow, India

Praveen Kumar Rai
Shalini Rai

Contents

Chapter 1
Climate Change and Negotiating Challenges: A Lesson for Success and Failure

Varun Chhachhar and Karishma Verma

1.1 Introduction

As we stand at the crossroads of a planet grappling with the profound impacts of climate change, this chapter embarks on a journey through the intricate web of challenges and triumphs in the realm of climate negotiations. From the foundational understanding of climate change to the evolution of global discourse and the critical analysis of negotiations, the chapter delves into the multifaceted facets of this crucial global issue.

Unravelling the very fabric of climate change, the chapter dissects its causes, exploring the anthropogenic factors that have fueled its rise and the impacts that reverberate across ecosystems. Beyond the scientific nuances, this chapter emphasizes the importance of law as an essential tool in minimizing the far-reaching implications of climate change.

Tracing the historical trajectory, the chapter further briefly highlights the emergence of climate crisis concerns, swiftly transitioning into a panorama of international endeavours. From seminal instruments such as the Stockholm Declaration and the creation of UNEP to the impactful formation of the IPCC, the exploration extends to the UNFCCC regime, Rio Declaration, and the labyrinth of COP conferences. Uncovering the key principles governing climate change negotiations, this section provides insights into the milestones that have shaped the global discourse on climate change.

The fourth part of the chapter critically examines the climate change negotiations. From the inception of the UNFCCC to the latest strides in COP28, the efficacy of these negotiations in mitigating the climate crisis is dissected. This critical analysis serves as a compass, guiding through lessons learned, successes celebrated, and

V. Chhachhar (✉) · K. Verma
Faculty of Law, University of Lucknow, Lucknow, India
e-mail: chhachhar_varun@lkouniv.ac.in

P. K. Rai and S. Rai (eds.), *Climate, Environment and Agricultural Development*, Advances in Geographical and Environmental Sciences,
https://doi.org/10.1007/978-981-97-8363-2_1

failures acknowledged, offering insights into the path ahead in the collective pursuit of a sustainable future.

1.2 Understanding the Concept of Climate Change

> "Climate is what you expect; weather is what you get."
> —Mark Twain

To comprehend the concept of climate change and fathom the consequences of the anthropogenic (human-induced) greenhouse effect, it's crucial to discern between "climate" and "weather." Weather denotes the varying atmospheric conditions that include temperature, wind, and precipitation, whereas climate encapsulates the average weather over a specific time span in a given region (Baede et al., 2001). The intricate web of the global climate encompasses the atmosphere, oceans, land surface, and diverse ecosystems rich in flora and fauna. In broad strokes, "climate" can be referred as the systematic summation of the Earth's land, atmosphere, and water system over time (Jager & Ferguson, 1992). The sun's energy, or radiation, powers this system. Once solar energy penetrates the atmosphere and Earth's surface, any un-reflected portion warms these elements. Consequently, the heated surface emits infrared radiation, or heat, back into space (Verheyen, 2005).

Climate change refers to the prolonged transformation in long-term weather patterns, which can occur over periods decades to millions of years (United Nations). It encompasses changes in average weather conditions and the frequency of weather events compared to the usual, resulting in extreme weather events. This phenomenon can be confined to a specific region or impact the entire globe. The American Meteorological Society provides a concise definition of climate change, describing it as any systematic alteration in the prolonged statistical patterns of climate elements, such as temperature, winds, or pressure, persisted over several decades or for longer period (AMS Glossary, 2024). Article 1.2 of The United Nations Framework Convention on Climate Change (UNFCCC 1992) articulates climate change as a deviation caused directly or indirectly by human activities altering the global atmosphere's composition beyond natural climate variability observed within similar time frames. The Intergovernmental Panel on Climate Change (IPCC), a United Nations body charged with evaluating climate science, takes a different approach, defining it as a shift in "either the mean state of the climate or in its variability, persisting for an extended period, typically decades or longer" (IPCC Glossary, 2018), In essence, the IPCC considers and uses the terms climate variability and climate change synonymously. Put simply, climate change refers to substantial variations in the Earth's temperature, precipitation, and wind that deviate significantly from historical norms which have a noticeable impact on ecosystems and socio-economic transactions.

1.2.1 Causes of Climate Change

Climate change results from two primary factors: natural processes and anthropogenic or human-induced activities. Natural factors encompass internal processes such as EL Nino and La Nina and external forcings, such as variations in solar emission or gradual shifts in Earth's orbital elements. Concurrently, human activities play a pivotal role in the elevated presence of greenhouse gases (GHGs) in the atmosphere. GHGs like carbon dioxide (CO_2), methane (CH4), and nitrogen dioxide (N_2O) trap the sun's heat, leading to the greenhouse effect (IPCC Glossary, 2018). The predominant driver of climate change lies in the escalating concentration of GHGs, intensifying the greenhouse effect (AR3, IPCC, 2001).

However, human activities, particularly the combustion of fossil fuels, deforestation and increased industrial activities have amplified this effect, leading to an upsurge in GHG concentrations and subsequent planetary warming (Global Climate Change, NASA). The genesis of this climatic metamorphosis can be traced back to the Industrial Revolution, a pivotal juncture where societies began liberally releasing greenhouse gases into atmosphere consequently warming the earth. From the pre-industrial era until now, human actions have led to a rise in Earth's average temperature by approximately 1 °C (1.8 °F), with the rate of increase now exceeding 0.2 °C (0.36 °F) per decade (AR5, IPCC, 2014). This warming is evidenced by elevated temperatures, rising sea levels, and an increase in extreme weather events, all of which impact ecosystems, livelihoods, and food security. When scientists delve into global climate change, their focus is on evolving patterns over many years, highlighting Earth's continuous temperature rise, commonly referred to as global warming.

Distinguishing between natural climate variability and changes induced by human behaviour becomes crucial, a distinction curated by the United Nations Framework Convention on Climate Change (UNFCCC). This convention predominantly addresses anthropogenic factors, underscoring the necessity for international collaboration to mitigate climate change stemming from human activities. This unambiguous consensus on human-induced warming is underscored by the Intergovernmental Panel on Climate Change (IPCC). The IPCC asserts, without room for reasonable dissent, that human influence has indisputably warmed the atmosphere, ocean, and land. Furthermore, their projections indicate a continued escalation in global surface temperature until at least mid-century under all considered emissions scenarios (AR2, IPCC, 1995). These scientific affirmations, grounded in comprehensive assessments, have long been acknowledged by the international community.

1.2.2 Impact of Climate Change

The IPCC has definitively stated that climate change is already affecting natural and human systems worldwide, with severe environmental and social repercussions that may unfold over centuries or millennia. Their evaluations highlight that continuous greenhouse gas emissions increase the likelihood of severe, widespread, and permanent effects. Climate change impacts numerous physical processes, leading to sea-level rise, ocean acidification, disturbances to rainfall patterns, and intensification of extreme weather events (Kaya et al., 2015). The toll on biodiversity, ecosystems, communities, their livelihoods, food security, and human well-being is clear, exacerbating conflicts and leading to forced migration (NOAA, 2021). Communities living in low-lying areas along the coast, remote islands, or flood-prone locations are especially susceptible to these threats. Developing nations, with restricted ability to deal with climatic effects, are at the forefront of this environmental conflict. States overwhelmingly regard climate change as a significant and urgent threat to human cultures and the earth, seeing it as one of the major concerns of our time.

1.2.3 Law as a Tool of Mitigation

Mitigating climate change and adapting to its impacts are twin imperatives of dealing with ongoing climate crisis (Mayer, 2022). Mitigation strives to address the underlying causes of climate change by curbing and controlling greenhouse gas (GHG) emissions. This involves addressing the sources of these gases, by reducing activities such as fossil fuel combustion, as well as their removal through initiatives like reforestation. In contrast, adaptation aims to facilitate coping mechanism for social and ecological consequences of climate change, encompassing disaster risk management, relocation and migration plans for populations, investments in infrastructure (AR5, IPCC, 2014). It is crucial to recognize that mitigation and adaptation are not mutually exclusive but supplementary goals.

The role of international law in mitigating climate change is pivotal. Existing legal frameworks, both general and specific, can be harnessed to mandate climate action. given the global nature of climate change, reliance on domestic law alone is insufficient. GHG emissions transcend national borders, affecting every country regardless of its emissions (Mayer, 2022). Effective mitigation necessitates coordinated efforts by significant emitters to regulate emissions within their jurisdictions. Although this cooperative strategy is crucial, it faces obstacles since different states may put their own interests ahead of the global good. States should work together since it benefits everyone, but certain states may be tempted to take advantage of others without contributing (Gardiner, 2006).

Addressing this collective action problem through the adoption of international rules proves to be a formidable task in the realm of international governance. Defining the modalities of international mitigation action requires navigating intricate diplomatic landscapes. Striking a balance between acknowledging national sovereignty and ensuring collective responsibility is a complex challenge that demands innovative solutions.

1.3 Navigating the Climate Crisis: The Evolution of Global Discourse

Scientists widely attribute the primary cause of climate change to the greenhouse effect, resulting from greenhouse gases (GHGs). Jean-Baptiste J. Fourier, a renowned French scientist in the early 19th Century, is credited with first proposing the concept of the greenhouse effect (Cowie, 2007; Giddens, 2009). It was the famous report titled 'The Limits to Growth' published in 1972 by the Club of Rome, which for the first-time highlighted climate change as a significant worldwide environmental risk due to the rising CO_2 levels in the atmosphere (Kaya et al., 2015).

1.3.1 Emergence of Climate Crisis Concern at Global Stage

In the initial years following its establishment, the United Nations showed minimal concern for environmental issues, let alone climate change. During the early twentieth century, climate change was predominantly perceived as a scientific phenomenon. It wasn't until the 1970s that environmental concerns gained prominence on the international agenda. Prior to this, discussions about environmental issues were primarily centred around the sufficiency of known natural resources to support the economic progress of states, particularly those that were underdeveloped (Dessler, 2022).

The UN Scientific Conference on Human Environment held in Stockholm (1972), laid the groundwork by adopting principles for environmental preservation, leading to the creation of the UN Environment Programme (UNEP) (General Assembly resolution 2997/27) and the formulation of 26 principles. Some of these principles later became the cornerstone of international environmental law.

1.3.1.1 Stockholm Declaration

In 1972, the UN Scientific Conference marked a pivotal moment in international environmental jurisprudence. During this event, a declaration outlining principles for the protection and improvement of the human environment was adopted, accompanied by an action plan suggesting recommendations for global environmental initiatives (Stockholm Declaration, 1972). The Stockholm conference, with its adoption of 26 principles focused on environmental preservation, stands as a crucial turning point. Additionally, a noteworthy outcome was the establishment of the United Nations Environment Programme (UNEP), further solidifying the conference's significance.

Principle 21 from the Stockholm Declaration became a linchpin for international environmental law, affirming both a state's sovereign right to exploit resources and its responsibility to prevent environmental harm beyond national borders (Stockholm

Declaration, 1972). This landmark declaration marked a shift, recognizing the interconnectedness of global environmental issues, including climate change (Verheyen, 2005).

In the ensuing two decades, global attention and action towards the atmosphere and climate intensified. Simultaneously, concerns about the ozone layer depletion led to the Vienna Convention for the Protection of the Ozone Layer in 1985 and the Montreal Protocol in 1987. These instruments were designed to mitigate the release of chlorofluorocarbons (CFCs) and halons, both contributors to ozone layer depletion and part of the broader category of greenhouse gases. In doing so, they indirectly contributed in addressing climate change (Mayer, 2022).

1.3.1.2 IPCC

The year 1988 emerged as a watershed moment, placing global warming and ozone layer depletion at the forefront. The Intergovernmental Panel on Climate Change (IPCC) was established in 1988 to examine greenhouse warming and global climate change.

Functioning as a vital international entity, the IPCC plays a crucial role in evaluating the scientific aspects of climate change and offering recommendations for adaptation and mitigation strategies. Comprised of three working groups: WG1 focusing on the physical aspects of climate change, WG2 on its impacts, and WG3 on mitigation- the IPCC consistently releases comprehensive reports on climate science and related issues (Verheyen, 2005). Since its first assessment report (AR1) in 1990, the IPCC has been instrumental in shaping global understanding and policies concerning climate change, with subsequent assessment reports (AR) published in 1995 (AR2), 2001 (AR3), 2007 (AR4), 2013 & 2014 (AR5), and 2022 (AR6).

In 1989, acknowledging climate change as a distinct and pressing concern, the UN General Assembly commenced thorough assessments and suggestions, establishing the foundation for the Framework Convention on Climate Change. Through resolution 44/207, the General Assembly supported the UNEP Governing Council's proposal to collaborate with World Meteorological Organization (WMO) in preparing for negotiations on a framework convention addressing climate change (Mayer, 2022; Verheyen, 2005).

The UNFCCC negotiations were significantly influenced by the IPCC's First Assessment Report (1990), which validated the concerns about climate change. This report played a pivotal role in shaping the language and discussions within the UNFCCC. Likewise, the 2015 Paris Agreement was greatly informed by the findings of the Fifth Assessment Report, concluded in 2014, highlighting the ongoing influence of IPCC assessments on global climate agreements (Schleussner et al., 2016).

1.3.2 Rio Earth Summit: UNFCCC Regime

In June 1992, Rio de Janeiro, Brazil hosted the UN Conference on Environment and Development, widely recognized as the Earth Summit or the Rio Summit. The gathering saw participation from representatives of 178 governments, resulting in the adoption of three significant non-binding instruments which are Agenda 21, Rio Declaration on Environment Development and The Statement of Forest Principles and two legally binding conventions which are United Nations Framework Convention on Climate Change (UNFCCC) and Convention on Biological Diversity (Rio Declaration, 1992).

1.3.2.1 UNFCCC, 1992

The evolution of the UNFCCC regime has been marked by challenging and lengthy international negotiations, striving for consensus amidst the conflicting expectations of various key stakeholders - European States advocating for robust mitigation efforts, while developing nations emphasizing the responsibilities of wealthier and developed counterparts, Conversely, the United States disclaiming such obligations, and oil-producing States resisting significant reductions in fossil fuel consumption; additionally, small island developing States calling for greater ambition, highlighting the diverse array of interests at play in these discussions (Mayer, 2018).

The UNFCCC stands as the cornerstone of international law concerning climate change, comprising treaties and decisions aimed at addressing this global challenge. Negotiated through diverse stakeholder expectations, it entered into force in 1994 (adopted in 1992) with an impressive 198 parties presently, making it one of the few conventions with near-universal membership. Serving as the foundation for global efforts to combat climate change, the UNFCCC's principles have provided foundation for landmark agreements such as the Kyoto Protocol (1997) and the Paris Agreement (2015), shaping international responses to climate change (Dessler, 2022).

The framework of UNFCCC can be categorised as following (UNFCCC, 1992):

(a) Preamble, definitions (Article 1), objectives (Article 2), and principles (Article 3);
(b) Obligations covering substantive actions, reporting, monitoring, and financial commitments, including those specific to developed and developing nations. These encompass mitigation and adaptation efforts, scientific collaboration, public awareness, financial aid, technology transfer, national policies, and the utilization of scientific and technical information (Article 4–6);
(c) Institutional arrangements for monitoring convention implementation (Article 7–10);
(d) Procedural guidelines concerning financial mechanism, annexes, entry into force, reservations, ratification, accession, protocols, dispute settlement mechanism and interim measures and miscellaneous provisions (Article 11–26).

The UNFCCC recognized that human actions are elevating greenhouse gas concentrations, leading to global warming and subsequent climate alterations. These changes pose risks to both natural ecosystems and human populations. The primary objective of the Convention, along with any associated legal frameworks, is to stabilize greenhouse gas levels in the atmosphere, preventing harmful disruptions to the climate system. This stabilization should happen within a timeframe that allows for natural ecosystem adaptation, safeguards food security, and promotes sustainable economic growth (Article 2, UNFCCC, 1992).

The preamble of the UNFCCC references various principles of international environmental law. These include acknowledging climate as a common concern of human-kind, affirming states' sovereign rights over their resources, emphasizing the avoidance of transboundary environmental harm, and advocating for intergenerational equity. It also highlights the differentiated response expected from developed and developing nations, recognizing potential economic and social costs for some countries and the necessity for energy consumption growth in developing countries for sustainable development.

The provision's location within a treaty determines its normative capacity. While the preamble provides interpretative context, it doesn't establish binding obligations (Vienna Convention on the Law of Treaties, 1969). Conversely, provisions within the operational part of the treaty may create enforceable rights and duties depending on the language (Rajamani, 2016). This is the reason why Article 3 of the UNFCCC, reaffirms principles from the preamble.

For instance, the first principle (Article 3.1) reiterates the notion of safeguarding the climate for both current and future generations, emphasizing the principle of equity and the concept of common but differentiated responsibilities and respective capabilities (CBDR-RC). Additionally, the second principle emphasizes giving due consideration to the specific needs and unique circumstances of developing countries, especially those most susceptible to the adverse effects of climate change, and those facing disproportionate burdens under the UNFCCC (Article 3.2). The remaining three principles, though not explicitly mentioned in the preamble, extend from it. The precautionary principle, embedded in the third principle, dictates that in the face of serious threats or irreversible harm, lack of full scientific certainty should not impede precautionary measures (Article 3.3). The fourth principle acknowledges the intrinsic link between development and climate change, recognizing the imperative of promoting sustainable development while understanding the importance of economic development in implementing climate change measures (Article 3.4). Furthermore, the fifth principle addresses the interplay between climate change and international trade, emphasizing that measures taken to combat climate change should not unfairly discriminate or hinder international trade (Article 3.5).

Climate law specialist Daniel Bodansky observes that certain principles were left out of the final UNFCCC draft, regardless of efforts by developing nations to push for more principles (1993). These include principles asserting equal rights to ocean sinks, mandating the growth of greenhouse gas (GHG) emissions for developing countries, opposing the imposition of environmental conditions on aid, and assigning primary responsibility to developed countries for addressing climate change due to their overconsumption and excessive emissions. Additionally, related principles regarding liability and compensation, linked to the 'main responsibility principle', were also absent from the UNFCCC (Bodansky, 1993).

To ensure fairness and equity in addressing climate change, the UNFCCC incorporates the principle of common but differentiated responsibilities (CBDR-RC), which is central to its framework. This principle recognizes that while all states share a collective responsibility to tackle climate change, developed nations bear a greater historical responsibility due to their significant contributions to the problem and their greater capacity to address it. Therefore, developed countries are expected to take the lead in addressing climate change and its adverse impacts, while developing nations are given more flexibility to reduce their emissions later (Article 3). This approach, similar to that employed in the Montreal Protocol, allows developed countries to pioneer emission reduction efforts, benefiting from economies of scale and technological advancements, which in turn reduces the cost of emission reduction for developing countries when they eventually embark on similar measures (Dessler, 2022).

Consequently, the UNFCCC categorizes states parties into three groups for specific commitments: Annex I parties consist of industrialised nations (OECD members in 1992) and the nations undergoing economic transformation (EIT parties). Their goal is to bring greenhouse gas emissions back to 1990 levels by the year 2000, and they are committed to implementing national policies to prevent climate change. Nonetheless, flexibility is provided, enabling EIT parties to select an alternative base year (Verheyen, 2005). Parties to Annex II, which is a subset of Annex I, promise to contribute financially to developing nations' attempts to reduce emissions and adapt. The remaining states, who are largely developing countries and are classified as non-Annex I parties, are not specifically obligated by the UNFCCC. Despite this, the UNFCCC stresses support for their requirements, including investment, insurance, and technology transfer, and acknowledges the vulnerability of some countries to climate change, such as tiny island states and those with dry areas (Article 4).

The UNFCCC establishes key institutions like the Conference of the Parties (COP) under article 7- A gathering of representatives from all participating parties convenes annually to make decisions regarding the implementation of the UNFCCC, including potential amendments or protocols as necessary, the Subsidiary Body for Scientific and Technological Advice (SBTA) under article 9, the Subsidiary Body for Implementation (SBI) under article 10, and a permanent Secretariat under article 8 to facilitate negotiations and oversee implementation.

The Convention's concluding articles cover a number of general and miscellaneous provisions. The Financial Mechanism is established in Article 11, and dispute resolution is covered in Article 14. Procedures for changing the Convention and updating its Annexes are described in detail in Articles 15 and 16, respectively. The protocol adoption procedure is described in Article 17. The UNFCCC's adoption and implementation are governed by Articles 19 through 23. Reservations to any provision of the Convention are forbidden by Article 24, and party withdrawal is covered by Article 25. Lastly, Article 26 attests to the six UN official languages' equal legitimacy.

1.3.2.2 Kyoto Protocol, 1997

The Kyoto Protocol was designed to overcome the lack of legally binding targets for emissions reduction and timelines. It was adopted in 1997 as the first subsidiary accord to the UNFCCC during COP3. Consistent with the goals of the UNFCCC, and based on the common but differentiated responsibilities principle, the protocol entered into effect on February 16, 2005 following Russia's ratification, so satisfying the requirement of participation from a minimum of 55 nations, who collectively account for more than 55% of worldwide emissions.

Article 3 stands as the core of the Kyoto Protocol, outlining commitments for 37 developed countries and the European Union (EU) to collectively reduce greenhouse gas emissions by a minimum of 5% below 1990 levels during the commitment period of 2008 to 2012. Notably, the European Union adopted an 8% reduction target, while the United States and Japan committed to 7% and 6% reductions, respectively. In line with the principle of "common but differentiated responsibilities" (CBDR) established in the Rio Declaration of 1992 (Principle 8) and reiterated in the UNFCCC (Article 4), developing countries were not mandated to accept specific emission targets (Clemencon, 2023).

After the initial commitment period of 2009–2012, all Parties with Quantified Emissions Limitation and Reduction Obligations (QELROs) under the Kyoto Protocol were assessed for compliance. While some Parties exceeded their domestic greenhouse gas (GHG) emissions targets, the Protocol provided avenues for fulfilment through three market mechanisms, known as flexibility mechanisms:

(a) Joint Implementation: This initiative enables countries committed to reducing emissions to earn emission reduction units (ERUs) by investing in projects that reduce emissions in other countries with emission reduction targets as outlined in Article 6 of the Kyoto Protocol.
(b) Clean Development Mechanism: This mechanism allows countries with emission reduction commitments to undertake projects in developing countries that reduce emissions, and these reductions are then counted toward their own emission reduction efforts as specified in Article 12.
(c) Emissions Trading: This mechanism permits countries with surplus emission units to sell them to countries that exceed their emission targets, creating a market for trading emission credits under Article 17 of the protocol.

By incorporating these flexibility mechanisms, the Protocol offers nations options to meet their emission limits in a cost-effective manner, allowing emissions reductions to be directed to where they can be achieved most economically (Dessler, 2022).

In 1998, the United States signed the Kyoto Protocol but refrained from ratifying it, which came as a significant setback to the treaty as at that time, US was the leading emitter of CHGs, contributing approximately a quarter of global CO_2 pollution, although this has since been surpassed by China. To garner wider participation, adjustments were made to the protocol's targets during the Bonn meeting in 2001, particularly to accommodate Japan, Canada, and Australia. Australia eventually ratified the Kyoto Protocol in December 2007 (Gupta, 2014).

Initially, developing countries were excluded from the Kyoto Protocol, with the expectation that they would join a subsequent agreement post-2012. However, the United States never ratified the Kyoto Protocol, and Canada withdrew before the first commitment period ended. Their reluctance to engage was partly due to objections from non-Annex I Parties regarding quantified GHG reduction commitments. In negotiations for the second commitment period of the Kyoto Protocol, efforts were made to ensure equal participation from all Parties. This led to the Durban Mandate, established at the 2011 COP17 in South Africa, aiming to develop a new protocol or legal instrument applicable to all Parties under the UNFCCC by 2015 (Osofsky & McAllister, 2012). Ultimately, this mandate paved the way for the 2015 Paris Agreement.

1.3.2.3 Paris Agreement, 2015

The Paris Agreement, which was adopted during COP21 on December 12, 2015 was formally implemented on November 4, 2016. The agreement has been ratified by almost all parties to the UNFCCC. Notably, the US initially ratified the treaty on September 3, 2016, during the Obama administration. However, following the Trump administration, the US withdrew from the pact on November 4, 2020. Subsequently, on January 20, 2021, the US re-joined the agreement during the Biden administration.

The agreement establishes a goal to keep the rise in global average temperature to less than 2 °C above preindustrial levels, with efforts aimed at achieving a 1.5 °C cap (Article 2, Paris Agreement, 2015). This 1.5 °C temperature cap was especially pushed for by small island developing states and least developed nations that are extremely vulnerable to the effects of climate change. The Paris Agreement employs a combination of top-down and bottom-up approaches. The bottom-up approach incorporates nationally determined contributions (NDCs) outlined in article 3 of the agreement, in which participating countries present their own plans with the shared goal of reaching the agreement's temperature target. However, concerns were raised about the potential self-serving nature of these contributions, prompting the inclusion of oversight mechanisms (top-down approach) such as the transparency system, global stocktake process, and compliance mechanism to ensure accountability (Mayer, 2022). Each party is obligated under Article 4 of the agreement to prepare, communicate, and update their NDCs every 5 years, outlining their climate action plans. These contributions are publicly recorded by the UNFCCC Secretariat, with the requirement that they progressively increase in ambition over time.

1.3.2.4 COP26: The Glasgow Climate Pact, 2021

A change from the negotiation dynamics seen at COP21 in Paris was evident during the COP26 conference, which met in Glasgow in November 2021. This conference created the foundation for a new global climate agenda and framework for ambition. Notably, COP26 served as the third meeting of the parties to the 2015 Paris Agreement and represented the inaugural occasion where nations collectively committed to enhancing the ambition of their carbon reduction commitments stipulated in the Paris Agreement.

The focal point of COP26 deliberations revolved around the concept of 'net zero' carbon emissions targets. The outcome of the conference was the Glasgow Climate Pact, which includes a strong affirmation of the necessity of achieving the 1.5 °C target. With this affirmation, the demand for a 45% decrease in CO_2 emissions by 2030 is reaffirmed. In addition, the agreement emphasises the need to "phase down" the use of coal and do away with wasteful fossil fuel subsidies while also pushing for an end to deforestation practices. The agreement also places a strong emphasis on mobilising climate funds to support developing countries as they make the shift to low-carbon and climate-resilient economies (Michaelowa, 2021).

1.3.2.5 COP 27, 2022

In 2022, the IPCC released two significant reports under its Sixth Assessment Report (IPCC AR6 2022): Working Group II's 'Impacts, adaptation, and vulnerability' and Working Group III's 'Mitigation of climate change.' These reports paint a dire picture of the future if greenhouse gas emissions continue unchecked. However, they also highlight that the solutions needed to combat climate change already exist; what is lacking is the scaling up of these solutions.

Against this backdrop, COP27 took place in Sharm El Sheik, Egypt, from November 6 to 18, 2022. Unlike COP26, which focused on setting targets, COP27 shifted its focus to implementation. One of its notable achievements was the establishment of a "loss and damage" fund (LDF) aimed at assisting vulnerable countries in coping with the escalating costs of climate-related damage. However, while the setup of the LDF is considered a significant accomplishment, questions remain to be negotiated in next conference regarding who will bear the financial burden and how this assistance will be delivered, particularly in countries like Pakistan, which are frequently affected by climate disasters (Masood et al., 2022; Clemencon, 2023). The phrase "Loss and damage" encompasses a wide range of impacts, including damage to infrastructure, displacement of communities due to rising sea levels, and loss of cultural heritage and livelihoods.

1.3.2.6 COP 28, 2023

The 28th session of the Conference of the Parties (COP28) under the UNFCCC) convened amidst the backdrop of mounting climate concerns, with 2023 poised to be the warmest year on record and climate impacts accelerating (Mckenzie, 2023). The conference, held in Dubai, concluded on December 13, 2023, following pivotal international negotiations.

Key developments at COP28 included the establishment of a fund to address losses and damages incurred by vulnerable countries due to climate impacts. Referred to as the Loss and Damage Fund (L&D Fund), its operational framework was agreed upon on the first day of the conference, with commitments exceeding USD 700 million towards its capital by the conference's end. Over the next four years, the World Bank will oversee the L&D Fund (Mckenzie, 2023).

Central to discussions at COP28 was the global transition towards renewable energy sources, signalling a decisive shift away from fossil fuels. Emphasizing the urgency of this transition, the Global Stocktake, a pivotal component of the Paris Agreement, highlighted the imperative of reducing fossil fuel dependency to limit global warming to 1.5 °C. Furthermore, enhancing climate finance, particularly for developing nations, emerged as a focal point. Discussions centred on devising more accessible and efficient funding mechanisms to support sustainability efforts in these countries. COP28 witnessed significant agreements and announcements, notably the recognition of the "beginning of the end of fossil fuels." This marked the first international deal explicitly addressing fossil fuels, emphasizing a transition away from them rather than a complete phase-out (Waskow et al., 2023). The nuanced agreement acknowledged the complexities of the global energy landscape, introducing the concept of "transitional fuels" such as gas and oil in the energy transition.

Another milestone was the inaugural Global Cooling Pledge, endorsed by 63 countries including the United States, Canada, and Kenya. This initiative aimed at substantially reducing emissions from cooling systems such as air conditioner and refrigerator by at least 68% by 2050 compared to 2022 levels, alongside establishing minimum energy performance standards by 2030. The culmination of COP28 saw the unveiling of the UAE Consensus, derived from the first-ever Global Stocktake (Waskow et al., 2023). This comprehensive assessment covered various climate issues, providing critical guidance for future climate action and informing the formulation of national climate commitments (NDCs) due in 2025.

1.4 Climate Change Negotiations: A Critique

At the global level, significant milestones like the inception of the UNFCCC in 1992, followed by the Kyoto Protocol and the Paris Agreement, were expected to catalyze substantial progress in combating climate change. These pivotal agreements were supplemented by numerous formal decisions, frameworks, and funding mechanisms like the Global Environment Facility and Green Climate Fund. Despite these efforts and a plethora of international meetings, global CO_2 emissions have continued to rise steadily over the past three decades, now standing at a staggering 60% higher than in 1990 (Stoddard et al., 2021).

Recent climate reports from various environmentalists underscore the inadequacy of current measures to address the looming challenges. The UNFCCC, originally aimed at stabilizing greenhouse gas concentrations, appears to have deviated from its core objective. This departure is evident as scientific evidence, notably from the

IPCC (), highlights the urgent need to limit global warming to 1.5 °C. However, negotiations on climate change have often fallen short of this goal.

Climate change adaptation has emerged as a critical priority for both developing and developed nations, reflecting the shortcomings of mitigation efforts, particularly by wealthier nations. The principle of equity and common but differentiated responsibilities, central to the UNFCCC, has faced challenges and neglect due to varying interests among participating countries. Another significant flaw in the current framework for climate change lies in its reliance on voluntary approaches within conventions, protocols, and agreements. The absence of legally binding mechanism and instances of withdrawal of parties from agreements further undermine the effectiveness of international climate governance mechanisms (Barua, 2023).

The Kyoto Protocol, while establishing firm commitments to reduce emissions, notably targets only developed countries, leaving out significant emitters like the United States (Russia joined later in 2005). This exclusion, coupled with lenient consequences for noncompliance, undermines its effectiveness. Furthermore, the Protocol's exemptions for greenhouse gas emissions fail to incentivize emerging big emitters such as China and India to curb their emissions. Both the United States and Australia initially refrained from joining the Kyoto Protocol due to concerns over the exemptions granted to these countries.

By the conclusion of the first commitment period of the Kyoto Protocol in 2012, a majority of industrialised nations had failed to achieve their respective targets (Dessler, 2022). The inability of the Protocol to effectively enforce carbon reductions among industrialised nations brought attention to deficiencies in global climate governance. In contrast to the obligations outlined in Articles 4.1 and 4.2 of the UNFCCC, the Quantified Emissions Limitation and Reduction Commitments (QELRCs) represent obligations to achieve specific outcomes rather than conduct. These commitments varied among nations, reflecting differences in energy needs and capacity for developing alternative energy sources. While the Kyoto Protocol aimed to reduce greenhouse gas emissions by 5% below 1990 levels for developed countries, no comparable commitments were imposed on developing nations. This discrepancy contributed to a net increase in global greenhouse gas emissions, driven by rapid growth in emerging economies (Mayer, 2018).

The Paris Agreement, touted as a landmark accord in the fight against climate change, falls short in several critical aspects. While it outlines processes that are legally binding, it lacks teeth when it comes to compelling nations to take concrete domestic actions to curb emissions. Instead of stringent mandates, it relies heavily on the submission and periodic updating of Intended Nationally Determined Contributions (INDCs), a mechanism that offers more flexibility than enforceable obligations (Clemencon, 2023). Furthermore, the agreement's provisions regarding the ambition of NDCs are disappointingly weak. Although it encourages developed nations to lead by example in implementing emission reduction targets, and nudges developing countries to gradually increase their ambition, these recommendations lack the force of binding commitments (Mayer, 2022). Consequently, there's a glaring absence of effective enforcement mechanisms, with no penalties stipulated for countries failing to meet their NDCs.

Additionally, while the Paris Agreement sets ambitious temperature targets, aiming to limit global warming to well below 2 °C, and ideally to 1.5 °C, the actual commitments made by signatory nations fall woefully short (Dessler, 2022). Despite being hailed as a success, the emission reduction pledges outlined in the agreement are insufficient to achieve these temperature goals, raising serious doubts about its effectiveness in tackling the climate crisis.

In contrast to the Kyoto Protocol, which at least targeted emissions reductions from developed countries, the Paris Agreement attempts a more inclusive approach by involving all nations. However, this inclusivity comes at the cost of enforceability, as the agreement lacks provisions to hold countries accountable for their commitments. Moreover, its failure to address greenhouse gas emissions from international shipping and aviation, a key provision in the Kyoto Protocol, highlights significant gaps in its scope and effectiveness (McGregor, 2021). Overall, while the Paris Agreement may signify progress, its fundamental weaknesses undermine its potential to drive meaningful climate action on a global scale.

Before COP26 held at Glasgow, major geopolitical changes concerning climate change occurred against the backdrop of the COVID-19 pandemic. In 2019, the UK made changes to the Climate Change Act that required a net zero emissions goal by 2050. In a similar vein, the EU increased its 40% reduction in GHG emissions to 55% by 2030. China, a major emitter, declared aims to attain carbon neutrality by 2060 and peak CO2 emissions by 2030, which is a substantial change from prior commitments (Michaelowa, 2021).

Notwithstanding these advancements, COP26 was criticised for a number of flaws. First of all, wealthy countries have not delivered on their ten-year commitment to donate $100 billion a year to help the least developed nations make the transition to sustainable energy. Second, the US and EU rejected the creation of the Glasgow Loss and damage fund, which was designed to compensate vulnerable countries for harm caused by climate change. Thirdly, there was leeway for insufficient action because national carbon-cutting pledges were not legally bound to the 1.5 °C trajectory outlined in the Paris Agreement. Projections show that by the end of the century, global warming may still surpass 2 °C even if all pledges are fulfilled (Dettmer, 2021).

Furthermore, despite the fact that the inclusion of fossil fuels in global climate agreements at COP26 was a historic moment, critics condemned the little attention paid to basic problems. The change in wording from 'phase out' to 'phase down' coal garnered attention, but the wider implications of this move were not taken into consideration (Lewis & Maslin, 2021). The COP26 conference brought to light structural shortcomings and insufficiencies in tackling the urgent problems caused by climate change on a worldwide level, despite some progress being made.

COP27 witnessed the establishment of a loss and damage deal, deemed a positive step, yet faced criticism for the slow progress in reducing greenhouse gas emissions. The absence of new targets or commitments posed a threat to the goal of limiting global temperature rise to 1.5 °C, set seven years prior in the Paris Agreement. Instead, the agreement merely called for new country pledges for COP28, delaying progress further. Disappointingly, the COP27 agreement fell short of advancing beyond the

Glasgow climate pact's commitment to phase down unabated coal power. Moreover, the COP27 negotiations were overshadowed by geopolitical tensions stemming from Russia's war against Ukraine, impacting gas supplies and prompting an expansion of domestic fossil fuel reserves. The event primarily served as a defensive action to prevent the dilution of progress made at Glasgow, with little enthusiasm from the Egyptian hosts to secure new agreements. Although a loss and damage fund was established, the aftermath revealed significant shortcomings in implementation. While celebrated by developing nations, the fund's current size falls short of addressing projected damages, with no clear commitment to increase funding levels or allocation mechanisms (Clemencon, 2023).

Following extensive negotiations throughout the year, the operationalization of the Loss and Damage Fund marked a significant milestone of COP28 held in Dubai. However, achieving this milestone came at the expense of substantial concessions from developing nations, highlighting the power dynamics at play within climate negotiations. While some countries contributed approximately \$700 million to the fund, this amount remains inadequate considering the projected climate-related damages vulnerable countries may face by 2030, estimated at \$580 billion (Waskow et al., 2023). The disparity between contributions and the scale of potential damages underscores the inadequacy of current efforts to address climate impacts.

Notably, all nations formally committed to transitioning away from fossil fuels and accelerating the adoption of renewable energy, with heightened emphasis on addressing climate vulnerabilities in cities and food systems (Mckenzie, 2023). Despite these pledges, the focus now shifts to translating COP28 goals into tangible national actions and ensuring sufficient financial resources for implementation. It is crucial to recognize that these commitments should serve as a baseline rather than a culmination in the fight against climate change.

1.5 Discussion

Almost three decades of painstakingly constructed political rhetoric, rumours of significant shifts, and pledges of ground-breaking technology breakthroughs have not been able to stop the unrelenting rise in emissions. In spite of unequivocal scientific evidence and global accords, CO_2 emissions have increased by about 60% since 1990. The complex causes and wide-ranging effects of climate change make it impossible to solve with straightforward methods. Ironically, the countries, companies, or individuals who bear the majority of the blame for emissions are frequently spared the immediate fallout, which lessens the urgency of taking swift action. On the other hand, the nations and marginalised populations that are least responsible for the emissions suffer the worst consequences. The foundation of economic progress is deeply entrenched in the combustion of fossil fuels and massive land-use shifts over centuries, which have produced intricate interdependencies that go beyond the economic realm to the material, social, cultural, and psychological spheres. The entrenched interests of dominant economic and political powers staunchly resist transformative change, perpetuating the status quo. The complexity of addressing climate change is further compounded by the fact that the roots of these issues are intertwined with larger historical legacies of colonialism, imperialism, and institutional inequalities.

Recent UN Climate Conferences, notably those in 2021, 2022 and 2023, held amidst the backdrop of the COVID-19 pandemic, have highlighted the difficulty in strengthening commitments to meet the ambitious targets set forth in the Paris Agreement. Despite the urgent need to limit global warming to 1.5 °C by 2030, negotiations have fallen short of adopting robust measures to achieve this goal. COP27, while marking a symbolic milestone with the establishment of a loss and damage fund, has underscored the challenges in translating international agreements into tangible action (Clemencon, 2023).

William Nordhaus, an economist at Yale University, asserts that voluntary agreements like the Paris Agreement inherently foster free-riding behaviours and are on bound to fail. He advocates for a shift towards negotiating a universal carbon price among governments, arguing that this approach would be more effective in reducing global emissions compared to the current emphasis on country-specific emission limits. Additionally, alternative proposals have surfaced, suggesting the development of new agreements tailored to address specific emissions or sectors, serving as complements to the Paris Accord.

According to analysis in international relations, the path to significant progress in climate action appears to be shifting away from global unity towards decentralized initiatives led by smaller groups and sectors. Recent examples supporting this perspective include commitments made by the Group of Twenty (G20) nations, collectively responsible for 80% of global greenhouse gas pollution. These decentralized efforts highlight a shifting approach to climate action, suggesting the potential for significant progress beyond the confines of global agreements.

The industrial sector stands as a significant contributor to carbon pollution, prompting numerous firms to pledge emissions reduction efforts or aspire to achieve carbon neutrality or negativity. This highlights the potential for enhanced climate change mitigation and adaptation through private sector involvement. Moreover, there has been a notable shift in momentum from global negotiations to engagement at national and subnational levels. This transition involves increased participation from societal and private sector actors, emphasizing the importance of ongoing collaboration among nations, non-governmental organizations (NGOs), private sectors, and local communities. Critical elements of this collaboration include fostering partnerships for technology transfer, sharing knowledge, and building capacity, particularly in developing countries.

Amid these changing dynamics, it is still widely recognized that policymakers remain essential in establishing and enforcing emissions targets.

1.6 Conclusion

In the face of the hottest year on record and the escalating climate crisis, there's a pressing need for bold, transformative actions to address the challenges ahead. While international climate negotiations have historically been crucial, a comprehensive analysis spanning three decades reveals a disconnect between these diplomatic efforts and actual national policies. Factors such as national political dynamics, normative shifts, economic considerations, and external events have largely shaped countries' climate policies independent of multilateral negotiations. Despite their declining relevance, these negotiations have sparked significant scientific cooperation, emphasized the equity dimension of the crisis, and facilitated the financing of climate action, particularly in vulnerable developing nations.

The Kyoto Protocol marked a pinnacle in international climate cooperation, establishing hard targets and deadlines that catalyzed capacity building across public and private sectors. Although emissions trading, a cornerstone of the protocol, hasn't fully delivered as anticipated, it spurred broad-based engagement and influenced climate action at subnational levels. Similarly, the Paris Accord of 2015, while non-binding, served as a critical milestone, prompting unilateral measures by major economic blocs like the United States and the European Union to enhance domestic climate policies.

Looking ahead, international climate negotiations will continue to play a vital role in coordinating research, addressing methodological challenges, and ensuring the scaling up of climate financing. While global agreements can send important signals, the ultimate fate of the planet rests on individual countries' willingness to fulfill their commitments and undertake decisive actions to combat the climate crisis. As the world grapples with this existential challenge, hope remains as long as nations prioritize the right policies, irrespective of international outcomes, to mitigate and adapt to climate change.

References

AMS Glossary. (2024). Glossary of meteorology. American Meteorological Society. https://glossary.ametsoc.org/wiki/Climate_change. Accessed 30 March 2024

Baede, A.P.M. et al. (2001). The climate system: An overview. In: Climate change 2001: The scientific basis, contribution of working group I to the third assessment report of the Intergovernmental Panel on Climate Change. Cambridge University Press.

Barua, A. (2023). UNFCCC: A broken and failed system. https://www.dhakatribune.com/opinion/op-ed/332913/unfccc-a-broken-and-failed-system. Accessed 30 March 2024

Bodansky, D. (1993). The United Nations framework convention on climate change: A commentary. *Yale Journal of International Law, 18* (2)

Clémençon, R. (2023). 30 years of international climate negotiations: Are they still our best hope? *The Journal of Environment & Development, 32*(2), 114–146. https://doi.org/10.1177/10704965231163908

Climate Change Impacts. NOAA (National Oceanic and Atmospheric Administration), USA. 2021. https://www.noaa.gov/education/resource-collections/climate/climate-change-impacts. Accessed 8 March 2024

Cowie, J. (2007). *Climate change: Biological and human aspects.* Cambridge University Press.

Dessler, A. E. (2022). *Introduction to modern climate change.* Cambridge University Press. https://doi.org/10.1017/9781108879125

Dettmer, J. (2021). COP26: Success or failure? https://www.voanews.com/a/cop26-success-or-failure/6349263.html. Accessed 29 March 2024

Gardiner, S. M. (2006). A perfect moral storm: Climate change, intergenerational ethics and the problem of moral corruption. *Perspectives on Environmental Values, 15*(3), 397–413. https://doi.org/10.3197/096327106778226310

Giddens, A. (2009). *The politics of climate change.* Polity Press.

Gupta, J. (2014). *The history of global climate governance.* Cambridge University Press.

IPCC. (1995). *Climate change 1995: Synthesis report. Contribution of working groups I, II and III to the fourth assessment report of the Intergovernmental Panel on Climate Change.* IPCC.

IPCC. (2001). *Climate change 2001: Synthesis report. A contribution of working groups I, II, and III to the third assessment report of the Intergovernmental Panel on Climate Change.* Cambridge University Press.

IPCC. (2014). *Climate change 2014: Synthesis report. Contribution of working groups I, II and III to the fifth assessment report of the Intergovernmental Panel on Climate Change.* IPCC.

IPCC. (2018). *Annex I: Glossary. In: Global Warming of 1.5°C, an IPCC special report on the impacts of global warming of 1.5°C above pre-industrial levels and related global greenhouse gas emission pathways, in the context of strengthening the global response to the threat of climate change, sustainable development, and efforts to eradicate poverty* (pp. 541–562). Cambridge University Press. https://doi.org/10.1017/9781009157940.008

IPCC. (2022). *Climate change 2022: Impacts, adaptation and vulnerability. Contribution of working group II to the sixth assessment report of the Intergovernmental Panel on Climate Change.* Cambridge University Press. https://doi.org/10.1017/9781009325844.

IPCC. (2023). *Climate change 2023: Synthesis report. Contribution of working groups I, II and III to the sixth assessment report of the Intergovernmental Panel on Climate Change.* IPCC. https://doi.org/10.59327/IPCC/AR6-9789291691647

Jager, J., & Ferguson, H. L. (eds) (1992). Climate change: Science, impacts and policy. *Environmental Conservation, 19* (2), 189–189. https://doi.org/10.1017/S0376892900030861

Kaya, Y., et al. (2015). *Climate change and energy: Japanese perspectives on climate change mitigation strategy, ICP series on climate change impacts, adaptation, and mitigation* (Vol. 4). Imperial College Press.

Kyoto Protocol to the United Nations Framework Convention on Climate Change, Kyoto. (1997). United Nations treaty series, Vol. 2303, No. 30822.

Lewis, S. L., & Maslin, M. A. (2021). Five things you need to know about the Glasgow climate pact. https://theconversation.com/five-things-you-need-to-know-about-the-glasgow-climate-pact-171799. Accessed 30 March 2024

Masood, E., et al. (2022). COP27 climate talks: What succeeded, what failed and what what's next. *Nature, 612*, 16–17. https://doi.org/10.1038/d41586-022-03807-0

Mayer, B. (2018). *The international law on climate change.* Cambridge University Press.

Mayer, B. (2022). *International law obligations on climate change mitigation.* Oxford University Press.

McGregor, D. (2021). A critical discussion on the strengths and weaknesses of the Paris Agreement (2015) on climate change. https://www.linkedin.com/pulse/critical-discussion-strengths-weaknesses-paris-2015-climate-mcgregor. Accessed 29 March 2024

Mckenzie, B. (2023). COP28: Key outcomes and where to next? https://www.bakermckenzie.com/-/media/files/insight/publications/2023/12/cop28-key-outcomes-briefing-paper.pdf. Accessed 30 March 2024

Michaelowa, A. (2021). The Glasgow climate pact: A robust basis for the international climate regime in the 2020s. *Intereconomics, 56*, 302–303. https://doi.org/10.1007/s10272-021-1004-7
Osofsky, H. M., & McAllister, L. K. (2012). *Climate change law and policy*. Aspen Publishing.
Paris Agreement. (2015). United Nations treaty series, No. 54113
Rajamani, L. (2016). Paris Agreement: Interplay between hard, soft and non-obligations. *Journal of Environmental Law, 28* (2), 337–358
Rio Declaration on Environment and Development. (1992). *Report of United Nations conference on environment and development*. United Nations Publication. A/CONF.151/26(Vol I)
Schleussner, C. F., et al. (2016). Science and policy characteristics of the Paris Agreement temperature goal. *Nature Climate Change, 6*, 827–835. https://doi.org/10.1038/nclimate3096
Stockholm Declaration. (1972). Report of the United Nations conference on the human environment. United Nations Publication. A/CONF.48/14/Rev.1
Stoddard, I., Anderson, K., Capstick, S., et al. (2021). Three decades of climate mitigation: Why haven't we bent the global emissions curve? *Annual Review of Environment and Resources, 46*, 653–689. https://doi.org/10.1146/annurev-environ-012220011104
United Nations Framework Convention on Climate Change. (1992). United Nations treaty series, Vol. 1771, No. 30822
Verheyen, R. (2005). *Climate change damage and international law: Prevention duties and state responsibility*. Martinus Nijhoff.
Vienna Convention on the Law of Treaties. (1969). United Nations treaty series, Vol. 1155, No. 18232
Waskow, D. et al. (2023). Unpacking COP28: Key outcomes from the Dubai climate talks, and what comes next. https://www.wri.org/insights/cop28-outcomes-next-steps. Accessed 31 March 2024

Chapter 2
Expected Changes in Aridity Land Degradation and Climate Risks to the Coastal State of Andhra Pradesh, South India and Road Map to Resilience

P. Dhanya, R. Jagannathan, and Sesha Srinivas Jonnalagadda

2.1 Introduction

Worsening aridity conditions may escalate serious concerns for sustainability and human wellbeing in the context of global climate change is been reported by United Nations Convention to Combat Desertification (UNCCD, 2016). Land degradation may exacerbate the already existing challenges of climate change (Evans & Geerken, 2004; IPCC, 2007a, 2012). Many research studies have carried to analyze the socio economic vulnerability of coastal areas of India, on the basis of exposure, sensitivity and adaptive capacity (Maiti et al., 2015; Mani Murali et al., 2013). Erratic rainfall that is 25% lower than usual is projected for semi-arid regions, if average global temperatures rise by 2 °C above pre-industrial period. According to De vente et al. (2017), management of green water for rain-fed farming is integral for eradicating hunger and achieving food security. Mainstreaming social resilience through national and state climate change adaptation framework is necessary at this juncture (Folke et al., 2010; UNCCD, 2012; EM-DAT9, 2019; IPCC, 2013a).

Cities and communities in the coastal areas of India are vulnerable to climate induced disasters. Knowing the seriousness of changing environmental scenarios, India's Coastal Regulation Zone (CRZ) act of 1991 was updated and reissued in 2011 to incorporate such threats to coastal zone management plans. It is estimated that population density in the coastal areas may increase three times and 50% of the world's population will be occupied within the vicinity of 100 km of coastal areas.

P. Dhanya (✉) · R. Jagannathan
Tamil Nadu Agricultural University, Coimbatore, Tamil Nadu, India
e-mail: dhanya.eptri@gmail.com

S. S. Jonnalagadda
Environment Protection Training and Research Institute, Gachibowli, Hyderabad, India

P. K. Rai and S. Rai (eds.), *Climate, Environment and Agricultural Development*, Advances in Geographical and Environmental Sciences,
https://doi.org/10.1007/978-981-97-8363-2_2

India has a very long coastline of 7500 kms, covering 16.7% of the world's population and a very high population growth rate. This makes India extremely vulnerable to environmental threats (McCarty et al., 2009). Mishra et al. (2017) suggested that India would be more vulnerable to the socio-economic impacts of extreme weather events such as floods, droughts, cyclones, heat waves etc. The densely populated coastal cities in India including the present study area, erstwhile Andhra Pradesh is highly prone to tropical cyclones (Table 2.1). Coastal states are anticipated to have rise in sea level which require more proactive management approaches (Praveen & Udayakumar, 2014). Additionally, the National Disaster Management (DM) act of 2005 and DM policy of 2009 and the revised National Disaster Management Plan of 2019, released by the National Disaster Management Authority (NDMA) emphasize that developing contemporary forecasting and early warning systems are prerogative for every states in India as a proactive disaster risk reduction strategy. Actions that aim in reducing disaster risks, stand as an efficient response and relief to the vulnerable sections of the society (IPCC, 2013b, 2014) in a proactive way (Chaturvedi et al., 2012). Ramachandran et al. (2015) warned that the land degradation for semi-arid areas especially, Tamil Nadu, south India may be increased during the end of twenty-first century as a result of rise in warming. This study is an attempt to understand the climate extremities especially the looming aridity conditions for peninsular India. Mishra et al. (2017) reported that population exposure to severe heat waves is projected to increase by about 15 and 92 times the current level by the mid and end-twenty-first century respectively under the 2 °C low-warming target. According to Ekardt (2020) the world climate has already become uncertain and unstable.

2.2 Background of the Study

The coastal cities in the erstwhile Andhra Pradesh are extremely vulnerable to tropical cyclones, storm surges and floods. The climate risk of the study is being aggravated by cyclones of moderate to severe intensity almost every year posing grave concern for sustainable living and developments. Data obtained from the Indian Meteorological Department (IMD) shows that since 1975, the state had faced more than 65 tropical cyclones. Cyclones are followed by heavy rains, flash floods and storm surges making the lives miserable in the whole of states. Even the recent Super Cyclonic Storm 'Amphan'-2020 and 'Nivar'-2021 were accompanied by rough seas storm surges, strong winds and unusual rains in many parts of coastal Andhra Pradesh, making it a multi hazard hot spot (Nanditha et al., 2020). NASA aids India in evaluating the tropical cyclones and gathers data on tropical cyclones. Global Precipitation Measurement mission or GPM core observatory satellites fly over Bay of Bengal and the Arabian Sea to track them and understand the rainfall intensity brought out by the cyclone (NASA, 2020). National Institute of Disaster Management (NIDM) and National Disaster Relief Force (NDRF) are equipped with ground level preparedness to face the cyclone repercussions.

Table 2.1 Major cyclones hit till date in the erstwhile Andhra Pradesh coast from 1975 to 2020

Sl. No	Cyclones	Year	Areas hit
1	Amphan	20 May 2020	Vizag
2	Fani	3 May 2019	Srikakulam
3	Phethai and Gaja and Titli	17 Oct to Dec 2018	E Godavari
4	Vardah, Roanu & Kyant	13 Dec 2016 May & Oct, 2016	Nellore, Chittor, Anantapuram & Vizag
5	Hudhud	13 Oct 2014	Vizag
6	Phailin	14 Oct 2013	Srikakulam
7	Helen, Lehar	23, 28 Nov 2013	Machilipatinam
8	Nilam	1 Nov 2012	Nellore
9	Laila	21 May 2010	Guntur, Prakasam
10	Jal	7 Nov 2010	Nellore
11	Khaimuk	17 Nov 2008	Kavali
12	Ogni	30 Oct 2006	Prkasam, Guntur and Krishna
13	CS-AP	19 Sep 2005	Kalingapatnam
14	VSCS	29 Nov 2000	Coastal Andhra
15	VSCS	15 Oct and Nov 1999	East Godavari, Visakhapatnam, Vizianagaram and Srikakulam
16	KKD	6 Nov 1996	E&W Godavari
17	TS-AP	8 May 1990	Prakasam, Guntur, Krishna, Machalipatinam
18	Kavali	8 Nov 1989	Nellore
19	SCS	Nov 1984	Nellore
20	Severe Cyclonic Storm	18 Oct 1980	Srikakulam to Krishna districts
	Severe Cyclonic Storm	13 May 1979	Nellore, Prakasam, Guntur, West Godavari to Visakhapatnam
21	TC-AP	20 Nov 1977	Nizampatinam Island of Diviseema

Source Elaborated by the authors

Achieving UN's Sustainable Development Goals (SDGs) by 2030 through state level action may have countless obstacles in multi hazard prone areas like Andhra Pradesh and Telangana. National governments alone cannot achieve the ambitious goals of the 2030 Agenda on disaster risk reduction and SDGs. States, regions, districts and cities must take their share of responsibilities and can contribute towards resilience building. A state which is already struggling with multiple natural and man-made disasters like severe cyclones, SARS-COVID-19 virus attack, droughts &

aridity, storm surges, gas leakage disasters and biological attacks by locust, safeguarding its populations and providing access to adequate food, safe drinking water, affordable and sustainable transport systems pose unlimited developmental challenges.

2.3 Materials and Methods

2.3.1 Study Area

The study area is the coastal area of souther state of India i.e. Andhra Pradesh which is the second longest mainland coastline with a total length of 974 km. Its coastal stretch extends between Ichchapuram of Sreekakulam District to Tada of Nellore District. It is located between 12°41' and 22° N latitude and 77° and 84°40' E longitude, and is bordered by Maharashtra, Chhattisgarh and Orissa in the north, the Bay of Bengal in the east, Tamil Nadu to the south and Karnataka to the west. The south-west monsoon accounts for about 70% of the annual rainfall in the State (SAPCC, Andhra Pradesh 2014). The study area i.e. Andhra Pradesh has a total population of 84,665,533 (as per the 2011 census of India) and density is 308 persons per sq. km. The Godavari, is the longest and the broadest river in South India, with a length of 1464 km length, of which about 772 km lies within the State, Census of India (2011).

2.4 Climate Change Modeling Approach

Abdus Salam International Centre for Theoretical Physics (ICTP), Italy has developed the Regional Climate Models (RCMs) viz., RegCM4.0 to simulate future climate scenarios under IPCC fifth assessment report based emission trajectories. The GCM boundaries used to drive the RCMs was Had GEM-ES model. The model simulation was run for 128 years (1971 to 2098) using the latest RCP 4.5 and RCP 8.5 scenarios. The weather variables maximum and minimum temperature are extracted from the RCM output. The future projections are based on a representative CO_2 concentration pathway that generates a radiative forcing of 4.5 Wm^{-2}, and 8.5 Wm^{-2} are chosen for the study. Chaturvedi et al. (2012); Taylor et al. (2012) and Kumar et al. (2010) have carried out enormous work on climate change simulations for India. RCP 4.5 represents moderate emissions scenario, presuming actions are being under taken to limit greenhouse gas emissions by 2050. Under RCP 4.5 scenario the atmospheric CO_2 concentration in the year 2050 would be 478 ppm, and by 2100, it would be 538 ppm. From the daily outputs of RegCM4, mean annual estimations were made using Perl programme for climate analysis. The annual estimations were done for each grid and then averaged for the state to find the increase or decrease in trend of climate parameters till the end of twenty-first century.

2.5 Sea Level Rise Simulator

IPCC's Assessment Report 5 has been used to project sea level rise for the coast of Andhra Pradesh under various emission scenarios (IPCC, 2014). A climate simulator model output is used to generate sea level projections for the twenty-first and twenty-second century based on low, medium and high emission pathways. The global SLR data were downloaded for the port areas of Visakhapatnam and processed to understand average future projections (Source: http://www.psmsl.org/data and https://tidesandcurrents.noaa.gov/sltrends/) to by the end of twenty-first and twenty second century. The sea level rise projection is employed based on GHG emission trajectories as per the Representative Concentration Pathways (RCPs) of IPCC AR5. Box and whisker plots were used to make visual regional comparisons of projected sea level rise for various time scales.

2.6 Aridity Assessment

Aridity Index is a numerical indicator of the degree of dryness of the climate at a given location of any region. "De Martonne" Aridity Index is used to evaluate present and future aridity trends usually in Telangana and Andhra Pradesh. The projected aridity was calculated for the near (2010–2040), mid (2041–2070), end century (2071–2098) periods. Spatio temporal aridity maps were calculated and prepared using Arc GIS platform.

De Martonne aridity index (DM_{Ia}) was calculated using the formula (Table 2.2).

$$DMI_a = \ldots\ldots\ldots P/(T + 10) \tag{2.1}$$

where P—annual precipitation sum [mm]; T—annual mean air temperature [°C].

Table 2.2 Aridity classes as per De' Martonne Index (DM_I) classification

Climate	Values of DMI
Dry	$DM_I < 10$
Semi-Dry	$10 \leq DM_I \leq 20$
Mediterranean	$20 \leq DM_I < 24$
Semi-Humid	$24 \leq DM_I < 28$
Humid	$28 \leq DM_I < 35$
Very Humid Extremely Humid	a.$35 \leq DM_I \leq 55$ b. $DM_I > 55$

2.7 Results and Discussion

2.7.1 Anticipated Sea Level Rise for Andhra Pradesh Coast

The results revealed that the study area may experience sea level rise in the later part of the twenty first and twenty second centuries. The present study area, state of Andhra Pradesh, situated in the south-east coastal areas of India shares coastal boundaries with Bay of Bengal in the east. The data from climate simulator model based on IPCC AR5 emission trajectories indicates a rise in sea level at Visakhapatnam (Vizag) coast under three different scenarios namely low, moderate, high strong emission scenarios. It is evident from the study that sea level may rise by 0.77 m, 0.89 m and 1.16 m under the RCP 2.6, RCP4.5, RCP 8.5 carbon pollution scenario respectively by 2100. It is also projected to further rise by 2.26 m, 2.64 m, 3.66 m by 2200 under RCP 2.6, 4.5 and 8.5 respectively (Table 2.2 & Fig. 2.1). It must be noted that only with aggressive cuts in carbon pollution, we will be able to limit the undesirable rise in sea level to be around 0.77 m as per RCP 2.6 (Table 2.3).

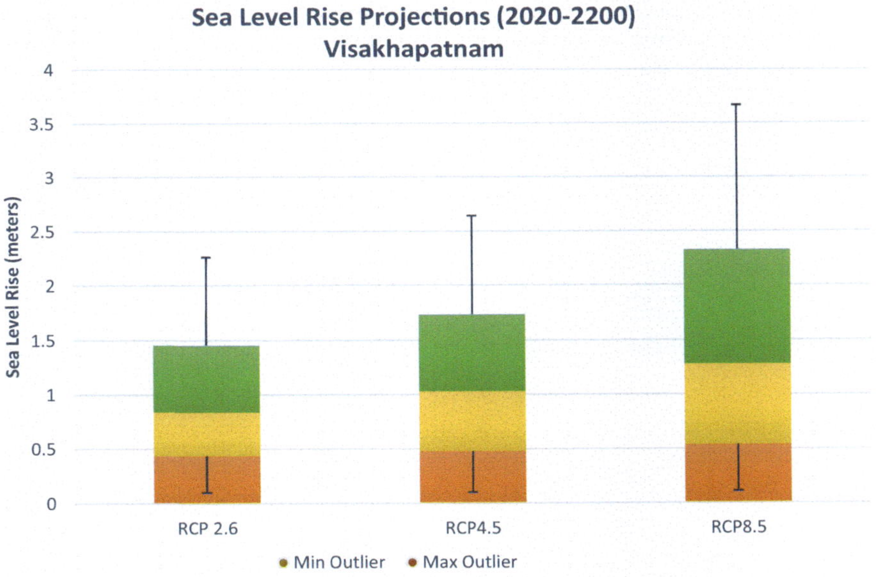

Fig. 2.1 Sea level rise projections for Visakhapatnam (2020–2200). *Source* https://tidesandcurrents.noaa.gov/sltrends/

Table 2.3 The median and interquartale ranges of sea level rise projections for Vizag 2020–2200, Andhrapradesh

Labels	RCP 2.6	RCP4.5	RCP8.5
Min	0.1	0.1	0.11
Q_1	0.435	0.48	0.54
Median	0.84	1.03	1.28
Q_3	1.455	1.735	2.325
Max	2.26	2.64	3.66
IQR	1.02	1.255	1.785

Source Elaborated by the authors

2.8 Aridity Assessment

Observed temperature and rainfall of the city of Andhrapradesh were collected to understand the basic climatology of the place and to compute aridity status. The observed climatology shows significant trend for average rainfall no significant trend for rainfall with an annual mean rainfall of 832.83 mm, however inter annual variations were noted.

De Martonne Aridity Index is presented in the Figs. 2.2 and 2.3 respectively under RCP 4.5 and 8.5 scenario. De Martonne Aridity Index is a widely used index to evaluate the present and future climate and aridity trends prevailing in a region. In Andhra Pradesh, apparently a wetting trend is noted in the near century period (2011–2040) with reference to the baseline period (1971–2000).

The temperature and rainfall changes as per the model HADGEM2-ES-RCP 4.5 trajectory show an increasing drying or aridity conditions for Visakhapatnam district. Under RCP 8.5 trajectory, wetter conditions are seen for the end century. Many districts in the interior Andhra Pradesh that have arid and semi-arid climate and districts such as Nellore, Cuddapah, Ongole, Kurnool, west Godavari districts are showing a wetting trend. Arid to semi-arid climatic conditions were prevailing throughout Visakhapatanam, West and East Godavari districts to Sreekakulam districts during the baseline period, however the analysis showed a wetting trend as the aridity values tend to decline in these areas in future. It shows that south east coastal districts may be more humid in future as shown in Figs.2.2 and 2.3. The projected aridity trend was in the range of IDM $28 \leq DM_I < 35$ the south east coastal districts and in the hilly areas during the mid-century (2010–2040) period, it is noted s IDM $35 \leq DM_I \leq 55$ in the coastal areas under RCP8.5 scenario.

2.9 Road Map to Resilience for Andhrapradesh

The journal Lancet countdown on health and climate has reported that India has been particularly affected by the rising frequency of heat wave events and lost about 75 billion hours of work, a significant part of it is in the agricultural sector. An

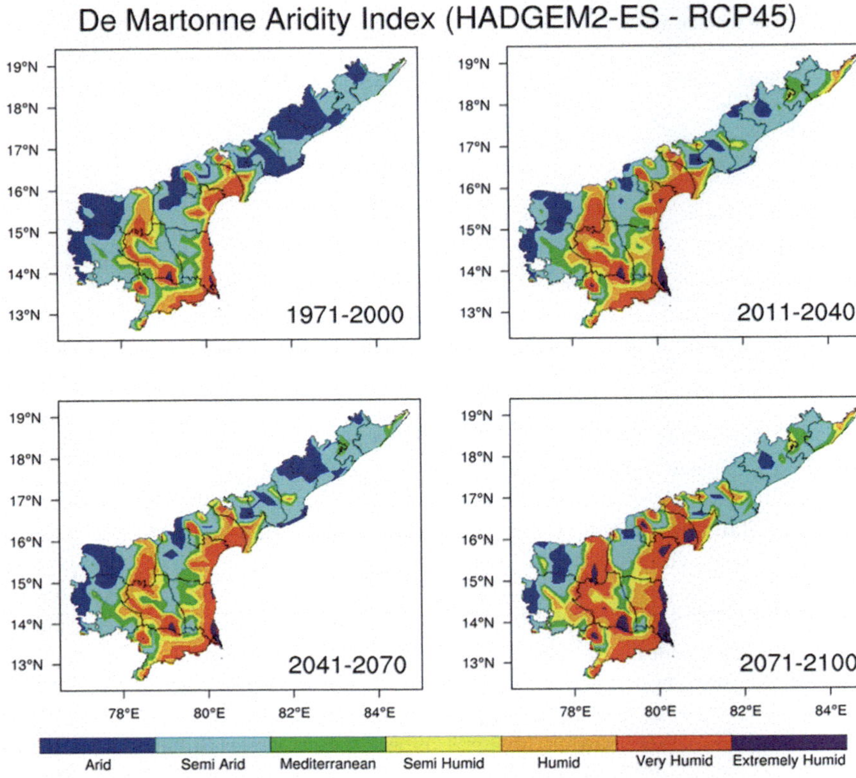

Fig. 2.2 Projected De Martonne Aridity index for Andhra Pradesh spanning three continuous four time slices (near-2010–2040, mid-2041–2070, End-2071–2100) under RCP 4.5

estimated loss of 153 billion hours of labor during 2017 due to rising temperatures around the globe is a reminder to dramatically curb greenhouse gas emissions. Lancet Countdown (2018) report recommends that Indian policy makers must take a series of initiatives to mitigate the increased risks to health, and the loss of labor hours due to a surge in heat wave events in the country over the 2012–2016 period (Watts et al., 2015, 2018). Severe heat wave conditions claims many lives in the states of Andhra Pradesh and Telangana. Guleria and Gupta (2018) have reported that severe heat wave conditions with variance more than 6 °C is seen in both the states during the summer, 2016, leading to 324 sunstroke deaths in Telangana State. Analysis carried out by Byers et al. (2018) suggests that impacts of sea level rise and storm surge could be more pervasive in sensitive systems such as low lying coastal areas and small islands and increasingly widespread by the 2070s (Brown et al., 2018) as temperatures rise from 1.5 °C to 2 °C, even when adaptation measures are considered, suggesting a transition to high risk for these areas. With 2.5 °C of warming, adaptation limits are expected to be exceeded in sensitive areas, and hence a transition to very high risk is

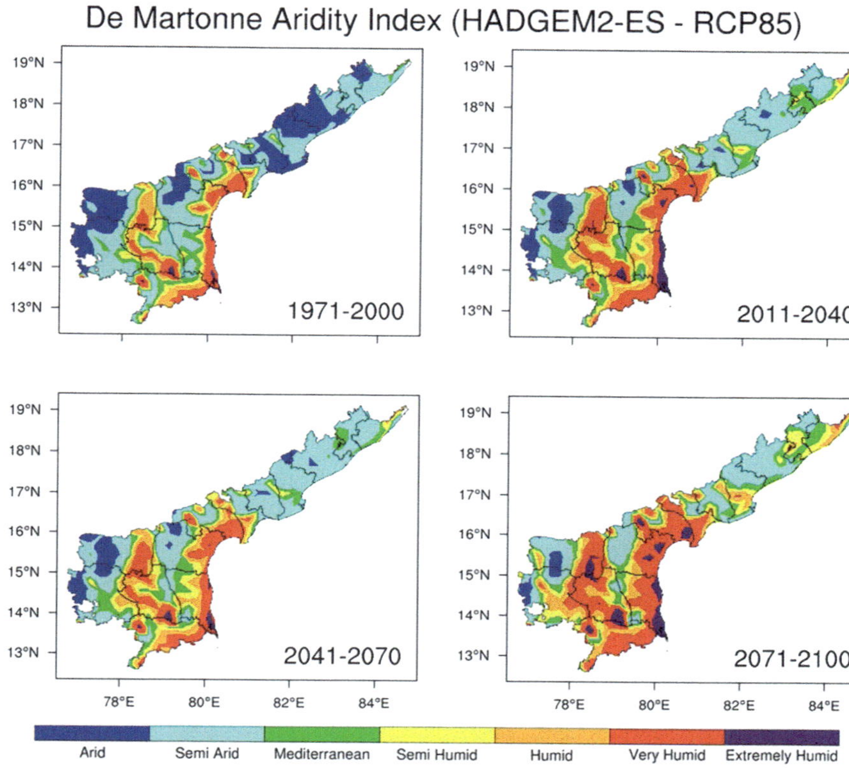

Fig. 2.3 Projected De Martonne Aridity index for Andhra Pradesh spanning three continuous four time slices (near- 2010–2040, mid-2041–2070, End-2071–2100) under RCP 8.5

projected. Nageswararao et al. (2020) has mentioned in a recent article that upsurge of heat waves in the tropical belt may be attributed primarily to increased temperature and humidity in the atmosphere under global warming. The most heavily affected areas in terms of casualties due to heat waves during pre-monsoon (March–May) are the central east coast of India (CECI; Odisha, Andhra Pradesh, and Telangana). East coastal areas of India are prone to cyclonic storms, water logging, flooding and storm surges, etc. (Ranjan et al., 2006; Shamsuddoh & Chowdhury, 2007).

The National Disaster Management Authority (NDMA) opinioned that climate change is causing the day and night time temperatures higher than normal in recent years (Pattanaik et al., 2017). The NDMA had held two planning workshops in 2017 and 2018 in Andhra Pradesh to address the issue of heat waves. Training and capacity building are essential as a preparedness plans to safeguard the communities from any disasters. Researchers around the world have valued the critical role of social networks in appraising adaptive capacity (Berkes & Ross, 2013; Folke, 2006). This is especially true for the rural communities in the global south, where often a

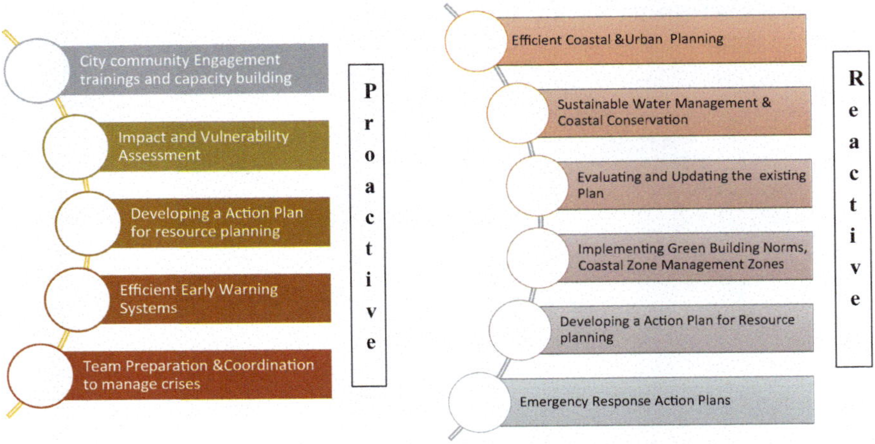

Fig. 2.4 Conceptual representation of road map to climate resilience. *Source* Author generated

lack of access to land and other natural resources, knowledge, and institutions are foremost obstacles to sustainable development (Etzold et al., 2012) (Fig. 2.4).

There should be a combination of proactive and reactive activities to attain resilience Fig. 2.4. Dhanya et al. (2015a) has mentioned various kinds of adaptation to achieve resilience such as proactive, reactive, passive and active. Reactive responses are driven by actual perceived climate induced disasters and proactive strategies can be drawn based on the risk assessments, forewarnings and forecasting future scenarios. Land degradation causes serious implications for biodiversity loss, the most critical global environmental threat (Dhanya et al., 2016). More than a third of the global species are facing extinction and an estimated 60% of the earth's ecosystems have been degraded in the past 50 years, posing adverse consequences for the services that depend on them (Joppa et al., 2016). As both the states are projected to experience severe warming in future, heat wave events and sea level rise may impact the major cities of these states such as Hyderabad and Visakhapatnam. Climate proofing of all the projects in water and agricultural, transport, rural development, disaster management sectors etc. is also required to build smarter cities and communities. As envisioned in the State Action Plan on Climate Change, Telanagana and Andhra Pradesh, mainstreaming climate and sea level rise forecasts into sectoral policies and processes are also essential for achieving green growth and resilient society. Strategic multi-stakeholder -adaptation and mitigation planning has to be carried out periodically as part of resilience building in the state. Context analysis is also an imperative constituent in the urban resilience process. It supports identification of most appropriate processes and means for resilience planning considering the geo-topographical, governance, socio-economic as well as climate elements, unique to an urban space.

2.10 Discussions

Many researchers have extensively reported on global warming and extreme climatic events, including aridity issues across the world (Coumou & Rahmstorf, 2012; Gavin et al., 2014; Golian et al., 2010; Kousari et al., 2011). Due to rising temperatures and increasing population growth, chronic water scarcity may be aggravated in the already resource poor areas of India, causing strong barriers for the healthy functioning of ecosystems (Patwardhan & Narayanan, 2003; Raji & Sreenivas, 2016; Rao, 2012). The capacity building is essential to cope with drastic changes that are happening in natural and cultural landscapes. The vulnerable Socio-Ecological Systems (SES) constantly require resilience building in order to withstand internal and external stresses from socio-economic and political changes (Wilson et al., 2013; Shifteh et al., 2013; Russo et al., 2013; Türkeş & Altan, 2013).

Policy makers face severe challenges in mitigating multiple environmental risks (Alugula et al., 2020; Komendantova et al., 2016). Multi-hazard and risk concepts are at its infancy stage in natural risk governance. In a region that is constantly at the mercy of the gamble of natural and cultural disasters, achieving SDGs and Nationally Determined Contribution (INDCs) are comprehensive trials. As shown in the present study, the positive impacts of future climate change such as more rainfall, wetting trend may put some solace to the southeastern part of the state. The water-stressed arable lands will be benefitted, modifying production patterns and economics in the mid and end of twenty-first century, heat related stress is projected if GHG emission is not curbed. The government of Telangana has undertaken two ambitious projects for addressing problems of drinking water and irrigation water supply namely Mission Bhageeratha and Mission Kakatiya respectively. As far as the newly formed state of Telangana is concerned, Mission Kakatiya (Our Village—Our Tank) may play a vital role in the revival & restoration of a chain of Tanks system was an aspiration for people during the Telangana Movement (Mission Kakatiya GoTS, Report, 2018). Under the Mission Bhageeratha, the policy of the Government is to provide drinking water in all the rural and urban areas by integrating the existing water supply systems in the rural and urban areas with the Grid. The inter-linking of Godavari and Krishna rivers are also proposed, to benefit drought-prone areas in Telangana and Andhra Pradesh. It is aimed at sustainable management of water resources in vulnerable districts of Prakasam, Nellore, Rayalseema districts in Andhra Pradesh governed by Krishna and Godavari river development board.

The commencement of SDGs implementation in India was followed after the establishment of NITI Aayog (successor to the erstwhile Planning Commission). The NITI Aayog (2019) has advised planning departments in the state governments to identify SDG focal points in all key sectors and prepare SDG inclusive budgeting for all departments giving special focus to disasters. By 2030, India is planning to lessen the adverse per capita environmental impact of cities, paying special attention to air quality improvement and municipal waste management. The Government of India's Smart Cities Mission, the Jawaharlal Nehru National Urban Renewal Mission, and the Atal Mission for Rejuvenation and Urban Transformation (AMRUT) are working towards addressing the challenges of improving urban and peri-urban spaces. The prime minister's Pradhan Mantri Awas Yojana also aims to achieve housing for all by 2022, NITI Ayog, (2019). Apart from all these, the state has planned afforestation activities in the dry land areas (not suitable for agriculture) before each Kharif season under the Mahatma Gandhi National Rural Employment Guarantee Act (MNREGA) scheme. The details of the dry land areas in the districts are given to the Zilla Panchayats on spatial maps for spatial planning. As part of resilience building, rejuvenation of existing lakes are also planned in the state. The state government has entered into multiple partnerships in the process of implementing the Sustainable Development Goals on issues encompassing women workforce participation rate and creation of an inclusive Disaster Risk Reduction framework, computing its own Multidimensional Poverty index etc. Authorities in these areas need to plan to build climate resilience. It is the need of the hour to transform existing systems to address multiple hazard risks without compromising on present development challenges. Proactive disaster management and accurate weather forecasting help in reducing the death toll and other impacts from heatwaves, cyclones and climate extremities.

2.11 Conclusions

It is accepted that there are various uncertainties associated both in modeling future climate change and in regional and local economic development and the wellbeing of people. Simulation results shows that the state of Andhra Pradesh is whirling with multi-hazard risks with respect to intensifying temperatures, rising sea levels, aridity, etc. Model predictions provided a more robust regional evaluation of the probable drying trends and sea-level rise. De Martonne Aridity Index for Andhra Pradesh, indicates apparently a slight wetting trend for southeastern parts in the near-century period (2011–2040) with reference to the baseline period (1971–2000. Enhanced research is required to understand how emerging cities and communities behave in the future scenario. Micro-level studies are the indispensable prerequisites for any developmental planning process due to spatial variability in climatic conditions. A research review carried out on the state government's ambitious programmes like Mission Kakatiya, Mission Bhageeratha and river linking projects indicates its huge potential in bringing water resilience to the state. Robust action plans, envisioned to tackle multi hazard risk reductions strategies may meet the water demands of the state. It is significant to have coordinated actions and inclusive growth in line with INDCs and SDGs for all interconnected sectors and departments in the state to accomplish sustainable water resources management.

Acknowledgements The first author would like to thank the Department of Science and technology (DST), Government of India for the support under the scheme under National Mission on Strategic Knowledge Management and Climate Change (SCCC-NMSKCCC) in the State Climate Change Centre in the Government of Telangana.

References

Berkes, F., & Ross, H. (2013). Community resilience: Toward an integrated approach. *Society and Natural Resources, 26*, 5–20. https://doi.org/10.1080/08941920.2012.736605

Boyaj, A., Dasari, H. P., Hoteit, I., & Ashok, K. (2020). Increasing heavy rainfall events in South India due to changing land use land cover. *Quarterly Journal of Royal Meteorological Society, 146*(732). https://doi.org/10.1002/qj.3826

Brown, S., Nicholls, R. J., Lázár, A. N., et al. (2018). What are the implications of sea-level rise for a 1.5, 2 and 3°C rise in global mean temperatures in the Ganges-Brahmaputra-Meghna and other vulnerable deltas? *Regional Environmental Change, 18*, 1829–1842. https://doi.org/10.1007/s10113-018-1311-0

Byers, E., et al. (2018). Global exposure and vulnerability to multi-sector development and climate change hotspots. *Environmental Research Letters, 13*(5), 055012. https://doi.org/10.1088/1748-9326/aabf45

Census of India. (2011). http://censusindia.gov.in/

Chaturvedi, R. K., Joshi, J., Jayaraman, M., Bala, G., & Ravindranath, N. H. (2012). Multi model climate change projections for India under representative concentration pathways (RCPs): A preliminary analysis. *Current Science, 103*, 791–802.

Coumou, D., & Rahmstorf, S. (2012). A decade of weather extremes. *Nature Climate Change, 2*, 491–496. https://doi.org/10.1038/NCLIMATE1452

De Vente, J., Chotte, J. L., Bernoux, M., Kust, G., Sanz, M. J., Ruiz, I., Almagro, M., Alloza, J. A., Vallejo, R., Castillo, V., Hebel, A., Akhtar-Schuster, M. (2017). *Sustainable land management contribution to successful land-based climate change adaptation and mitigation*. A report of the science-policy interface. United Nations Convention to Combat Desertification (UNCCD).

Dhanya, P., & Ramachandran, A. (2015a). Farmers' perceptions of climate change and the proposed agriculture adaptation strategies in a semi-arid region of South India. *Journal of Integrative Environmental Sciences, 13*(1), 2016. https://doi.org /https://doi.org/10.1080/1943815X.2015.1062031

Dhanya, P., Ramachandran, A., & Palanivelu, K. (2016). The urgent call for land degradation vulnerability assessment for conserving land quality in the purview of climate change: Perspective from South Indian coast. Land management practices for soil conservation in climate change scenarios. *AIMS Agriculture and Food, 1*(3) 330–341. https://doi.org/10.3934/agrfood.2016.3.330

Ekardt, F. (2020). *Politics and governance of sustainability, sustainability, Environmental humanities; transformation, governance and ethics law* (p. 225). Springer Nature.

EM-DAT9. The international disaster database. https://www.emdat.be/. Accessed on December 2019.

Etzold, B., Jülich, S., Keck, M., Sakdapolrak, P., Schmitt, T., & Zimmer, A. (2012). Doing institutions. A dialectic reading of institutions and social practices and its relevance for development geography. *Erdkunde, 66*, 185–195. https://doi.org/10.3112/erdkunde.2012.03.01

Evans, J., & Geerken, R. (2004). Discrimination between climate and human-induced dry land degradation. *Journal of Arid Environments, 57*, 535–554.

Folke, C. (2006). Resilience: The emergence of a perspective for social-ecological systems analyses. *Global Environmental Change, 16*, 253–267. https://doi.org/10.1016/j.gloenvcha.2006.04.002

Folke, C., Carpenter, S. R., Walker, B., Scheffer, M., Chapin, T., & Rockström, J. (2010). Resilience thinking: Integrating resilience, adaptability and transformability. *Ecology and Society*, 15(4),20. [online] http://www.ecologyandsociety.org/vol15/iss4/art20/

Gavin, A. S., Drew, T. S., & Kostas, T. (2014). Reconciling warming trends. *Nature Geoscience, 7*, 158–160. https://doi.org/10.1038/ngeo2105

Golian, S., Saghafian, B., Sheshangosht, S. (2010). Comparison of classification and clustering methods in spatial rainfall pattern recognition at Northern Iran. *Theoretical and Applied Climatology, 3*(110), 319–329. Ground water resources. *Global Environmental Change*, 16(1), 388–399.

Guleria, S., & Gupta, A. K. (2018). *Heat wave in India documentation of state of Telangana and Odisha (2016)* (p. 124). National Institute of Disaster Management.
IMD Data obtained through https://www.imdtvm.gov.in/index.php?option=com_content&task=view&id=29&Itemid=43
IPCC. (2007a). Climate Change 2007. *The physical science basis, contribution of working group I to the fourth assessment report of the Intergovernmental Panel on Climate Change* (p. 167). Cambridge University Press.
IPCC. (2012). *Emergent risks and key vulnerabilities*, Final draft, Chapter 19 [online] http://ipccwg2.agov/AR5/images/uploads/WGIIAR5-Chap19_FGDall.pdf
IPCC. (2013a). *Climate change 2013: The physical science basis. Contribution of working group I to the fifth assessment report of the Intergovernmental Panel on Climate Change.* Cambridge University Press.
IPCC. (2013b). *Terrestrial and inland water systems*, Final draft, Chapter 4 (online) http://ipccwg2
IPCC. (2014). Climate change 2014: Impacts, adaptation, and vulnerability. In C. Field, V. Barros, K. Mach & M. Mastrandrea (Eds.), *Contribution of working group II to the fifth assessment report of the intergovernmental panel on climate change.* Cambridge University Press.
Joppa, L. N., O'Connor, B., Visconti, P., Smith, C., Geldmann, J., Hoffmann, M., & Burgess, N. D. (2016). Filling in biodiversity threat gaps. *Science, 352*, 416–418.
Komendantova, N., Scolobig, A., Garcia-Aristizabal, A., Monfort, D., & Fleming, K. (2016). Multi-risk approach and urban resilience. *International Journal of Disaster Resilience in the Built Environment, 7*, 114–132. https://doi.org/10.1108/IJDRBE-03-2015-0013
Kousari, M. R., Ekhtesasi, M. R., Tazeh, M., & AsadiZarch, M. A. (2011). An investigation of the Iranian climatic changes by considering the precipitation, temperature, and relative humidity parameter. *Theoretical Applied Climatology, 103*(3–4), 321–335.
Kumar, T. S., Mahendra, R. S., Nayak, S., Radhakrishnan, K., & Sahu, K. C. (2010). Coastal vulnerability assessment for Orissa state, East coast of India. *Journal of Coastal Research, 26*, 523–534.
Maiti, S., Jha, S. K., Garai, S., et al. (2015). Assessment of social vulnerability to climate change in the eastern coast of India. *Climatic Change, 131*, 287–306. https://doi.org/10.1007/s10584-015-1379-1
Mani Murali, R., Ankita, M., Amrita, S., & Vethamony, P. (2013). Coastal vulnerability assessment of Puducherry coast, India, using the analytical hierarchical process. *Natural Hazards and Earth System Sciences, 13*(12), 3291–3311.
McCarty, J. P., Wolfenbarger, L. L., Wilson, J. A. (2009). Biological impacts of climate change. In *Encyclopedia of life sciences (ELS)*. Wiley https://doi.org/10.1002/9780470015902.a0020480
Mishra, V. et al. (2017). Heat wave exposure in India in current, 1.5°C, and 2.0°C worlds. *Environmental Research Letters, 12*(12), 124012. https://doi.org/10.1088/1748-9326/aa9388
Mission Kakatiya Report. (2018). https://missionkakatiya.cgg.gov.in/, https://tweb.telangana.gov.in/initiatives/mission-kakatiya/
Nageswararao, M. M., Sinha, P., Mohanty, U. C., et al. (2020). Occurrence of more heat waves over the central east coast of India in the recent warming era. *Pure and Applied Geophysics, 177*, 1143–1155. https://doi.org/10.1007/s00024-019-02304-2
Nanditha, J. S., van der Wiel, K., Bhatial, U., Stone, D., Selton, F., & Mishra, V. (2020). A seven-fold rise in the probability of exceeding the observed hottest summer in India in a 2 °C warmer world. *Environmental Research Letters, 15*, 044028.
NASA's Report, *Amphan cyclone.* (2020). https://blogs.nasa.gov/hurricanes/2020/05/21/amphan-northern-indian-ocean-5/
NITI Ayog Report. (2019). *Localising SDGs, the early lessons from India.* http://niti.gov.in/writereaddata/files/LSDGs_July_8_Web.pdf
Pattanaik, D. R., Mohapatra, M., Srivastava, A. K., & Kumar, A. (2017). Heat wave over India during summer 2015: An assessment of real time extended range forecast. *Meteorology and Atmospheric Physics, 129*, 375–393. https://doi.org/10.1007/s00703-016-0469-6

Patwardhan, A., & Narayanan, K. (2003). Assessment of vulnerability of Indian coastal zones to climate change, conference on water resources, coastal zones and human health.

Praveen, K. & Udayakumar. (2014). Effects of sea level change on vulnerable east coast of India. *Research Journal of Marine Sciences*, *2*(1) 1–5, 2321–1296, International Science Congress Association 1.

Raji Reddy, D., & Sreenivas, G. (2016). Climate change impact on agriculture in India. *International Journal of Economic Plants, 3*(1), 23–26.

Ramachandran, A., Praveen, D., Jaganathan, R., & Palanivelu, K. (2015). Projected and observed aridity and climate change in the east coast of South India under RCP 4.5. *The Scientific World Journal, 2015*, 11, Article ID 169761. https://doi.org/10.1155/2015/169761

Ranjan, P., Kazama, S., & Sawamoto, M. (2006). Effects of climate change on coastal fresh ground water resources. *Global Environmental Change, 16*(1), 388–399.

Rao, V. U. M. (2012). Impacts of climate change on Indian agriculture. http://www.cseindia.org/userfiles/VUM-CSE-Delhi-16-11-12.pdf. Accessed 6 July 2015.

SAPCC, Andhra Pradesh. (2014). State Action Plan on Climate Change (2014). http://moef.gov.in/wp-content/uploads/2017/08/Andhra-pradesh.pdf. Accessed on 30th August 2017.

Sea Level Rise projections obtained from https://tidesandcurrents.noaa.gov/sltrends/& http://www.psmsl.org/data. Accessed on October 25 2017.

Shamsuddoha, M., & Chowdhury, R. K. (2007). Climate change impact and disaster vulnerabilities in the coastal areas of Bangladesh. *Report prepared by Coast Trust and Equity and Justice Working Group (EJWG).*

Russo Some'e, B., Ezani, A., & Tabari, H. (2013). Spatiotemporal trends of aridity index in arid and semi-arid regions of Iran. *Theoretical and Applied Climatology, 111*, 149–160. https://doi.org/10.1007/s00704-012-0650-x

Simone, R. et al. (2015). Top ten European heat waves since 1950 and their occurrence in the coming decades. *Environmental Research Letters, 10*(12), 124003. https://iopscience.iop.org/article

Taylor, K. E., Stouffer, R. J., & Meehl, G. A. (2012). An overview of CMIP5 and the experiment design. *Bulletin of the American Meteorological Society, 93*(1). https://doi.org/10.1175/BAMS-D-11-00094.1

Türkeş, M., & Altan, G. (2013). An ecological biogeography analysis of the impacts of climatic variations and forest fires on the natural environment, nature protection areas and biota in the Muğla district. *EgeCoğrafyaDergisi, 22*(2), 57–76.

United Nations Convention to Combat Desertification (UNCCD). (2012). http://www.unccd.int/ActionProgrammes/india-eng2001.pdf, report. Accessed on September 23 2014.

United Nations Convention to Combat Desertification, UNCCD-SPI. (2016). Land in balance. The scientific conceptual framework for land degradation neutrality. Science-Policy Brief 02. United Nations Convention to Combat Desertification, Science-Policy Interface (UNCCD-SPI). http://www.unccd.int/Lists/SiteDocumentLibrary/Publications/10_2016_spi_pb_multipage_eng.pdf

Watts, N., Adger, W. N., Agnolucci, P., et al. (2015). Health and climate change: Policy responses to protect public health. *Lancet, 2015*(386), 1861–1914.

Watts, N., Amann, M., Ayeb-Karlsson, S., et al. (2018). The Lancet countdown on health and climate change: From 25 years of inaction to a global transformation for public health. *Lancet, 2018*(391), 581–630.

Wilson, S., Pearson, L. J., Kashima, Y., Lusher, D., & Pearson, C. (2013). Separating adaptive maintenance (resilience) and transformative capacity of social-ecological systems. *Ecology and Society, 18*(1), 22. https://doi.org/10.5751/es-05100-180122

Chapter 3
Climate Change and Agriculture in Rajasthan: Present Perspective and Future Strategies

Alok Chauhan and S. K. Kulshrestha

3.1 Introduction

The rapid changes in global climate over recent decades have indeed had profound impacts on various aspects of our world. These changes have exposed vulnerabilities in social, economic, and environmental systems, placing our living and physical environments at significant and multifaceted risks (Solomon, 2007; Stocker et al, 2013). In India, climate change is indeed exerting significant impacts on various weather parameters, particularly affecting the monsoon timing, temperature patterns, and overall atmospheric conditions. These changes have profound implications for both natural ecosystems and human livelihoods across the country (Fulekar & Kale, 2010). Climate change poses significant risks to smallholder farmers in developing economies, where agriculture is often a primary source of livelihood for rural populations. The impacts of climate change on agriculture exacerbate existing challenges and introduce new risks that threaten food security, economic stability, and overall well-being of these vulnerable communities. (Singh, 2009). Climate indeed plays a crucial role in determining the type of agriculture that can be practiced in a region. Changes in temperature, rainfall patterns, and extreme weather events can have significant impacts on agriculture, affecting crop productivity and soil quality. In countries like India, where a large portion of agriculture is rainfed, climate change poses a serious concern. Even small changes in climate parameters can lead to fluctuations in agricultural production. It is important for farmers and policymakers to adapt to these changing climatic conditions to ensure food security and sustainable agriculture

A. Chauhan (✉) · S. K. Kulshrestha
Department of Geography, School of Humanities and Social Sciences, Vardhman Mahaveer Open University, Rajasthan Kota 324010, India
e-mail: achauhan@vmou.ac.in

S. K. Kulshrestha
e-mail: skulshrestha@vmou.ac.in

P. K. Rai and S. Rai (eds.), *Climate, Environment and Agricultural Development*, Advances in Geographical and Environmental Sciences,
https://doi.org/10.1007/978-981-97-8363-2_3

practices (Mayowa et al., 2015; Sa'adi et al., 2019). As per the Intergovernmental Panel on Climate Change (IPCC, 2018), greenhouse gas (GHG) accumulation owing to increased anthropogenic emissions has caused 1.0 °C of global warming above pre-industrial levels, which is likely to reach 1.5 °C between 2030 and 2052, causing greater frequency of extreme weather events (droughts, floods, and heat waves). In India, a number of efforts have been made to investigate the relationship between climate change and agriculture in India (Abrol & Sangar, 2006; Aggarwal, 2008; Begum & Mahanta, 2017; Birthal et al., 2014; Dinar et al., 1998; Gupta et al., 2014; Mall et al., 2006; Rao et al., 2016; Singh et al., 2020; Tripathi & Mishra, 2017). Climate change impacts like drought and desertification can have severe consequences on agriculture, particularly in states like Rajasthan in India. The varying climate in Rajasthan, from arid in the west to sub-humid in the east and south, makes the region particularly sensitive to climate change. Agriculture in arid and semi-arid regions heavily relies on rainfall, making it more vulnerable to changing climate patterns. It is crucial for policymakers and stakeholders in these regions to implement strategies to mitigate the impacts of climate change on agriculture and ensure the sustainability of farming practices. Agricultural practices in Rajasthan can indeed be challenging due to the harsh and dry climate prevalent in most parts of the state (Chand & Raju, 2009). The heavy reliance of agriculture in Rajasthan on monsoon rainfall indeed makes the region more vulnerable to the impacts of climate change. Variability in monsoon patterns, including changes in timing, intensity, and distribution of rainfall, can significantly affect crop yields and agricultural productivity in Rajasthan (Singh, 1994; Singh & Kumar, 2013). There are, however, very few researches on the patterns and trends of rainfall as well as associated extreme events that are relevant to India's dryland ecology, specifically the state of Rajasthan.

The state of Rajasthan has a monsoon climate, with the south west monsoon providing the majority of the state's yearly rainfall. Agriculture in many sections of the state is threatened by the irregular and insufficient rainfall. According to Pingale et al. (2014), the rainfall distribution in Rajasthan's dryland environment is extremely irregular and varies greatly from area to region and year to year. Furthermore, the state has been known for having the highest likelihood of drought and rainfall deficits (Rathore, 2005). Rainfall is extremely significant to the area because it replenishes groundwater. The state's water resources and agriculture may be impacted by long-term rainfall-related climatic changes (Yadav et al., 2018). In order to better understand the rainfall and associated extreme events for this state, several researches have been made (Machiwal & Jha, 2017; Meena et al., 2019; Pingale et al., 2014; Poonia et al., 2014; Rao et al., 2011; Singh et al., 2012; Singh & Kumar, 2016; Srivastava & Pradhan, 2018). These studies, however, are limited to the examination of a specific district, western arid region, rainfall event, and brief time frame of data. A more comprehensive region-by-region plan is urgently needed because the rainfall patterns in Rajasthan differ, just as the state's climate varies. As a result, a single contingency plan for agriculture sustainability cannot be formulated for the entire state. Based on research findings, Rajasthan is located in regions with the highest degree of climate sensitivity, maximum vulnerability, and lowest degree of adaptability. The State already has limited water resources, which are distributed very unevenly in terms of both time and space. Additionally, the State has the highest chance of experiencing a drought nationwide. Therefore, a threat like climate change necessitates prompt and cogent policy responses and actions that will help the state become less vulnerable to expected climate impacts and more resilient to them. There hasn't been a study that looks at temperature and rainfall variations on the basis of agro-climatic zones in detail across the Rajasthan state's dryland ecology in northwestern India. For this, a thorough examination of the unpredictability of recent rainfall and temperature change is necessary. Given these details, the primary goal of this research is to investigate the variability of temperature and rainfall in all the agro-climatic zones of Rajasthan during 1990–2019.

The study also examines the change in agriculture pattern of the state caused by change in the climate. The study also suggests the ways to mitigate the impacts of climate change.

3.2 Objectives

The objectives of the study are:

(i) To analyze the present climate change in Rajasthan,
(ii) To assess the interrelationship between climate change and agriculture in Rajasthan and
(iii) To suggest some adaption and mitigation options to deal with the consequences of climate change.

3.3 Study Area

The Indian state of Rajasthan is located between 23°3' and 30°12' north latitude and 69°30' and 78°17' east longitude in the northwest of the country. It makes up 10.41% (342,239 km2) of the nation's total land area. It is the biggest state in India and the one where the percentage of area covered by deserts is highest. 33 districts, comprising 7 divisions, 244 tehsils, 249 panchayat sammitees (Blocks), 9168 g panchayats, and 43,264 inhabited villages make up the state (Fig. 3.1).

Physiographically, the state may be divided into four primary regions: (i) the western desert, which consists of stony plains, sandy plains, and barren hills; (ii) the Aravalli hills, which travel from south-west to northeast and from Gujarat to Delhi. (iv) the eastern plains' rich alluvial soils and (iii) the plateau in the southeast. The Mahi, Chambal, and Banas are the three principal rivers in the state. Sandy plains make up over 62% of the state's total area, therefore, it is known as the Desert State of India. The diagonally spreading Aravalli hills across the state define the eastward geomorphic and climatic border of the desert. The state's main water divide is the Aravalli Range.

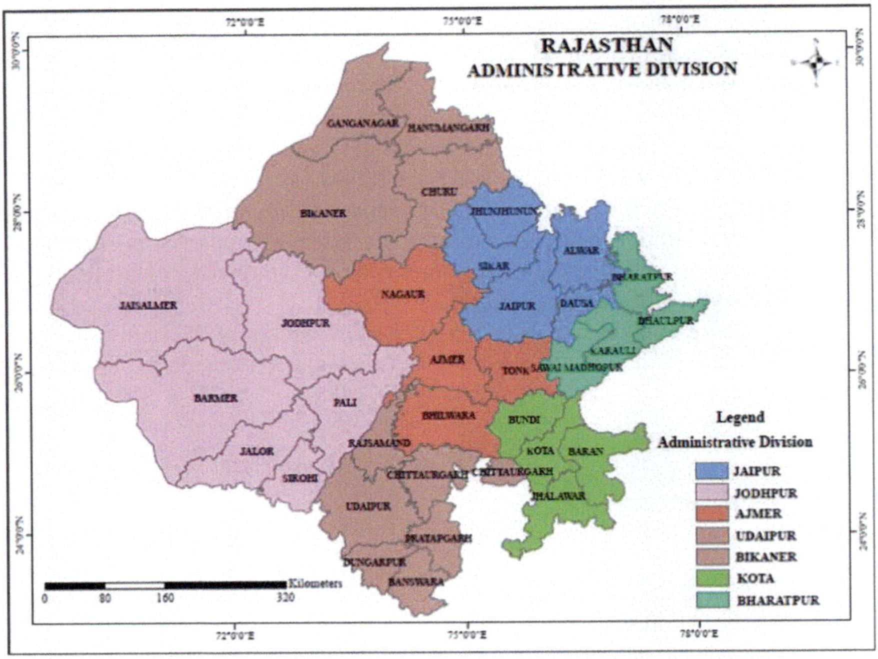

Fig. 3.1 Rajasthan-Administrative divisions

High wind speeds, low humidity, daily and annual temperature extremes, and sporadic, uneven rainfall are the main features of the climate. According to Climatological tables of observatories in India, 1991–2020 (Indian Meteorological Department), the state receives 58.3 mm of rain on average annually. Rainfall is the primary source of irrigation for agriculture in Rajasthan, where canals irrigate 24.47% of the area and wells and tube wells irrigate 73% of the land (State Agriculture Plan (SAP) and State Agriculture Infrastructure Development Plan (SAIDP)—Rajasthan State—2017–2019). The state is the biggest producer of mustard, isabgol, cluster beans, cumin, and fenugreek in India, as well as the second-largest producer of maize. It also produces the most pearl millet, or Bajra (Department of Agriculture, Government of Rajasthan). To support the agriculture sector's sustainable development, the state must methodically address the numerous obstacles it faces.

Topography and climate have a major impact on agricultural methods. Climate variables like as rainfall, temperature, humidity, wind speed, and sunshine length have a big impact on cropping patterns in a given area. The Indian Council of Agricultural Research has classified the nation into 126 agro climatic zones, of which the state is divided into ten (Table 3.1 and Fig. 3.2). Agro-climatic factors such as rainfall, temperature patterns, terrain, soil properties, cropping patterns, and irrigation accessibility have all been taken into consideration when classifying these zones.

Table 3.1 Agro-climatic zones of Rajasthan

Zone	Agro-climatic zone	District	Rainfall (mm)	Climate	Soil	Important crops grown
I a	Arid western plains zone	Jodhpur Barmer	100–370	Arid	Aridisol, Entisol	Pearl-millet, Green Gram, Moth Bean, Cluster Bean, Ber, Lasoda and Ker
I b	Irrigated north western plain zone	Hanumangarh Sriganganager	100–350	Arid	Aridisol, Entisol	Wheat, Gram, Mustard, Cotton, Kinnow, Citrus
I c	Hyper arid and partially irrigated western plain	Bikaner Jaisalmer Churu	100–350	Arid	Aridisol, Entisol	Pearl millet, green gram, moth bean, cluster bean, gram, ber, datepalm and aonla
II a	Transitional plain of inland drainage	Sikar Nagaur Jhunjhunu	300–500	Semi-Arid	Aridisol, Entisol	Pearl millet, green gram, moth bean, cluster bean, groundnut, wheat, barley, ker, lasoda
II b	Transitional plain of Luni basin	Jalore Sirohi Pali	300–500	Semi-Arid	Aridisol, Entisol, Hills and inceptisols	Cumin, green gram, moth bean, cluster bean, wheat, barley, cotton, gram, pomegranate, aonla
III a	Semi-arid eastern plain	Jaipur Ajmer Tonk Dausa	500–700	Semi-Arid	Entisols and alluvium soils	Groundnut, jowar, mustard, gram, green gram, moth bean, wheat and barley, ber, citrus
III b	Flood prone eastern plain zone	Alwar Bharatpur Dholpur	500–700	Semi-Arid	Inceptisols, Entisols and Vertisols	Mustard, pigeon pea, gram, wheat, barley, mango, citrus

(continued)

Table 3.1 (continued)

Zone	Agro-climatic zone	District	Rainfall (mm)	Climate	Soil	Important crops grown
IV a	Sub humid southern plain & Aravalli hill zone	Bhilwara Chittorgarh Rajsamand Udaipur	500–900	Sub-Humid	Inceptisols, Entisols, Vertisols and Hills	Groundnut, jowar, maize, gram, green gram cotton, mustard, black gram, and soybean, papaya, guava
IV b	Humid southern plain	Banswara Dungarpur	525–1000	Humid	Vertisols, Alfisols and hills	Cotton, soybean linseed, maize paddy, wheat, gram, mustard, guava
V	Humid south eastern plain	Kota Bundi Jhalawar Baran S. Madhopur Karauli	650–1000	Humid	Vertisols	Sorghum, maize, cotton, paddy, sugarcane, soybean, wheat, gram, mustard. Guava, citrus

Source Department of Agriculture, Govt. of Rajasthan-2024 https://agriculture.rajasthan.gov.in/agriculture/#/sm/jankalyan-category-and-entry-type/48843/5/72/382

3.4 Research Methodology

The study is totally based on secondary data. Data (1990–2019) in respect of rainfall, maximum temperature, minimum temperature and rainy days obtained from India Meteorological Department (IMD). Data on crop production growth rate has been taken from 58 Years Agricultural Statistics of Rajasthan (1956–57 to 2013–14), Agricultural Statistics-2019–20, Statistical Year Book of Rajasthan 2023, Directorate of Economics and Statistics Department of Planning Rajasthan, Jaipur. Information about Ground water position has been sourced from Central Ground Water Board, Western Region, Jaipur. The climate variability is analyzed on both temporal and spatial dimensions. Moving averages method has been employed to determine the trend of climatic parameters over past 30 years. The spatial change in climate parameters has been deduced through combination of interpolation and change detection technique using GIS software. Similarly, the changes in climate regions as classified by Thornthwaite are determined using the same set of software. The data thus obtained have also been represented and analyzed through tables, graphs and charts.

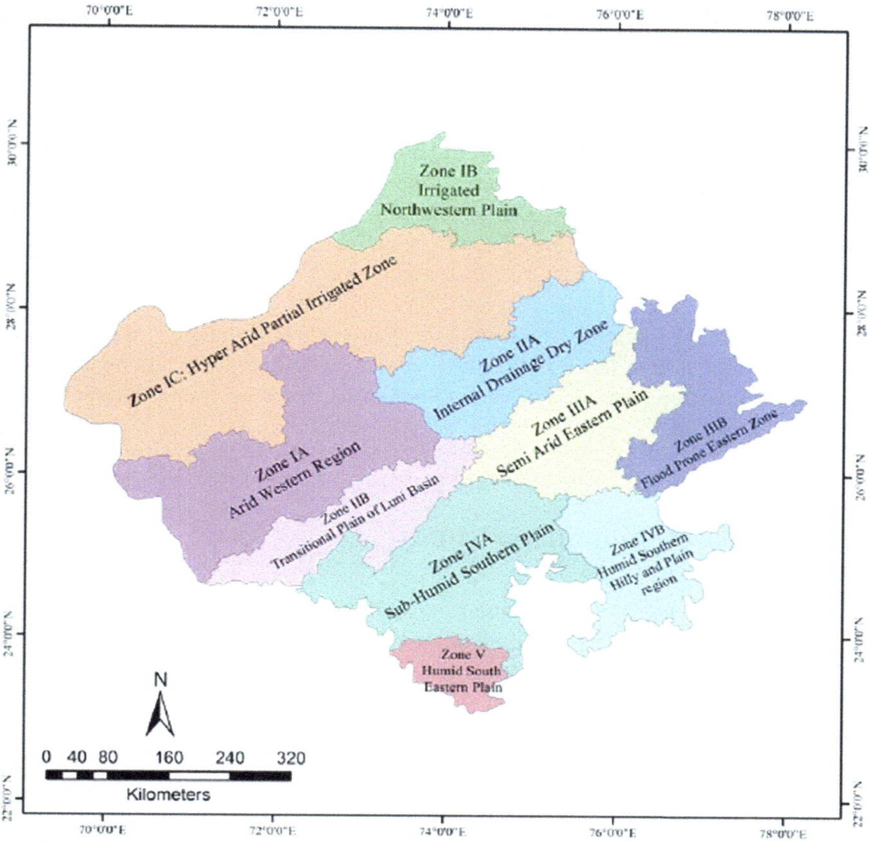

Fig. 3.2 Agro-climatic zones of Rajasthan

Regression, correlation coefficient of variance techniques have also been used wherever necessary. For such analysis SPSS-28 software has been used. For calculating crop productivity, crop combination and crop diversification appropriate formulas have been used which have been discussed in the respective section of the study.

3.5 Analysis and Findings

Change in temperature and rainfall variability in all the agro-climatic zones of the state during 1990–2009 has been discussed in the following paragraphs in sequel.

3.5.1 Change in Temperature

Decadal mean temperature of Rajasthan according to agro-climatic zone has been presented in Table 3.2. It is clear from the table that average decadal temperature is continuously increasing in all the agro-climatic zones since 1990. Highest increase I temperature was observed in Arid western plain-zone I A with an increase of 2.7 °C during 1990–1999 to 2010–2019. During the same period least change in temperature 0.5 °C was registered in Sub-humid southern plains-IVA. During the period of last thirty year lowest increase (0.1 °C) in temperature was registered in Sub-humid southern plains-zone IVA.

It is evident that the temperature has risen in each of the state's agro-climatic zones. This implies that during last thirty years, the temperature has been rising. The state's increasing trend in temperature across its major agro-climatic regions is shown in Fig. 3.3.

3.6 Analysis of Rainfall Variability

Rainfall variability refers to the fluctuations in the amount and distribution of rainfall over time and across a geographical area. It is a crucial aspect of climate that directly impacts water availability and is instrumental in understanding climate change patterns. By studying rainfall variability, researchers can gain insights into climate trends and develop strategies for adapting to future changes in precipitation patterns. Variability in rainfall influences water resources, affecting the occurrence of floods and droughts, and altering the nature of rainfall itself. In regions like Rajasthan, where agriculture relies heavily on rainfed practices, variability in rainfall

Table 3.2 Decadal mean temperature of Rajasthan: Agro-climatic zone wise (1990–2019)

Agro-climatic zone	1990–1999	2000–2009	2010–2019
Arid western plain- I A	25.9	28.1	28.6
Irrigated north western plain-I B	25.1	26	27.3
Hyper arid partial irrigated zone-I C	26.2	27.4	27.9
Internal drainage dry zone-II A	24.2	26	26.6
Transitional plain of Luni basin-II B	24.4	26	26.5
Semi arid eastern plains-III A	25.1	26.4	26.8
Flood prone eastern plain-III B	25.9	26.6	26.8
Sub-humid southern plains-IVA	24.7	24.8	25.2
Humid southern plains-IV B	24.8	26.2	26.7
Humid south eastern plain-V	25.9	27	27.1

Source Calculated by the authors from Climatological Normals (1981–2010) and Climatological Tables (1991–2020) Indian Meteorological Department

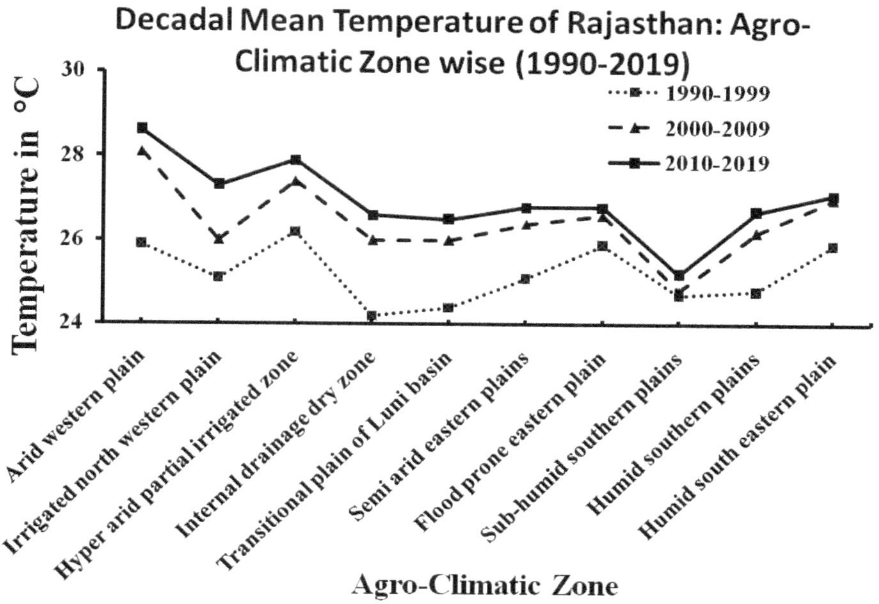

Fig. 3.3 Agro-climatic zones wise decadal average temperature (1990–2019)

can significantly impact crop growth and yields, particularly for crops with shallow root systems. Therefore, managing and adapting to rainfall variability is essential for sustainable agricultural practices and livelihood security in such areas.

Here, in this section rainfall variability of the state has been analyzed. This may be considered as the large scale variability analysis. The time span taken is of 30 years starting from 1990 up to 2019. The coefficient of variation (CV) which is expressed in percentage is defined as below:

$$\text{C.V.} = \text{Standard deviation } (\sigma) \times 3.100/ \text{ Normal Annual Rainfall (N)}$$

where, $N = 58.3$ and the standard deviation and CV are tabulated in Table 3.3.

From 1990 to 2019, there has been a decline in the amount of rainfall in the state (IMD, 2020). Zone IVB has shown the greatest drop followed by Zone V and IIIB. There is a drop in both the amount of rainfall and the number of rainy days. Zones IB and IC experienced the least reduction in rainfall between 1990 and 2019 (IMD, 2020). The state's CV was 13.83% during 1990 to 1999 and increased to 15.18% during 2010–2019. CV of zone-IA, which was 20.02%, was at its greatest in the 30 years between 2010 and 2019. During 1990–1999 and 2000–2009 the highest increase in variability of rainfall was recorded in zone-IA and lowest in zone-IIIA (Table 3.3). Hyper arid partial irrigated zone-I C registered highest increase in rainfall variability during 2000–2009 and 2010–2019. Zone wise coefficient of variation of rainfall has been depicted in Fig. 3.4.

Table 3.3 Decadal rainfall variability in Rajasthan: Agro-climatic zone wise (1990–2019)

Agro-climatic zone	1990–1999	2000–2009	2010–2019
Arid western plain-I A	16.64	18.89	20.02
Irrigated north western plain-I B	13.42	14.33	15.43
Hyper arid partial irrigated zone-I C	15.23	17.24	19.64
Internal drainage dry zone-II A	13.1	14.97	15.13
Transitional plain of Luni basin-II B	17.63	17.89	18.62
Semi arid eastern plains-III A	14.75	13.26	12.87
Flood prone eastern plain-III B	10.15	11.62	11.95
Sub-humid southern plains-IVA	15.44	16.23	16.89
Humid southern plains-IV B	12.5	11.68	10.78
Humid south eastern plain-V	9.5	10.31	10.52

Source Calculated by the authors from Climatological Normals (1981–2010) and Climatological Tables (1991–2020) Indian Meteorological Department

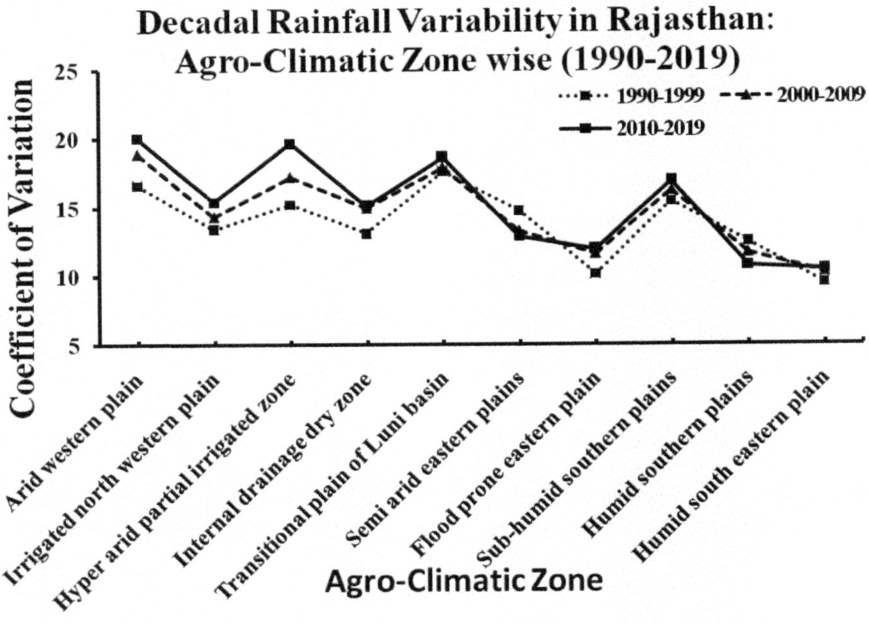

Fig. 3.4 Coefficient of variation of rainfall (1990–2019)

It is noted that during the course of the observed 30 years Humid south eastern plain zone-V registered the lowest coefficient of variation of rainfall. Additionally, it is noted that during the course of the observed 30 years, the CV values varied from as high as 20.02% to as low as 9.50%.

In Rajasthan, rainfall exhibits significant variability both annually and spatially, posing challenges to water availability and agricultural productivity. The state's average annual rainfall is typically meager at 58.4 mm. However, between 2010 and 2019, the highest average annual rainfall occurred in 2011, contrasting sharply with the lowest in 2017. In 2019, the average annual rainfall increased notably, reaching 829.4 mm, which was 29.5% above the normal annual rainfall. This variability underscores the unpredictable nature of rainfall in the region, impacting water resources and agriculture, where rainfed practices dominate. Adaptation strategies are crucial to mitigate the effects of such fluctuations on livelihoods and sustainability in Rajasthan.

Rainfall variability, spanning from annual to daily scales, is a fundamental aspect of climate that goes beyond total precipitation amounts. Low rainfall values alone do not always signify drought, nor do high values guarantee flooding. Understanding this variability helps characterize regional climates and provides insights into climate change impacts. Studying rainfall patterns and variability is essential for developing strategies to adapt to future climate changes. This knowledge informs adaptation efforts related to water resource management, flood and drought preparedness, and understanding shifts in rainfall patterns. By addressing current challenges posed by rainfall variability, societies can better prepare for and mitigate the impacts of future climate changes on various facets of life and the environment.

3.7 Climate Change

Thornthwaite's classification of climatic regions in Rajasthan from 1990 to 2019 has been analyzed by mapping shifts in climate zones based on several indices. These indices include potential evapotranspiration (PET), heat index (I), moisture index (MI), and thermal index (TI). PET, calculated from mean monthly temperatures and day length, measures potential water loss from the soil and vegetation. The heat index (I) considers monthly mean temperatures adjusted for daylight hours, influencing climate categorization. The moisture index (MI), derived from PET and precipitation (P), assesses water balance, while the thermal index (TI) averages daily temperatures to gauge thermal conditions. These indices collectively inform Thornthwaite's classification, which divides regions based on humidity into five distinct categories, providing insights into climate shifts and implications for adaptation strategies in Rajasthan over the studied period.

In Rajasthan there are three regions of climate as per Thornthwaite classification. The Thornthwaite-defined humidity provinces have moved eastward due to changes in precipitation patterns and a general increase in temperature throughout the state. Between 1990 and 2020, the western portion of Rajasthan, as well as a few isolated areas in the eastern and southern regions, was covered by an arid climatic zone (Fig. 3.5). The Aravali region was in the semi-arid zone. The sub-humid zone covered the southern and southeast regions of Rajasthan. When comparing 1990 to 2019, the climatic zones moved. The semi-arid areas of Pali, Jalore, and the sub-humid zone

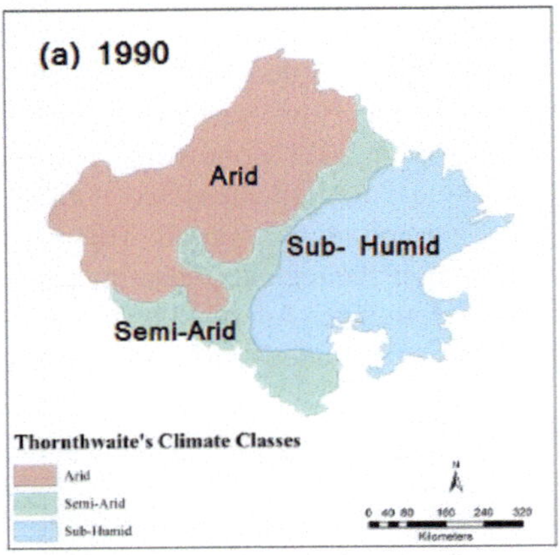

Fig. 3.5 Change in climatic zones (1990–2020)

of Sirohi were all covered by the arid zone. Both the north- and south-western halves are currently located in a semi-arid area. The introduction of canal irrigation in the state's northwest is primarily responsible for the change. In the northern section, the sub-humid zone stretched along the Aravali range (Fig. 3.5).

3.8 Assessment of Agricultural Variability

Change in temperature and rainfall mediated by climate change and anthropogenic impacts may threaten the sustainability of agriculture. Agriculture is highly dependent on the spatial and temporal distribution of rainfall in Rajasthan. Here, attempt has been made to assess the impact of climate change on agriculture by comparing various parameters of agricultural production and development, e.g., Productivity, Crop combination Crop diversification and Agriculture efficiency.

3.8.1 Agricultural Productivity

By analyzing the data obtained from 58 Years Agricultural Statistics of Rajasthan (1956–57 to 2013–14) and Agricultural Statistics-2019–20, it has been observed that productivity of major kharif crops has declined from 1990 to 2019. Crop efficiency has been used to analyze the temporal change in productivity. Crop efficiency has been obtained through dividing yield of a particular crop by area under that particular crop. It has been noticed that decline in kharif pulses have been declined invariably across the state, a sharp decline was observed in IIB (2006.03 to 851.47), IVA (1053.66 to 324.70) and IVB (2121.63 to 13.60). Besides, productivity of Maize have also shown

Table 3.4 Crop efficiency of major Kharif crops

Climatic-zone	Sorghum		Pearl millet		Maize		Kharif pulses	
	1990–91	2019–20	1990–91	2019–20	1990–91	2019–20	1990–91	2019–20
I A	101.26	92.18	3217.67	2563.84	3.44	0.12	216.03	130.49
I B	596.83	0.63	962.96	1172.53	45.39	0.36	63.23	20.95
IC	63.13	11.02	1661.83	1212.92	0.18	0	182.14	44.28
II A	485.73	176.88	3060.68	5014.09	13.43	1.05	228.97	235.9
II B	678.21	687.31	2465.01	3104.23	274.86	209.77	2006.03	851.47
III A	1458.28	1507.26	2335.38	5017.26	860.56	380.6	514.98	213.69
III B	1724.17	1248.65	5070.62	12,583.65	227.45	142.7	335.9	330.69
IVA	617.62	738.23	514.13	183.21	4886.95	6145.67	1053.66	324.7
IV B	75.78	31.03	39	2.46	3280.95	4526.96	2121.63	13.6
V	9123.19	850.76	120.47	68.24	1030.02	1891.39	491.38	155.47

Source Calculated by researcher based on data from agricultural statistics of Rajasthan, Directorate of Economics & Statistics, Department of Planning, Govt of Rajasthan, 1990–2020

a downward trend in almost all region, exception being IVA, IVB and V, which have registered a quite decent rise in productivity owing to suitable physical conditions of the region. Pearl Millet is the only crop registering growth in productivity in IB, IIA, IIB and IIIA. Productivity in IIIB has doubled between 1990 to 2020. Region wise trends show that productivity of all major crops have declined in IA and IC. In IVB and V, only productivity of maize crop has increased significantly. In IIB and IIIA, productivity of sorghum and pearl millet have increased (Table 3.4).

Zone IA and IV B have witnessed a small decline in productivity of all major rabi crops. Region IB has registered increase in productivity of wheat, barley and mustard. A decrease in productivity of Wheat, Gram and Mustard has been noticed in zone IIA, IIIA and IVA (Table 3.5).

3.8.2 *Crop Combination*

Cultivating a crop in total isolation is quite rare. In the context of Rajasthan, monoculture is extremely uncommon. Crops are planted in rotation to preserve soil fertility, make prudent use of the limited amount of water available, and make effective use of farm labor. Since crops are complementary to one another, they are also grown in conjunction. Physical, economic and cultural determinants generally operate in combination in a region and they influence the crop distribution, the production associations are necessarily dependant on them. The crop combination indices provide information about the current crop combinations as well as how crop combinations are changing in the area. Crop combination allows for the evaluation of crop compatibility and consistency in relation to the importance of the local environment. Better

Table 3.5 Crop efficiency of major rabi crops

Climatic-zone	Wheat		Barley		Gram		Mustard	
	1990–91	2019–20	1990–91	2019–20	1990–91	2019–20	1990–91	2019–20
IA	88.65	83.16	110.94	105.41	116.44	88	132.67	103.51
IB	101.18	118.31	53.47	96.66	119.9	56.23	50.62	109.38
IC	88.17	92.47	61.79	84.56	102.01	68.03	130.52	81.37
IIA	98.22	105.72	125.98	100.62	56.33	126.14	57.59	96.9
IIB	83.67	64.22	97.77	88.26	62.68	89.88	71.5	85.84
IIIA	105.14	98.74	98.88	100.3	94.6	88.03	94.75	74.49
IIIB	121.7	120.42	100.37	107.6	133.93	122.98	55.83	106.39
IVA	112.87	98.93	105	97.25	79.79	97.83	114.77	100.04
IVB	92.69	78.65	107.12	91.25	118.23	112.01	134.81	102.81
V	82.71	114.38	113.67	103.09	91.09	125.88	131.95	114.28

Source Calculated by researcher based on data from agricultural statistics of Rajasthan, Directorate of Economics & Statistics, Department of Planning, Govt of Rajasthan, 1990–2020

agricultural infrastructure development will allow for the introduction of new crop varieties to replace traditional, less profitable crop combinations.

The crop combination formula used in the study is Weaver's (1954) formula, which is based on standard deviation computation. For net sown area under major crops with area greater than 5%, the standard deviation is computed. While calculating the standard deviations the mean used is equal to ideal per cent of area under each category viz. 100 per cent in case of monoculture, 50 per cent in case of double cropping, etc. The standard deviation calculated for different cases are compared and the case having minimum value is considered as the prevailing crop combination in the region.

The findings indicate that pearl millet dominated the arid zone throughout the kharif season season which was replaced by Moth at the end of twentieth century. However, as monsoonal rainfall fluctuated more, the tendency changed back towards pearl millet production. The Turkish bean, or guar, has become more popular in zones IA and IC recently as a result of the farmer's increased income from its cultivation. Cotton farming has been chosen for zone IB with efficient irrigation techniques (Singh & Chauhan, 2011).

Maize has long been the ideal crop to dominate the southern zone IVA. The region's maize crop is supported by soil conditions. The area benefits from less rainfall unpredictability, which makes maize farming easier. Soybean farming predominates in zone V. This area, which is a plateau, borders Madhya Pradesh, the state that produces the most soybeans. In zone IIA in the central tract, the crop combination has changed from monoculture to the cultivation of six crops.

In the rabi season, mustard has replaced the gram as the dominant crop in terms of crop combination. Since the gram crop depends on monsoonal rainfall, this shift can be attributed to the increasing variability of monsoonal rainfall. More irrigation facilities have led to an increase in mustard cultivation as well.

In zones V and IC, there has also been an increase in the cultivation of spices. The cultivation of cumin predominates in zone IC, while coriander cultivation predominates in zone V. Farmers have also been drawn to cultivate mustard and spices due to their higher, more lucrative prices. The cowpea has been a popular choice in Rajasthan's central region as well. The predominant crop combination in Rajasthan is Wheat-Mustard. The central region of Rajasthan has shown the greatest shifts in crop-combination.

3.8.3 Crop Diversification

Crop variety lowers the risk associated with climatic whims. The cultivators plant a variety of crops in order to generate income despite the unfavorable weather. Additionally, diversification enables farmers to produce the majority of their domestic needs on their land. Crop diversification's extent demonstrates the influence of institutional, technological, social, and physical factors. Understanding the competition between various agricultural activities for space requires an understanding of the geographical pattern of agriculture diversification. Farmers choose diversification when the local environment is favorable for the cultivation of a range of crops.

The Gibbs and Martin (1962) Index of Diversification is useful for measuring the extent of diversification. The formula is:

$$\mathit{Diversification} = 1 - \frac{\varepsilon x^2}{(\varepsilon x)^2}$$

where x is the per cent of total cropped area occupied by each crop.

Different scenarios for the rabi and kharif seasons has been analyzed. In the kharif season, the diversification increased at the end of twentieth century and presently it is declining. On the other hand, since the 1990s, crop diversification for the rabi season has consistently risen. The southern region of Rajasthan has seen the biggest shifts in diversification for the kharif season. During the kharif season, there has been a marked decrease in diversity in zones IIIB and V. The cause is related to the expansion of fallow land used for the production of more lucrative rabi crops. Farmers like to cultivate kharif crops for subsistence during the kharif season. A significant rise in crop diversification has been observed in the middle tract of zones IIA and IIB during the rabi season. Farmers employ crop diversification as a technique to respond to climatic variability. Additionally, there have been improvements in the IC zone, where agricultural diversification has gone from extremely low to high. The zone IVB has a consistency and the crop diversification has always been moderate in the region.

3.8.4 Agriculture Efficiency

The overall performance of different crops in relation to their output per hectare is known as agriculture efficiency. The following formula was employed by Sapre and Despande (1964) to measure the efficiency of agriculture:

$$E_i = \frac{I_{ya} \times C_a + I_{yb} \times C_b + \cdots + I_{yn} \times C_n}{C_a + C_b + \cdots + C_n}$$

where, E_i = Agriculture Efficiency Index
I_{ya} = Yield Index of Crop *a*, calculated by following formula:

$$I_{ya} = \frac{Y_C}{Y_r} \times 100$$

C_a = Percentage of Cropland under crop *a*.

Throughout the state, the kharif season has seen an increase in overall agricultural efficiency. During the first half of the study period, the rabi season's agricultural efficiency increased but in the second half it decreased. For the rabi season, all zones have moderate to low agricultural efficiency. The rationale is that as crop diversity has increased, less land is planted to each crop, and some crops perform better in certain zones than others. These factors lower crop efficiency, which in turn lowers the efficiency of agriculture as a whole.

3.8.5 Adaption and Mitigation Options for Sustained Agriculture

Societal vulnerability to the risks associated with climate change may exacerbate ongoing social and economic challenges, particularly for poor rural people and societies dependent on natural resources that are sensitive to climate change. Thus, in order to solve the issues related to global warming and climate change, society will need solid understanding to pursue solutions for both mitigation and adaptation. The authors recommended several measures to mitigate the adverse impacts of climate change on agriculture in the study area:

(i) Reduction and/or sequestration of greenhouse gases: Implementing practices that reduce emissions and enhance carbon sequestration in agricultural systems can mitigate climate change impacts.

(ii) Biodiversity conservation and ecosystem functioning: Promoting biodiversity conservation and maintaining healthy ecosystems supports resilient agricultural systems that are better able to withstand climate variability.

(iii) Enhancing the yield of livelihood goods and services to local people: Improving agricultural productivity through sustainable practices and technologies ensures food security and livelihood stability for local communities.
(iv) Reduction in poverty and vulnerability, and improving resilience and adaptive capacity: Supporting measures that reduce poverty, enhance resilience, and improve adaptive capacity of farmers to climate impacts strengthens their ability to cope with changing conditions.
(v) Local empowerment and capacity development: Empowering local farmers through training, access to information, and participatory decision-making processes enables them to adopt climate-smart practices effectively.
(vi) Synergy with objectives of international instruments and conventions: Aligning mitigation and adaptation strategies with global climate goals and agreements enhances effectiveness and promotes international cooperation.
(vii) Coherence with local strategies for sustainable development: Integrating climate adaptation and mitigation measures into local sustainable development plans ensures that interventions are context-specific and contribute to long-term resilience and sustainability.

These measures collectively aim to build resilience, enhance sustainability, and mitigate the adverse impacts of climate change on agriculture, thereby safeguarding livelihoods and food security in the study area.

For climate change resilient agriculture focus should be given to water management, management of drylands, Management of forests and promotion of agroforestry, diversified agriculture, pasture/grazing lands, traditional farming systems, storage of food grain and seed bank, fodder bank, network of mega-shelterbelts, sequential restoration of dunes in Thar desert, solar, wind and biomass-based energy, management of protected areas and biodiversity and scientific research to generate policy options.

3.9 Conclusion

It is evident from the study that the temperature has risen in each of the state's agro-climatic zones. From 1990 to 2019, there has been a decline in the amount of rainfall in the state (IMD, 2020). Rainfall in large parts of the State is not only inadequate but also varies sharply from year to year and place to place. The Thornthwaite-defined humidity provinces have moved eastward due to changes in precipitation patterns and a general increase in temperature throughout the state. It has been observed that productivity of major kharif crops has declined from 1990 to 2019. Zone IA and IV B have witnessed a small decline in productivity of all major rabi crops. The findings indicate that pearl millet dominated the arid zone throughout the kharif season season which was replaced by Moth at the end of twentieth century. On the other hand, since the 1990s, crop diversification for the rabi season has consistently risen. The study found that variations in the total amount of rainfall as well as changes in the timing of the rainfall have a negative impact on agriculture. An increase in temperature in Rajasthan is predicted to significantly decrease the yield of pearl millet, a staple crop grown during the kharif season. A significant factor influencing agricultural output in Rajasthan's dry regions is climate change and variability linked to frequent extreme weather events, in addition to ecological, technological, and socioeconomic factors. The state's area and productivity have consistently increased due to the application of contemporary farming technologies and inputs. The pattern of cropping is evolving during the Rabi and Kharif seasons. A vast network of mega-shelterbelts should be developed to check the eastward expansion of arid zone over semi-arid zone. Cropping pattern and agricultural practices should be followed according to changing climatic characteristics of the state.

References

Abrol, I. P., & Sangar, S. (2006). Sustaining Indian agriculture–conservation agriculture the way forward. *Current Science, 91*(8), 1020–1025.

Aggarwal, P. K. (2008). Global climate change and Indian agriculture: Impacts, adaptation and mitigation. *Indian Journal of Agricultural Sciences, 78*(10), 911–919.

Begum, A., & Mahanta, R. (2017). Adaptation to climate change and factors affecting it in Assam. *Indian Journal of Agricultural Economics, 72*(3), 446–455.

Birthal, P. S., Negi, D. S., Kumar, S., Aggarwal, S., Suresh, A., & Khan, M. (2014). How sensitive is Indian agriculture to climate change? *Indian Journal of Agricultural Economics, 69* (902–2016–68357), 474–487.

Chand, R., & Raju, S. S. (2009). Instability in Indian agriculture during different phases of technology and policy. *Indian Journal of Agricultural Economics, 64*(2), 283–288.

Dinar, A., Mendelsohn, R., Evenson, R., Parikh, J., Sanghi, A., Kumar, K., McKinsey, J., & Lonergan, S. (1998). *Measuring the impact of climate change on Indian agriculture.* World Bank Technical Paper 402. Retrieved January 08, 2024, from https://econpapers.repec.org/paper/fthwobate/402.htm

Fulekar, M., & Kale, R. (2010). Impact of climate change: Indian scenario. *University News, 48*(24), 14–20.

Gibbs, J., & Martin, W. (1962). "Index of diversification," taken from quantitative techniques in geography: an introduction. In R. Hammond & P. S. McGullagh (Eds.), Clarendon Press, Oxford. p. 21.

Gupta, S., Sen, P., & Srinivasan, S. (2014). Impact of climate change on the Indian economy: Evidence from food grain yields. *Climate Change Economics, 5*(2), 1450001(1–29).

Intergovernmental Panel on Climate Change-IPCC. (2018). Summary for policymakers. In V. Masson-Delmotte, P. Zhai, H. O. Pörtner, D. Roberts, J. Skea, P.R. Shukla, A. Pirani, W. Moufouma-Okia, C. Péan, R. Pidcock, S. Connors, J. B. R. Matthews, Y. Chen, X. Zhou, M. I. Gomis, E. Lonnoy, T. Maycock, M. Tignor & T. Waterfield (Eds.), Global warming of 1.5°C. World Meteorological Organization. Retrieved on October 22, 2023, from https://www.ipcc.ch/sr15/chapter/spm/

Machiwal, D., & Jha, M. K. (2017). Evaluating persistence, and identifying trends and abrupt changes in monthly and annual rainfalls of a semi-arid region in western India. *Theoretical and Applied Climatology, 128*, 689–708.

Mall, R. K., Singh, R., Gupta, A., Srinivasan, G., & Rathore, L. S. (2006). Impact of climate change on Indian agriculture: A review. *Climatic Change, 78*, 445–478.

Mayowa, O. O., Pour, S. H., Shahid, S., Mohsenipour, M., Harun, S. B., Heryansyah, A., & Ismail, T. (2015). Trends in rainfall and rainfall related extremes in the east coast of peninsular Malaysia. *Journal of Earth System Science, 124*, 1609–1622.

Meena, H. M., Machiwal, D., Santra, P., Moharana, P. C., & Singh, D. V. (2019). Trends and homogeneity of monthly, seasonal, and annual rainfall over arid region of Rajasthan, India. *Theoretical and Applied Climatology, 136*, 795–811.

Pingale, S. M., Khare, D., Jat, M. K., & Adamowski, J. (2014). Spatial and temporal trends of mean and extreme rainfall and temperature for the 33 urban centers of the arid and semi-arid state of Rajasthan, India. *Atmospheric Research, 138*, 73–90.

Poonia, S., Rao, A. S., Singh, D. V., & Choudhary, S. (2014). Rainfall characteristics and meteorological drought in Hanumangarh district of arid Rajasthan. *Annals of Plant and Soil Research, 16*, 279–284.

Rao, V. B., Arai, E., Franchito, S. H., Shimabukuro, Y. E., Ramakrishna, S. S. V. S., & Naidu, C. V. (2011). The Thar, Rajputana desert unprecedented rainfall in 2006 and 2010: Effect of climate change? *Geofísica Internacional, 50*, 363–370.

Rao, C. A. R., Raju, B. M. K., Rao, A. V. M. S., Rao, K. V., Rao, V. U. M., Ramachandran, K., Venkateswarlu, B., Sikka, A. K., Rao, M. S., Maheswari, M., & Rao, Ch. S. (2016). A district level assessment of vulnerability of Indian agriculture to climate change. *Current Science, 110*(10), 1939–1946.

Rathore, M. S. (2005). *State level analysis of drought policies and impacts in Rajasthan, India*, International Water Management Institute, (Working paper 93, Drought Series Paper No. 6). Colombo, Sri Lanka. Retrieved January 14, 2024, from https://www.iwmi.cgiar.org/Publications/Working_Papers/working/WOR93.pdf

Sa'adi, Z., Shahid, S., Ismail, T., Chung, E. S., Wang, X. J. (2019). Trend analysis of rainfall and rainfall extremes in Sarawak, Malaysia using modified Mann-Kendall test. *Meteorology and Atmospheric Physics, 131*, 263–277.

Sapre, S. G., & Deshpande, V. D. (1964). The inter-district variations in agricultural efficiency in Mahrashtra state. *Indian Journal of Agricultural Economics, 19*, 242–252.

Singh, B., Rajpurohit, D., Vasishth, A., & Singh, J. (2012). Probability analysis for estimation of annual one day maximum rainfall of Jhalarapatan area of Rajasthan, India. *Plant Archives, 12*, 1093–1100.

Singh, J., & Chauhan, A. (2011). Regional disparities in levels of development in Rajasthan. *Transactions-Institute of Indian Geographer, 33*(2), 231–245.

Singh, N. P., Bhawna, A., & Singh, S. (2020). *Impact of climate change on agriculture in India: Assessment for agro-climatic zones* (Policy Paper 35). ICAR-National Institute of Agricultural Economics and Policy Research (NIAP), Retrieved January 10, 2024, from https://niap.icar.gov.in/pdf/pp35.pdf

Singh, R. B. (2009). Integrated water resources management in the vulnerable Indian environment. *Paper Presented at the IAHS-AISH Publication, 330*, 187–193.

Singh, R. B. (1994). Desert region in India: Resource management issues and research strategy. In T. Miyazaki & A. Tsunekawa (Eds.), *Towards solving the global desertification problem (2)*. National Institute of Environmental Studies.

Singh, R. B., & Kumar, A. (2013). Impact of changing monsoonal pattern of rainfall on dryland agriculture: A case of Nagaur district. *Journal of Water and Land Use Management, 13*(1), 1–14.

Singh, R. B., & Kumar, A. (2016). Agriculture dynamics in response to climate change in Rajasthan. *The Delhi University Journal of the Humanities and the Social Sciences, 3*, 115–138.

Solomon, S. (Ed.). (2007). *Climate change 2007—The physical science basis: Working group I contribution to the fourth assessment report of the IPCC* (Vol. 4). Cambridge University Press. Retrieved November 14, 2023, from https://www.ipcc.ch/report/ar4/wg1/

Srivastava, K., & Pradhan, D. (2018). Real-time extremely heavy rainfall forecast and warning over Rajasthan during the monsoon season (2016). *Pure and Applied Geophysics, 175*, 421–448.

Stocker, T. F., Qin, D., Plattner, G. K., Alexander, L. V., Allen, S. K., Bindoff, N. L., Bréon, F. M., Church, J. A., Cubasch, U., Emori, S., & Forster, P. (2013). Technical summary. In *Climate change 2013: The physical science basis. Contribution of working group I to the Fifth assessment report of the intergovernmental panel on climate change* (pp. 33–115). Cambridge University Press. Retrieved on November 28, 2023, from https://www.ipcc.ch/site/assets/uploads/2018/02/WG1AR5_TS_FINAL.pdf

Tripathi, A., & Mishra, A. K. (2017). Knowledge and passive adaptation to climate change: An example from Indian farmers. *Climate Risk Management, 16*, 195–207.

Yadav, S. K., Gautam, S., & Rawat, S. (2018). Rainfall variability estimation for western Rajasthan, India. *International Journal of Current Microbiology and Applied Sciences, 7*, 4344–4348.

Chapter 4
Bridging the Gap: Exploring Adaptation Strategies for Climate-Resilient Agriculture and Enhanced Livelihoods in Odisha

Chadetrik Rout and Arabinda Sharma

4.1 Introduction

Odisha, located on India's east coast, is a treasure trove of rich cultural history, a thriving economy, and vast natural resources. Odisha is rich in mineral resources, notably iron ore, coal, and bauxite, which contribute considerably to India's industrial sector. The state has achieved significant progress in industries like as steel, electricity, and manufacturing, attracting investments and supporting economic growth. However, in recent years, the state has faced several economic and environmental issues as a result of changing climatic events, including agricultural susceptibility, fisheries losses, infrastructure damage, erosion and coastal degradation, deforestation, water scarcity, and so on.

The state has a tropical climate characterized by high temperatures and copious rains (Mahala, 2019). The state has seen considerable climate change trends over the years, which have influenced various parts of its ecology, economy (Bahinipati & Venkatachalam, 2016) and agricultural production (Das and Mishra, 2017). The rise in average temperatures is one of the most visible signs of climate change in Odisha (Patra et al., 2012). Heatwaves are growing more common and powerful as temperatures rise, presenting dangers to human health and agricultural output (Das and Mishra, 2017; Patz et al., 2005). Furthermore, warming in the Bay of Bengal

C. Rout
Department of Civil Engineering, Maharishi Markandeshwar Engineering College, Maharishi Markandeshwar (Deemed to be University), Mullana, Ambala 133207, Haryana, India

A. Sharma (✉)
School of Geography, Gangadhar Meher University, Amruta Vihar, Sambalpur 768 004, Odisha, India
e-mail: arbind_78@rediffmail.com

P. K. Rai and S. Rai (eds.), *Climate, Environment and Agricultural Development*, Advances in Geographical and Environmental Sciences,
https://doi.org/10.1007/978-981-97-8363-2_4

has resulted in more severe cyclones and storms, causing significant damage to property and life (Vivekanandan et al., 2016). Changes in rainfall patterns are another important component of climate change in Odisha (Patra et al., 2012). While the state receives a significant quantity of rain throughout the monsoon season, there has been a considerable shift in rainfall distribution and intensity (Prabhakar et al., 2019). Droughts and floods have ensued, affecting agricultural productivity and lives (Roy et al., 2022). Furthermore, sea level rise is a major problem for Odisha, particularly in its coastal districts (Khristodas et al., 2022). The state is subject to coastal erosion and floods, affecting coastal resident's infrastructure, ecosystems, and livelihoods (Senapati & Gupta, 2014).

To protect Odisha's economy and environment from the harmful effects of changing climatic events, a comprehensive strategy incorporating sustainable practices, community resilience development, and policy interventions is required (Pradhan, 2018). The state government of Odisha has been actively involved in adopting adaptation and mitigation measures in response to these climate change trends (Sharma et al., 2018). At this outset, the article aims at exploring the effect of climate change on Odisha's economy particularly the agricultural produces and to summarize the comprehensive adoption and mitigation initiatives needed to protect the environment, economy, and wellbeing of its population through improved resilience to climate change.

4.2 Economic Impact of Climate Change

Climate change isn't just an environmental threat; it's already impacting economic activities across the globe. Its impact on economic activities is multifaceted and varies depending on the sector and location. The state is also experiencing several concerns from the climate change front that are having a substantial influence on its economy (Hota & Behera, 2016). The summary of the major economic impacts of climate change are presented in Table 4.1.

It is obvious that climate change is disrupting established economic systems and creating new challenges. However, Table 4.1 provides a simplified overview and may not capture the full complexity of the impacts. Further research and analysis are necessary to understand the specific effects in different geographical regions and economic sectors. The vulnerability of Odisha is especially because of its long coastline, low-lying areas, and agricultural reliance.

The impact of climate change on the people of Odisha's economic and food security is a complicated problem with far reaching ramifications. The increasingly frequent and violent cyclones, floods, and droughts as a result of climate change further entail deep social problems like poverty, food insecurity, malnutrition, and migrations. It is because Odisha's population relies primarily on locally produced food, and agricultural productivity reduction has a direct influence on food security. Climate change induced environmental risk has also a substantial impact on displacement and migration trends. These events frequently result in temporary or permanent

Table 4.1 Summary of impact of climate change on different economic activities

Economic activity	Effect of climate change	Examples
Agriculture	- Reduced crop yields due to droughts, floods, and heatwaves - Increased pest and disease outbreaks -Water scarcity impacting irrigation, changes in growing seasons	- Decrease in global agricultural output - Salinization of coastal lands - Droughts impacting grain production, warmer temperatures increasing insect populations - Salinization affecting coastal farms, shorter growing seasons in mountainous regions
Fisheries	- Ocean acidification harming marine life and disrupting food chains - Changes in fish migration patterns impacting catches - Coastal ecosystem degradation reducing fish populations	- Coral bleaching impacting coral reefs and fish habitats, fish moving to cooler waters impacting traditional fishing grounds - Sea level rise affecting coastal spawning grounds, increased competition for depleted fish stocks
Tourism and Recreation	- Extreme weather events damaging infrastructure and tourist destinations - Reduced snow and ice cover impacting winter sports - Sea level rise and coral bleaching harming beaches and marine beauty	- Lost revenue and job losses in tourism-dependent regions - Shorter ski seasons in mountain regions, damaging coastal resorts
Infrastructure	- Damage from floods, storms, and heatwaves requiring costly repairs - Sea level rise threatening coastal infrastructure and property - Changes in water availability impacting hydropower and industrial water use	- Billions of rupees in annual damage from extreme weather events - Relocation costs for threatened coastal infrastructure, floods damaging roads and bridges, rising sea levels inundating coastal infrastructure - Droughts impacting hydropower generation, more frequent repairs due to extreme weather events
Energy	- Disruptions to renewable energy sources (e.g., wind, solar) - Increased demand for cooling due to heatwaves - Potential for instability in energy grids	- Extreme weather events impacting wind farms and solar panels - Increased energy consumption for air conditioning, heatwaves overloading power grids

(continued)

Table 4.1 (continued)

Economic activity	Effect of climate change	Examples
Insurance	- Rising insurance premiums due to increased climate risks - Difficulty obtaining insurance for high-risk areas - Potential for insurance market instability	- Hurricanes and floods lead to higher insurance costs for coastal properties - Wildfires impacting insurance availability in certain regions, potential for large-scale losses affecting insurance companies
Overall Economy	- Disruptions and losses from extreme weather events - Reduced productivity and efficiency - Increased costs and risks from adaptation and mitigation - Opportunities for green technologies and innovation	- Trillions of dollars in annual economic losses by the end of the century - Potential for job creation in clean energy sectors

migration, both within the state and to other areas of the nation. Furthermore, shifting weather patterns and environmental deterioration have made traditional livelihoods such as farming and fishing less sustainable (Patel et al., 2019). This lack of livelihood possibilities adds to individuals migrating from rural to urban regions in pursuit of alternate sources of income.

Climate change also poses several other socioeconomic challenges such as earning level (Moharaj & Rout, 2021; Sam et al., 2020), water resources including drinking and irrigation water (Kumar et al., 2021; Sam et al., 2020), health wellbeing (Mishra, 2017), inequality among different section of the society (Mishra, 2017) and community's natural resilience (Sam et al., 2020).

4.3 Climate Change and Agriculture

Agriculture and climate are intricately linked, with the success of agricultural activities heavily dependent on various climate factors (Mertz et al., 2009). From the first seed sown to the final harvest, climate plays a crucial role in every step of agricultural activities. The spatio-temporal variations in climatic variables such as temperature, precipitation, sunlight, wind etc. have several direct and indirect influences on the crop type, its location, farming practices, quality and quantity of the agricultural produce (Belal et al., 2014). The summary of the potential impact of climate change on agricultural activities is presented in Table 4.2.

It is evident that climate plays a critical role in determining the success of agricultural activities (Yildiz, 2019). Understanding the intricate relationships between climate and agriculture is crucial for developing effective adaptation strategies and

Table 4.2 Summary of effect of climate change on agriculture

Impact category	Description	Examples
Reduced Crop Yields	• Droughts, floods, and heatwaves disrupt growing seasons. • Erratic rainfall patterns affect pollination and germination. • Increased pest and disease outbreaks damage crops	• Droughts impacting grain production in Africa. • Floods destroying rice paddies in Asia. • Heatwaves reducing yields of corn and soybeans in North America
Water Scarcity	• Changing rainfall patterns reduce water availability for irrigation. • Rising temperatures increase evaporation rates. • Depletion of groundwater resources	• Limited irrigation leading to failed harvests in arid regions. • Competition for water resources leading to conflicts among farmers. • Salinization of coastal lands reducing arable land
Pests and Diseases	• Warmer temperatures and altered ecosystems favor pest and disease outbreaks (Newton et al., 2011). • Increased pesticide use leads to environmental pollution and resistance development	• Tropical insects expanding ranges and attacking new crops. • Fungal diseases thriving in warmer, wetter conditions. • Rising costs of pest control for farmers
Soil Degradation	• Erosion from extreme weather events reduces soil fertility. • Unsustainable agricultural practices deplete organic matter and nutrients	• Soil erosion washing away topsoil in hilly regions. • Desertification expanding in arid areas due to land degradation. • Reduced crop yields due to poor soil quality
Livestock Productivity	• Heat stress reduces animal growth and milk production. • Water scarcity limits access to drinking water and forage. • Changes in weather patterns disrupt animal breeding cycles	• Heatwaves causing livestock deaths in tropical regions. • Droughts forcing farmers to sell off livestock due to lack of water. • Increased risk of diseases and infections in stressed animals
Limited Positive Impacts	• CO_2 fertilization can, in some cases, enhance photosynthesis and crop yields in specific regions. • Longer growing seasons in some areas may allow for additional crop cycles or new crops	• Rice yields potentially increasing in parts of Asia with CO_2 fertilization. • Introduction of heat-tolerant crops in some regions with extended growing seasons

building resilient agricultural systems in the face of a changing climate. Thus, addressing climate dependency in agriculture requires a multi-pronged approach involving research, technological advancements, policy changes, and community-based adaptation strategies.

4.4 Reason for Vulnerability of Odisha's Agriculture

Odisha, a state in eastern India, faces significant threats to its agricultural activities due to climate change. Its dependence on rain fed agriculture, coupled with its geographical location, makes it particularly vulnerable to the following factors:

Increased Droughts and Erratic Rainfall:

- Odisha receives most of its rainfall during the monsoon (June–September), but the timing and intensity are increasingly unpredictable. Prolonged dry spells and delayed monsoon onset can severely impact crop yields, especially for drought sensitive crops like rice and pulses (Gautam & Bana, 2014).
- Excessive rainfall during short periods can lead to flash floods, damaging crops and infrastructure.

Sea Level Rise and Coastal Erosion:

- Odisha's long coastline is vulnerable to sea level rise, which threatens agricultural land and disrupts freshwater sources. Saline intrusion into rivers can contaminate irrigation water and reduce its suitability for agriculture (Sarkar et al., 2021).
- Coastal erosion damages agricultural infrastructure like embankments and irrigation canals, further impacting productivity.

Rising Temperatures and Heat Stress:

- Increasing temperatures put crops under heat stress, reducing yields and impacting the growth cycle.
- Heatwaves can lead to wilting and stunted plant growth, further affecting production.
- Increased evaporation rates exacerbate water scarcity, compounding the challenges faced by farmers.

Extreme Weather Events:

- Odisha is prone to cyclones, floods, and other extreme weather events. These events can devastate crops, destroy infrastructure, and disrupt food supply chains.
- Rising sea levels pose a threat to coastal agricultural lands and salinize freshwater sources, impacting irrigation.
- Cyclones, for instance Fani which hit the state in 2019, caused widespread damage to crops and infrastructure, highlighting the vulnerability of coastal communities.

Water Scarcity:

- Climate change is expected to increase water stress in Odisha, putting pressure on already limited water resources. This can lead to competition for water between agriculture, industry, and domestic use, impacting agricultural production.
- While Odisha receives a good amount of rainfall, its distribution is uneven. This leads to water scarcity during crucial planting and growing seasons.
- Overexploitation of groundwater resources further exacerbates the problem, putting irrigation-dependent agriculture at risk.
- Salinization of coastal aquifers due to rising sea levels further diminishes freshwater availability.

Soil Degradation:

- Unsustainable farming practices and soil erosion from extreme weather events degrade soil fertility and productivity.
- Loss of organic matter reduces soil's ability to retain water and nutrients, impacting crop yields.
- Deforestation and soil erosion contribute to land degradation, further reducing agricultural potential.

Limited Resources:

- Smallholder farmers, who form a significant part of the agricultural workforce, often lack access to resources and technologies for adaptation.

Socio-Economic Vulnerabilities:

- Odisha is one of India's poorest states, with many farmers lacking access to resources and technology to adapt to climate change.
- Limited access to irrigation and drought-resistant seeds further increases vulnerability.

4.5 Effect of Climate Change on Major Agricultural Produce of Odisha

Odisha, boasts a rich agricultural heritage, with over 70% of its population engaged in farming. The state's diverse climate and topography give rise to a variety of major agricultural crops, each playing a vital role in the local economy and food security (Swain, 2014). However, climate change casts a long shadow over Odisha's diverse agricultural landscape, impacting the cultivation of each major crop in distinct ways:

4.5.1 Rice

Unquestionably the king of crops in Odisha, rice is the staple food for over 45 million people in the state. It thrives in the humid subtropical climate and abundant rainfall, particularly in the coastal plains and river deltas. Odisha is the 6th largest rice producer in India, contributing significantly to the national food basket. Popular rice varieties include Basmati, Swarna, and Non-Basmati types.

Climate Change Effect

- Droughts and erratic rainfall: Disrupted planting and harvesting cycles, leading to yield losses and food insecurity.
- Heat stress: Reduced yields and quality due to damage during crucial flowering and grain-filling stages.
- Salinization: Coastal rice fields threatened by rising sea levels, reducing land availability and impacting production.

4.5.2 Pulses

Legumes like lentils, chickpeas, and pigeon peas are vital sources of protein and dietary fiber for Odisha people. Odisha is a major producer of pulses in India, contributing to the national goal of self-sufficiency in pulses production. These are grown across the state, particularly in upland areas with well-drained soil. Pulses contribute about 10% of the state's agricultural output and are often intercropped with rice for improved soil fertility and nutrient balance. Pulses like lentils, chickpeas, and pigeon peas are an essential source of protein and dietary fiber for the.

Climate Change Effect

- Unpredictable rainfall: Erratic patterns affect planting and harvesting, impacting yields and production.
- Nutrient deficiencies: Soil degradation and erosion due to extreme weather events reduces soil fertility lead to nutrient deficiencies, affecting pulse quality and yields.
- Market volatility: Climate-induced fluctuations in production cause market instability, impacting farmers' income.

4.5.3 Maize

Maize, or corn, is another important crop in Odisha, grown mainly for animal feed and industrial purposes like starch production. It is also gaining popularity as a food grain due to its high nutritional value and shorter growing season. Maize cultivation is concentrated in the western and southern parts of the state, where the climate is suitable for its growth.

Climate Change Effect

- Droughts and heatwaves: Similar to rice, maize suffers from droughts, water and heat stress, affecting yields and posing challenges for farmers.
- Pests and diseases: Warmer temperatures favor the spread of pests and diseases, impacting production and farmer income.
- Soil erosion: Heavy rainfall events can lead to soil erosion, impacting maize yields.

4.5.4 Oilseeds

Groundnut, mustard, and sesame are important oilseed crops in Odisha, providing edible oil for domestic consumption and generating income for farmers. They collectively account for around 5% of the state's agricultural production.

Climate Change Effect

- Droughts and heatwaves: Reduced yields and pose challenges for farmers dependent on these crops.
- Pests and diseases: Warmer temperatures favor the spread of pests and diseases like fall armyworm, leading to crop losses and increased pesticide use.
- Soil erosion: Heavy rainfall events can lead to soil erosion, impacting soil fertility and reducing oilseed yields.

4.5.5 Fruits and Vegetables

Odisha boasts a variety of vegetables, including potatoes, tomatoes, okra, brinjal, and leafy greens. These add vital vitamins and minerals to the diets of Odias and contribute significantly to the state's agricultural income. Odisha is known for its delicious fruits like mangoes, bananas, citrus fruits, and papayas, contributing significantly to the state's economy and exports. The state has a diverse range of fruit-growing regions, each with its own specialties. For example, the coastal belt is famous for its sweet mangoes, while the hilly regions are known for their citrus fruits.

Climate Change Effect

- Extreme weather events: Floods and cyclones damage crops and disrupt production (i.e. Mangoes and vegetable are vulnerable to cyclones and hailstorms), leading to shortages and price fluctuations.
- Heat stress: Certain fruits and vegetables are sensitive to heat stress, causing wilting and reduced yields.
- Changing rainfall patterns: Can affect fruit quality and quantity, impacting market value and farmer income.
- Water scarcity: Certain fruits (i.e.banana) and vegetables have significant water requirements make them vulnerable to droughts, impacting production and farmer income.
- Pests and diseases: Rising temperatures and rainfall changes can favor pests and diseases, requiring increased pesticide use and impacting fruit quality.

4.5.6 *Jute*

A major fiber crop, jute is used for making ropes, sacks, and other agricultural products. It plays a significant role in Odisha's rural economy and provides livelihood for many farmers.

Climate Change Effect

- Droughts and floods: Disrupt jute cultivation, leading to reduced fiber quality, production and market value which in turn affect farmer income.
- Pests and diseases: Warmer temperatures favor the spread of pests and diseases, requiring increased pesticide use and impacting fiber quality.
- Waterlogging: Heavy rainfall can lead to waterlogging, damaging jute plants and reducing yields.

4.5.7 *Sugarcane*

Odisha is a major producer of sugarcane in India, with the sugar industry playing a crucial role in the state's economy. Sugarcane cultivation is concentrated in the coastal districts of the state.

Climate Change Effect

- Droughts and heatwaves: Reduce sugarcane yields and sugar content, impacting production and profitability.
- Soil degradation: Salinization and erosion due to extreme weather events reduce soil fertility, impacting sugarcane growth.
- Water scarcity: Droughts pose challenges for irrigation-dependent sugarcane cultivation, impacting production costs and yields.

4.5.8 *Cash Crops (Cashew and Coconut)*

Cashew, coconut, and coffee are important cash crops grown in Odisha, particularly in the coastal and hilly regions. They provide valuable export earnings and contribute to the diversification of the state's agricultural sector.

Climate Change Effect

- Cashew: Rising temperatures (heat stress) and water scarcity (drought) affect fruit setting and nut development, impacting cashew production and its quality
- Coconut: Coastal coconut plantations are threatened extreme weather events flood and storm surges damage trees and infrastructure and reducing yields.
- Sea level rise and salinization also threatens coastal plantations of both cashew and coconut, reducing land availability and impacting production.

4.5.9 Spices

Turmeric, ginger, and chilies are important spice crops grown in Odisha. They add flavor to Odia cuisine and generate income for farmers, particularly in the hilly regions of the state.

Climate Change Effect

- Droughts and erratic rainfall: Affect spice production and quality, leading to market fluctuations and impacting farmer income.
- Soil degradation: Extreme weather events can lead to soil erosion and degradation, reducing soil fertility and impacting spice yields.
- Pests and diseases: Warmer temperatures can favor the spread of pests and diseases, requiring increased pesticide use and impacting spice quality.

4.6 Making Odisha's Agriculture Climate Resistant

4.6.1 Adaptation of New Techniques

4.6.1.1 Climate-Resilient Agricultural Practices

Climate resilient agriculture methods include:

- **Creating ponds, check dams, and water harvesting facilities:** Building ponds, check dams, and water harvesting facilities helps to alleviate the impact of unpredictable rainfall, as observed in Odisha (Sivakumar et al., 2021).
- **Natural agricultural promotion:** The NITI Aayog project focuses on encouraging natural agricultural practices, minimizing reliance on external resources, and increasing resilience (Anantha et al., 2020).
- **World Bank assistance:** Projects funded by the World Bank aim to improve agricultural resilience to climate change in Odisha, in line with the state's vision (Jogesh & Dubash, 2014).
- **Adoption of climate resilient technology:** Farmers, notably women and those from disadvantaged backgrounds, have adopted resilient agricultural technology, increasing productivity and income (Jena et al., 2023).

4.6.1.2 Innovative Technology Implementation

Several new technological implementations targeted at enhancing agricultural yield have adopted in Odisha. Among the prominent instances are:

- **Weather forecasting and advisory services:** Efforts have been made to give timely weather predictions and advisories to farmers (Ray et al., 2017). Mobile

applications provide critical information on weather trends, allowing farmers to make smart crop management decisions.
- **Drones:** Drones outfitted with imaging technologies are used to monitor crop health, analyze soil conditions, and detect problem areas (Nayak, 2022). This aids precision agriculture by allowing for focused treatments.
- **Mobile based extension services:** Farmers now have access to agricultural specialists, market information, and best practices through mobile applications and platforms (Dominic et al., 2022). These platforms provide advice on numerous areas of farming, allowing for more informed decisions.
- **Solar-powered irrigation systems:** Solar powered irrigation systems have been created to alleviate water scarcity challenges (Giri et al., 2020). These systems use renewable energy sources to efficiently irrigate crops, which is especially important in locations with limited access to power.
- **Farmers' biometric attendance:** Some regions have used biometric attendance systems to track farmers' attendance at government-sponsored agricultural training sessions, assuring greater diffusion of technology breakthroughs and farming methods.

4.6.2 Policies and Efforts of the Government

The state government has launched many initiatives to boost agricultural productivity in the face of climate change:

- **Initiatives for collaboration:** Odisha, in collaboration with USAID, initiated a project to increase tree coverage, with an emphasis on carbon sequestration and agricultural climate resilience (Mishra & Nayak, 2011).
- **Climate resilient agriculture initiative:** The goal of this initiative is to increase productivity while also empowering farmers to adapt to climatic pressures affecting crops and aquaculture (Patel and Kumar, 2022; Rao et al., 2016).

The stated initiatives including importance of working together provide farmers with the information and resources to the to boost agricultural output in the face of climatic chnage.

4.6.3 Resilience and Mitigation in Communities

4.6.3.1 Community Based Adaptation Strategies

To address climate related difficulties, communities in Odisha have used a variety of CBA techniques (Bahinipati & Sahu, 2012; Jyotishi & Manjula, 2021; Khristodas et al., 2023; Sharma et al., 2018), including:

- Farmers are moving to drought and flood resistant crop types. They are promoting water conservation measures such as rainwater gathering and drip irrigation, as well as agroforestry activities to boost soil health and decrease erosion.
- Communities have established early warning systems for cyclones and floods, which utilize traditional knowledge and contemporary technology to offer timely alarms.
- Community seed banks protect traditional and varied agricultural types, assuring seed availability during calamities.
- Coastal communities are actively engaged in mangrove protection to guard against storms and storm surges.
- To lessen reliance on agriculture, communities are investigating other occupations such as fishing, handicrafts, and eco-tourism.
- Training programmes provide communities with knowledge about climate change adaptation and sustainable practices.
- Community based organizations collaborate with government agencies, non-governmental organizations, and academic institutes to exchange information and resources.

These CBA solutions in Odisha illustrate communities' resilience and adaptation in the face of climate change. They are aiming towards sustainable agricultural production and general community well-being by blending traditional wisdom with modern understanding.

4.6.3.2 Collaboration with Local Stakeholders

Various projects are being implemented in Odisha in collaboration with local stakeholders to combat climate change (Jogesh & Dubash, 2014; Prasad & Sud, 2019; Seidler et al., 2018; Sharma et al., 2018):

- **Community engagement programmes:** Educating farmers about climate change, its effects, and possible adaptation methods through workshops, seminars, and awareness campaigns to adopt climate resilient agricultural techniques.
- **Partnerships with NGOs and community based organizations:** Collaboration with non-profits and community based organizations to undertake climate change adaptation and mitigation methods such as afforestation efforts, waste management programs, or renewable energy initiatives.
- **Government partnerships:** Involving local governments to create and implement climate change resilience and sustainability policies. This might include collaborative initiatives to reduce catastrophe risk, create infrastructure, or manage water resources.
- **Capacity building and knowledge sharing:** Facilitating knowledge sharing on climate-smart activities, technology advances, or data gathering methods for climate monitoring among scientific professionals, policymakers, and local stakeholders.

- **Indigenous knowledge integration:** Recognising and incorporating traditional knowledge systems of indigenous groups in Odisha, can offer useful insights into sustainable methods suited to local conditions.

These collaborations seek to promote a bottom-up approach, ensuring that solutions are context specific, inclusive, and aligned with the needs and viewpoints of Odisha's local communities.

4.6.3.3 Mitigation Strategies for Long Term Livelihoods

The consequences are especially troubling given the country's reliance on agriculture as a key source of income for a huge section of its population.

Strategies:

- **Drought resistant crops:** Promoting the production of drought-resistant crop types can assist farmers in adapting to changing rainfall patterns and reducing the effect of droughts (Towolawi et al., 2024).
- **Water conservation practices:** Using effective irrigation systems and water conservation practices can assist farmers in optimizing water consumption and mitigating the consequences of water shortage (Hamdy et al., 2003; Ward & Pulido-Velazquez, 2008).
- **Crop diversification:** Encouraging farmers to diversify their crops reduces the danger of relying on a single crop. Farmers can better resist climate related issues by growing a diversity of crops (Lin, 2011).
- **Agroforestry:** Including trees in agricultural systems may offer shade, increase soil fertility, and decrease soil erosion. Agroforestry methods can also aid with carbon sequestration, which can help to prevent climate change (Lasco et al., 2014).
- **Soil conservation:** Promoting soil conservation measures such as conservation tillage, mulching, and contour farming to preserve soil fertility and minimize erosion (Rwakaningiri, 2014).
- **Capacity building:** Providing farmers with climate resilient agricultural techniques and training can help them adapt and prosper in the face of climate change (Bhusal et al., 2020; Chouksey et al., 2021).
- **Weather forecasting and advisory services:** Improving the transmission of weather-related information and advisories to assist farmers in making educated crop selection and timing decisions (Kumar et al., 2020).

By addressing the impact of climate change on agriculture, Odisha can improve its long-term agricultural production and guarantee the lives of its farming population by implementing these mitigation techniques and encouraging sustainable agricultural practices.

4.6.3.4 Assistance to Vulnerable Populations

Odisha is implementing various policies and efforts to adapt to climate change, including vulnerable population aid, to address the country's significant climate-related challenges.

- **State action plan on climate change (SAPCC):** Odisha has a SAPCC that contains strategies and action plans to reduce and adapt to the consequences of climate change. This strategy frequently contains protections for disadvantaged groups (Kumar, 2018).
- **Disaster management policies:** Because the state is subject to natural catastrophes, disaster management policies include efforts to help vulnerable people during and after such incidents (Ray-Bennett, 2018).
- **Social welfare programs:** The government may implement special social welfare programs for disadvantaged groups, including those affected by climate change, offering rehabilitation, relocation, income support, and access to necessities.
- **National and International assistance:** Odisha may also get assistance from national and international institutions or funds devoted to climate change adaptation and vulnerable people (Prasad & Sud, 2019; Walch, 2019).
- **Water body programs:** The primary goal is programs targeted at reclaiming, revitalizing, and improving water bodies in water-stressed areas, particularly in Western Odisha (Swain & Swain, 2015).

4.7 Conclusion

Climate change has had a substantial influence on agriculture in Odisha, hurting people's lives significantly. Rising temperatures, irregular rainfall patterns, and extreme weather events have disturbed conventional farming techniques, resulting in lower crop yields and increased farmer vulnerability. Odisha's agriculture dependent population faces issues such as water shortages, soil degradation, and the advent of new pests and illnesses, all of which influence their capacity to sustain their livelihoods. Small scale farmers face the brunt of the impact since they lack the means to adequately respond to these changes. Adaptation options including cultivating drought resistant crops, using water management techniques, and adopting sustainable agricultural practices have shown promise, but they require significant assistance and money. Climate resilient agriculture efforts and community based programs that provide farmers with access to technology, education, and financial help are critical for reducing these negative consequences. To address the problems posed by climate change in agriculture in Odisha, a holistic strategy combining politicians, local people, scientists, and international collaboration is required. Collaboration, creativity, and timely responses are required to protect people's livelihoods and maintain food security in the face of climate change.

References

Anantha, K. H., Garg, K. K., & Dixit, S. (2020). Building resilience to climate change in agriculture: Integrated natural resource management and institutional measures. *Global Climate Change: Resilient and Smart Agriculture*, 109–136.

Bahinipati, C. S., & Sahu, N. (2012). Mangrove conservation as a sustainable adaptation to cyclonic risk in Kendrapada District of Odisha, India. *Asian Journal of Environment and Disaster Management, 4*(2), 183–202.

Bahinipati, C. S., & Venkatachalam, L. (2016). Role of climate risks and socio-economic factors in influencing the impact of climatic extremes: A normalisation study in the context of Odisha, India. *Regional Environmental Change, 16*, 177–188.

Behera, B., & Tandi, S. (2019). Impact of climate change on agriculture: A study of western Odisha. *Asian Journal of Multidimensional Research, 8*(2), 145–156.

Belal, A. A., El-Ramady, H. R., Mohamed, E. S., & Saleh, A. M. (2014). Drought risk assessment using remote sensing and GIS techniques. *Arabian Journal of Geosciences, 7*, 35–53.

Bhusal, A., Khatri, L., Bhandari, B., Sherpa, L., & Neupane, G. (2020). Climate-smart agriculture (CSA): Training manual.

Chouksey, R., Singh, K. C., Singh, C., & Birle, Y. (2021). Adaptation of farmers regarding climate resilient technologies in Rewa block of Rewa District in Madhya Pradesh. *Indian Journal of Extension Education, 57*(1), 26–31.

Dalei, N. N., Gupta, A., & Anand, N. (2020). Empirical nexus between global temperature, local weather and agriculture: Evidence from the Indian state of Odisha. *Energy, Environment and Globalization: Recent Trends, Opportunities and Challenges in India*, 143–156.

Das, M. K., & Mishra, P. (2017). Impacts of climate change on agricultural yield: Evidence from Odisha, India. *International Journal of African and Asian Studies, 38*, 58–65.

Dominic, D. M., Mitta, N., Sulaiman, V., & R., Bharti, P., & Sharma, S. (2022). *Enhancing capacities of digital extension and advisory services in Odisha*. Policy Brief.

Gautam, R. C., & Bana, R. S. (2014). Drought in India: Its impact and mitigation strategies–A review. *Indian Journal of Agronomy, 59*(2), 179–190.

Giri, N. C., Rana, D., Mish, S. P., & Pani, B. B. (2020). Efficacy of solar powered water pumps for Rural farmers in Odisha; India. *PalArch's Journal of Archaeology of Egypt/egyptology, 17*(9), 2215–2224.

Hamdy, A., Ragab, R., & Scarascia-Mugnozza, E. (2003). Coping with water scarcity: Water saving and increasing water productivity. *Irrigation and Drainage: THe Journal of the International Commission on Irrigation and Drainage, 52*(1), 3–20.

Hota, P., & Behera, B. (2016). Opencast coal mining and sustainable local livelihoods in Odisha, India. *Mineral Economics, 29*, 1–13.

Jena, P. R., Tanti, P. C., & Maharjan, K. L. (2023). Determinants of adoption of climate resilient practices and their impact on yield and household income. *Journal of Agriculture and Food Research*, 100659.

Jogesh, A., & Dubash, N. K. (2014). Mainstreaming climate change in state development planning: An analysis of the Odisha climate change action plan. *Working paper/Centre for Policy Research (CPR), Climate Initiative.*

Jyotishi, A., & Manjula, M. (2021). Alternative approaches to measure sustainability in a subsistence economy: Empirical insight from shifting cultivation in Odisha. *Environment, Development and Sustainability in India: Perspectives, Issues and Alternatives*, 263–286.

Khristodas, P. M., Palanivelu, K., Ramachandran, A., Anushiya, J., Prusty, B. A. K., & Guganesh, S. (2022). Assessment of climate-induced sea-level rise scenarios and its inundation in coastal Odisha, India. *Applied Ecology and Environmental Research, 20*(4), 3393–3409.

Khristodas, P., Palanivelu, K., Ramachandran, A., & Prusty, B. (2023). Prediction of climate change-induced sea level rise in Chilika-Puri coast of Odisha, India: With special prominence on adaptation action strategy framework. *Applied Ecology & Environmental Research, 21*(5), 4425–4444.

Kumar, P., Dasgupta, R., Dhyani, S., Kadaverugu, R., Johnson, B. A., Hashimoto, S., & Mishra, B. K. (2021). Scenario-based hydrological modeling for designing climate-resilient coastal water resource management measures: Lessons from Brahmani river, Odisha, Eastern India. *Sustainability, 13*(11), 6339.

Kumar, U., Werners, S., Roy, S., Ashraf, S., Hoang, L. P., Kumar Datta, D., & Ludwig, F. (2020). Role of information in farmers' response to weather and water related stresses in the lower Bengal Delta, Bangladesh. *Sustainability, 12*(16), 6598.

Kumar, V. (2018). *Coping with climate change: An analysis of India's state action plans on climate change.* Centre for Science and Environment.

Landsea, C. W. (2000). Climate variability of tropical cyclones: Past, present and future. *Storms, 1*, 220–241.

Lasco, R. D., Delfino, R. J. P., & Espaldon, M. L. O. (2014). Agroforestry systems: Helping smallholders adapt to climate risks while mitigating climate change. *Wiley Interdisciplinary Reviews: Climate Change, 5*(6), 825–833.

Lin, B. B. (2011). Resilience in agriculture through crop diversification: Adaptive management for environmental change. *BioScience, 61*(3), 183–193.

Mahala, S. C. (2019). *Geology, chemistry and genesis of thermal springs of Odisha, India.* Springer International Publishing.

Mertz, O., Mbow, C., Reenberg, A., & Diouf, A. (2009). Farmers' perceptions of climate change and agricultural adaptation strategies in rural Sahel. *Environmental Management, 43*, 804–816.

Mishra, B., & Nayak, B. K. (2011). Community forest management in Odisha: Macro and micro views. *A report of XIM Bhubaneswar, Connecting Business and Society*, 1–23.

Mishra, P. K. (2017). Socio-economic impacts of climate change in Odisha: Issues, challenges and policy options. *Journal of Climate Change, 3*(1), 93–107.

Mohapatra, S., Sharp, B., & Sahoo, D. (2022). How changes in climate affect crop yields in Eastern India. *Climate Change Economics, 13*(02), 2250001.

Moharaj, P., & Rout, S. (2021). Changing climatic conditions and agricultural livelihoods: An impact study in Jagatsinghpur District, Odisha. *Social Change, 51*(2), 160–179.

Nayak, A. K. (2022). Advanced techniques for precision farming in rice. *Climate resilient technologies for rice based production systems in eastern India.*

Nayak, S. R. (2014). Forced displacement and resistance: A study of Lanjigarh project, Odisha. *Educational Quest-an International Journal of Education and Applied Social Sciences, 5*(1), 1–10.

Newton, A. C., Johnson, S. N., & Gregory, P. J. (2011). Implications of climate change for diseases, crop yields and food security. *Euphytica, 179*, 3–18.

Patel, J., & Kumar, G. S. (2022). Climate resilient agriculture practices-the future of farming. *Plant Science Today, 9*(sp3), 63–70.

Patel, S. K., Mathew, B., Nanda, A., Pati, S., & Nayak, H. (2019). A review on extreme weather events and livelihood in Odisha, India. *Mausam, 70*(3), 551–560.

Patra, J. P., Mishra, A., Singh, R., & Raghuwanshi, N. S. (2012). Detecting rainfall trends in twentieth century (1871–2006) over Orissa State, India. *Climatic Change, 111*, 801–817.

Patz, J. A., Campbell-Lendrum, D., Holloway, T., & Foley, J. A. (2005). Impact of regional climate change on human health. *Nature, 438*(7066), 310–317.

Prabhakar, A. K., Singh, K. K., Lohani, A. K., & Chandniha, S. K. (2019). Assessment of regional-level long-term gridded rainfall variability over the Odisha State of India. *Applied Water Science, 9*, 1–15.

Pradhan, S. (2018). *Understanding the role of politics of scale and power relations in local governance of climate change adaptation: Case study from coastal Odisha, India* (Doctoral dissertation, University of Reading).

Prasad, R. S., & Sud, R. (2019). Implementing climate change adaptation: Lessons from India's national adaptation fund on climate change (NAFCC). *Climate Policy, 19*(3), 354–366.

Rao, C. S., Gopinath, K. A., Prasad, J. V. N. S., & Singh, A. K. (2016). Climate resilient villages for sustainable food security in tropical India: Concept, process, technologies, institutions, and impacts. *Advances in Agronomy, 140*, 101–214.

Ray, M., Patro, H., Biswasi, S., Dash, S. R., & Dash, A. C. (2017). Economic assessment of weather based agromet advisories in Keonjhar district, Odisha. *Vayu Mandal, 43*(1&2), 38–48.

Ray-Bennett, N. S. (2018). Disasters, deaths, and the Sendai goal one: Lessons from Odisha, India. *World Development, 103*, 27–39.

Roy, P., Pal, S. C., Chakrabortty, R., Chowdhuri, I., Saha, A., & Shit, M. (2022). Climate change and groundwater overdraft impacts on agricultural drought in India: Vulnerability assessment, food security measures and policy recommendation. *Science of the Total Environment, 849*, 157850.

Rwakaningiri, F. (2014). The role of soil conservation practices in improving soil fertility.

Sam, A. S., Padmaja, S. S., Kächele, H., Kumar, R., & Müller, K. (2020). Climate change, drought and rural communities: Understanding people's perceptions and adaptations in rural eastern India. *International Journal of Disaster Risk Reduction, 44*, 101436.

Sarkar, B., Islam, A., & Majumder, A. (2021). Seawater intrusion into groundwater and its impact on irrigation and agriculture: Evidence from the coastal region of West Bengal, India. *Regional Studies in Marine Science, 44*, 101751.

Seidler, R., Dietrich, K., Schweizer, S., Bawa, K. S., Chopde, S., Zaman, F., Sharma, A., Bhattacharya, S., Devkota, L. P., & Khaling, S. (2018). Progress on integrating climate change adaptation and disaster risk reduction for sustainable development pathways in South Asia: Evidence from six research projects. *International Journal of Disaster Risk Reduction, 31*, 92–101.

Senapati, S., & Gupta, V. (2014). Climate change and coastal ecosystem in India: Issues in perspectives. *International Journal of Environmental Sciences, 5*(3), 530–543.

Sharma, V., Reddy, B., & Sahu, N. (2018). Sustainable rural livelihoods approach for climate change adaptation in Western Odisha, Eastern India. In *Climate change adaptation and development* (pp. 161–174). Routledge.

Sivakumar, M. V. K., Vittal, K. P. R., & Eslamian, S. (2021). Indian experiences in water harvesting systems. *Handbook of water harvesting and conservation: Case studies and application examples* (pp. 325–339).

Subrahmanyam, V. P. (1988). Hazards of floods and droughts in India. In *Natural and man-made Hazards: Proceedings of the international symposium held at Rimouski, Quebec, Canada, 3–9 August, 1986* (pp. 337–356). Springer Netherlands.

Swain, M. (2014). Sources of growth and instability in agricultural production in Western Odisha, India. *Asian Journal of Agriculture and Development, 11*(1362-2016-107732), 51–70.

Swain, M., & Swain, M. (2015). Role of institutional support system in drought management: The case of Western Odisha. *Journal of Rural Development, 34*(3), 343–369.

Towolawi, A. T., Oguntoke, O., Bada, B. S., & Adejuwon, J. O. (2024). Water management and crop-resistant varieties adaptation due to climate change by farmers. *Theoretical and Applied Climatology*, 1–18.

USEPA. (2017). *Climate impacts on agriculture and food supply.* United States Environmental Protection Agency.

Verma, O. (2021). Climate change and its impacts with special reference to India. In C. Water (Ed.), *And climate change in the Himalayas: A geospatial approach* (pp. 39–55). Springer International Publishing.

Vivekanandan, E., Hermes, R., & O'Brien, C. (2016). Climate change effects in the Bay of Bengal large marine ecosystem. *Environmental Development, 17*, 46–56.

Walch, C. (2019). Adaptive governance in the developing world: Disaster risk reduction in the State of Odisha, India. *Climate and Development, 11*(3), 238–252.

Ward, F. A., & Pulido-Velazquez, M. (2008). Water conservation in irrigation can increase water use. *Proceedings of the National Academy of Sciences, 105*(47), 18215–18220.

Yildiz, I. (2019). Review of climate change issues: A forcing function perspective in agricultural and energy innovation. *International Journal of Energy Research, 43*(6), 2200–2215.

Chapter 5
Application of New Techniques in Fodder Farming for Enhancing the Pasture Grass for Yaks in the Arunachal Himalayas

Pranjit Sarma, Bhagwan B. Ghute, Arvind P. Naik, and Dinesh D. Thorat

5.1 Introduction

The State of Arunachal Pradesh lies between 26° 28″ to 29° 30″ N latitudes and 91″ 30″ to 97″ 30″ E longitudes at an average elevation of 3048 m from MSL with a total area of 83,743 km^2. The state has international boundaries all around except in the south, where it has a border with the states of Assam and Nagaland. The entire state forms part of the Lower Himalayas. Many of the regions in the state experience snowfall, and snowmelt feeds the downstream river system. The area is ascribed as one of the important biodiversity hotspots in the world (Myers et al., 2000; Wambulwa et al., 2021). The state is home to various endangered species. The state is enriched with a plethora of natural resources, including abundant medicinal plants and wildlife habitat. Physiographically, this region has steep mountains, gorges, canyons, and waterfalls. The study on the lower part of Arunachal Himalaya, which has a temperate climate and tropical vegetation, was carried out in the year 2019–20. The experimental plot had an average annual rainfall ranging from 18.78 to 15.36 mm, whereas the average temperature varies between 16.42 °c to 10.63 °c.

P. Sarma
Department of Geography, Guwahati University, Guwahati, Assam 781014, India

B. B. Ghute (✉)
Department of Geology, Toshniwal Arts, Commerce, and Science College, Hingoli, Maharashtra 431542, India
e-mail: bhagwangeo@gmail.com

A. P. Naik · D. D. Thorat
Department of Dairy Science, Toshniwal Arts, Commerce, and Science College, Hingoli, Maharashtra 431542, India

P. K. Rai and S. Rai (eds.), *Climate, Environment and Agricultural Development*, Advances in Geographical and Environmental Sciences,
https://doi.org/10.1007/978-981-97-8363-2_5

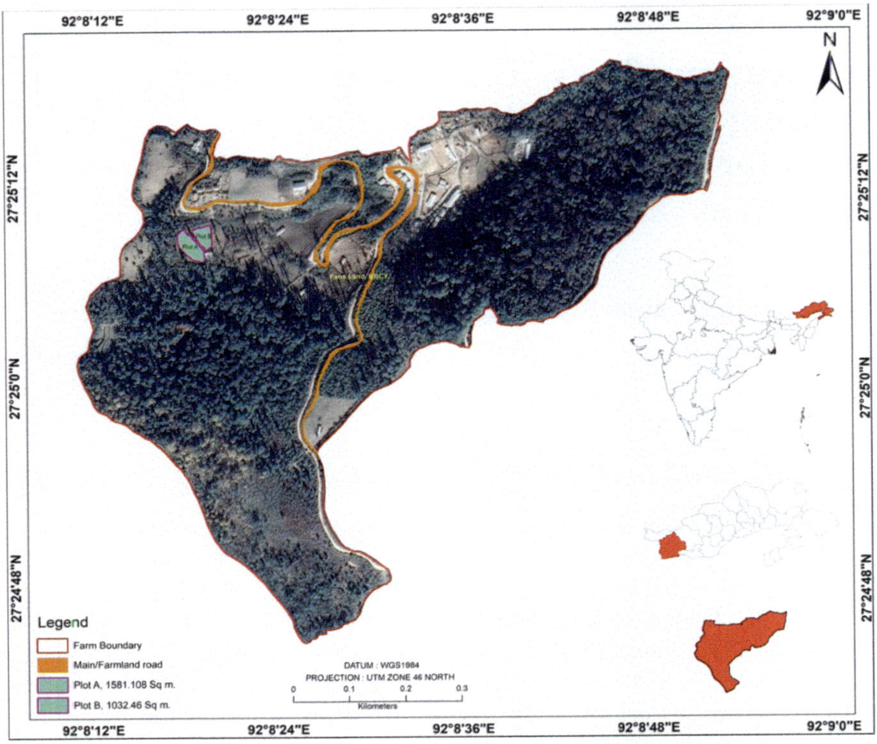

Fig. 5.1 Base map showing the pasture plots

The region is characterized by Clay Mountain red-yellow soil developed on a sandy rock (Fig. 5.1).

5.2 Materials and Methods

5.2.1 Methodology

An experiment was done to record the growth and development in PLOT-A designated with plastic mulch and PLOT-B without mulch, where sowing the high-yielding verities (HYV) seeds by scientific and systematic methods was carried out. Structural beds were constructed at an equivalent distance (6×0.91 m) with a $2''$ interval between two beds for sowing the seeds in PLOT-A and PLOT-B for further analysis (Fig. 5.2). The spacing of holes was planned at 40 cm intervals at a frequency of 13–45 for Mulch, while for sectors without mulch, the hole interval was chosen at 40 cm with

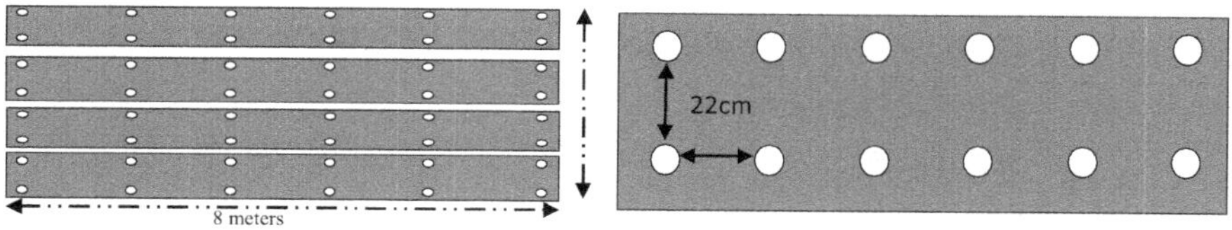

Fig. 5.2 Sketch profile of established Mulching beds

a frequency of 14–45. Arranged the beds like lucernes-dactylics-lucernes, dactylics-lucernes-dactylic, frenchbeans-dactylic-frenchbeans, dactylic-frenchbeans-dactylic, only dactylic, lucerne-lolium-lucerne, lolium-lucerne-lolium, lolium-lolium, only lolium beds were prepared for sowing.

5.2.2 *Sample Collection*

Fifth natural growth was sampled from different beds covered with (Mulch and without mulch) on the southern aspects of NKD farm at an elevation of 2623 m from AMSL. From the plots, samples were collected at particular intervals throughout 2019 and 2020 from May to September. The different attributes like plant height (cm), the number of sward density/plant, and the number of fruits/plants were determined manually for both variants, the one with the mulch and the other without mulch. The length of French bean fruits planted in between the beds covered with mulch and without mulch was measured. Proximate analyses were carried out from the samples from mulch and without mulch (Table 5.1).

Table 5.1 Range of the chemical composition of herbage biomass

Chemical composition	Biomass type	Mulch (%)	Without Mulch (%)
Moisture	Dactylics glomerata	29.86–87.00	24.11–81.37
	Lolium perenee	29.42–82.95	24.88–110.83
Ash	Dactylics glomerata	6.59–11.24	7.69–10.09
	Lolium perenee	7.97–12.07	7.37–12.07
Fats	Dactylics glomerata	5.62–6.09	4.33–8.76
	Lolium perenee	5.06–9.62	6.27–10.08
Crude fibre	Dactylics glomerata	27.34–36.12	38.11–28.37
	Lolium perenee	27.04–29.99	26.80–48.35
Crude protein	Dactylics glomerata	5.02–12.02	4.59–6.59
	Lolium perenee	5.02–6.76	4.59–10.5

5.3 Management of Livestock and Their Pasture Field

Sowing of fodder grass for animals is essential to mitigate the scarcity of yak fodder on the slope of Himalaya at Nyukmadung Yak farm, which is located in the sloppy mountainous area of the lower Himalayas at an elevation of 2631 m MSL. PLOT-A is spread over an area of 1581.108 m^2 while PLOT-B is spread over an area of 1032.46 m^2 in Nyukmadung of the lower Himalayas. So, at the aspect of Himalaya, two plots were selected for the practices of fodder crops designated as PLOT-A and PLOT-B, where dactylics *glomerata* and lolium *perenee* were sown on beds covered with mulch and beds without mulch. Attempts were made to assess the feasibility of growing vegetables in the Himalayan aspect (Wambulwa et al., 2021). As an experiment, French beans between the fodder beds were planted in both mulch and non-mulch plots. Observation shows that the growth and yields of French beans in mulching beds are much higher than in non-mulching beds (Fig. 5.3). Observations further show that plant height ranges from 15 to 48 cm with a density of 2–4 on mulch, whereas plant height and density are recorded from 0.9 to 9 cm and 1 to 2, respectively. The length of French bean fruits ranged from 3 to 19 cm, where the production were from 0.7 to 1.5 kg during the practice period from July to August, 2019.

5.3.1 Combination of Fodder Crops Analysis

For the evaluation of degradation of grasslands on Himalayan slope, identification and scientific management of pasture crop land is most essential for the conservation of yak husbandry, amidst the Brokpa community. In some parts of the region yak is the main source of transportation for carrying the goods and commodities from one place to another. Off late there are concerted efforts to reduce the pressure on grass lands with improved fodder crop practices specially uses of plastic for pasture crop development. The development of pasture crop at the slope of lower Himalaya was given more emphasis. Salix and maize (green forage) practises were promoted extensively for preservation during scarcity days mainly in winter season. For the

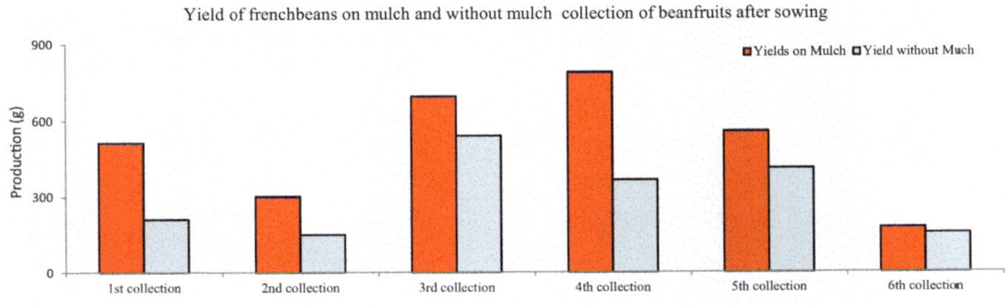

Fig. 5.3 Yield of frenchbeans in different time on mulch

first time lucernes medicagosativa, with protein content of 10% were introduced on the mulching beds.

From the combination beds of lucernes-dactylics-lucernes on the beds with mulch, the plant height ranged from 88 to 109 cm, whereas the plant density varies from 75 to 85.The plant height from the same line ranged from 28 to 58 cm and the plant density ranged from 18 to 65. On another bed which was constructed on mulch with the combination of dactylics-frenchbeans-dactylics shows the plant height ranges from 6 to 108 cm whereas the plant density ranged from 2 to 215 on mulch. Hence, the plant height varied from 4 to 15 cm and plant density ranged from 1 to 5 on without mulch. Again, on other mulching bed covered with frenchbeans-dactylics-frenchbeans shows the growth performance of plant height ranges from 12 to 105 cm while the plant density ranges from 1 to 125 on mulch whereas the plant height ranges from 2 to 98 cm and plant density ranges from 1 to 78 on without mulch. The beds covered with dactylics-lucernes-dactylics, the plant height ranges from 65 to 120 cm and plant density ranges from 60 to 102 on mulch whereas the plant height ranges from 51 to 79 cm whereas the plant density ranges from 25 to 45 on without mulch. At the top of the mulching plot covered with lolium-lolium on the both side of bed recorded the plant height ranges from 81 to 89 cm where the plant density varies from 65 to 75 on mulch, whereas the plant height ranges from 48 to 51 cm and plant density varies from 25 to 40 on without mulch. The application of plastic culture to the establishment of plant for fodder production is more essential during scarcity in winter period (Bahadur et al., 2018). Several studies have proved that the plastic mulch improves nutrient availability and regulates soil temperature and moisture content and enhancing the total yields. Experiment shows that the growth of herbage biomass on plastic mulch without fertilizer N had perform healthy with high sward density then without plastic.

5.3.2 Lucerne (Medicagosativa) on the Pasture Plot

Lucerne is one of the forage crops cultivated with the aim of producing forage in farms leaning animal production?. It has superior forage qualities and high yields that can be consumed by livestock readily (Yolcu et al., 2006). It is important for good quality meat and milk production. Lucerne is rich in vitamins, minerals and protein and also the productivity of crude protein from the unit area is high. Lucerne hay has significantly high digestibility coefficients for crude protein, crude fibre, organic matter and fat compared to grass (Cochran et al., 1986; Sommer et al., 2005). To meet the increasing demand of fodder crops, *medicagosativa* (Fig. 5.5) were introduced on the four mulching beds at the pasture plot on the aspect of Himalaya. Continuous monitoring was carried out to record their growth performance and development in various climatic conditions. Detailed study on the height, density, and seeding time for the Lucerne crop was done from experimental site. Before cut the plant height was ranging from 2.3 to 12.4 cm whereas the plant density ranges from 7 to 19 on mulch. After first cut the plant height ranges from 3 to 40 cm whereas the plant

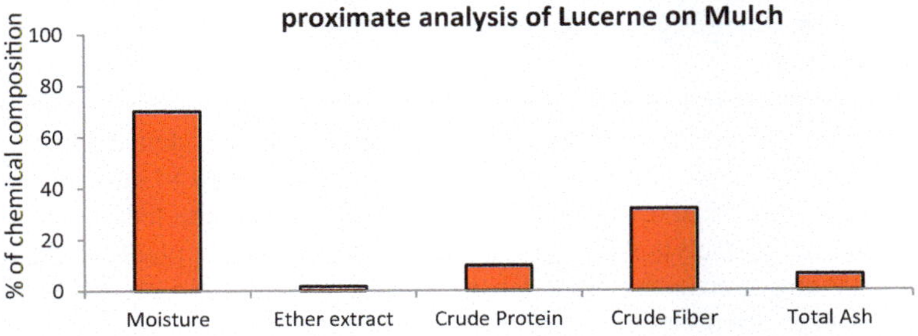

Fig. 5.4 Chemical composition of lucerne on mulch

Fig. 5.5 Shows the growth of medicago sativa L on beds cover with mulch

density ranges from 7 to 26 on mulch and recorded the total yield of 3 kg. Satisfactory growth has been observed without use of nitrogen fertilizer in the pasture plot during summer season. *Lucerne* is suited to naturally alkaline and free-draining soils. It is severely restricted on soils with pH below 6.2. Application of lime will not be enough to correct naturally acidic soils. Chalk and limestone soils are most suitable for Lucerne while the crop does not tolerate waterlogged soils *Lucerne* is slower to introduce than ryegrass.

5.3.3 Results

The chemical composition of lucerne was carried out to compare the crude protein (9.84%), Moisture (70.06%), Total Ash (6.28%), Ether extract (1.78%) and Crude fibre (32.05%) were laboratories (Fig. 5.4).

5.3.4 Proximate Analysis

Proximate analysis comprises of determination of Moisture, Ash, Fats, Crude Fibre and Protein. Thus it is important to extract the nutritional value from the biomass through proximate analysis (Karatepe & Küçükbayrak, 1993). The moisture content of dactylics *glomerata* ranges from **87.00%** to **29.86%** on mulch whereas it ranges from **81.37%** to **24.11%** for the beds without mulch.

On the other hand the moisture content of lolium *perenee* ranges from **82.95%** to **29.42%** on mulch whereas it varies from **110.83%** to **24.88%** on without mulch. Some climatic factors determine moisture content of that area. Ash content from dactylics *glomerata* ranges from **11.24%** to **6.59%** on mulch whereas it ranges from **10.09%** to **7.69%** on without mulch (Fig. 5.4).

Analysis of total Ash content of lolium *perenee* ranges from **12.07%** to **7.97%** on mulch whereas it ranges from **12.07%** to **7.37%** on without mulch.

Estimation of fats is very vital for health of animals to survive in high altitude areas hence the quantity of fats was analysed from fodder crops like dactylics *glomerata* and it was found that the fat quantity ranges from **6.09%** to **5.62%** on mulch whereas it ranges from **8.76%** to **4.33%** on without mulch. Simultaneously estimation of fats from lolium *perenee* was found to be in the range of **9.62%** to **5.06%** on mulch whereas it ranges from **6.27%** to **10.08%** on without mulch.

Again, crude fibre which is considered the most important elements for animal health for nutrition found in the fodder crops. Experiment shows that the fibre content in dactylics *glomerata* varies from **36.12%** to **27.34%** on mulch whereas it varies from **28.37%** to **38.11%** on without mulch whereas the percentage of fibre on lolium perenee varies from **27.04%** to **29.99%** on mulch whereas it ranges from **26.80%** to **48.35%** on without mulch.

Determination of protein is another important element for animal's health, shows the significant increases of protein in fodder crops. Vitamins including protein are more essentials for reproduction. Significant variation was observed after laboratory experiments about protein from the fodder. Protein content from dactylics *glomerata* ranges from 5.02% to 12.02% on mulch whereas it ranges from **4.59%** to **6. 59%** on without mulch hence, the protein contents of lolium *perenee* ranges from 5.02% to 6.76% on mulch whereas it ranges from **4.59%** to **10.5%** on without mulch (Fig. 5.6).

5.4 Impact of Climatic Phenomena on the Production of Herbage Biomass on Mulch and Without Mulch (*Dactylics Glomerata and Lolium Perenee*)

Dactylics is a genus of Eurasian and North African plants known as orchard grass, it is a cool season perennial grass which was sowing last year at the pasture plot located on the Himalaya. Animal husbandry is the prime occupation for the people who are the native dwellers in mountainous region. Due to pressure on local environment on the global hotspot region by direct activities to fulfilled their needs inverse effect have been marked on the environment and on yak habitat as well. An attempt is made to record the yield of herbage biomass in various climatic condition of

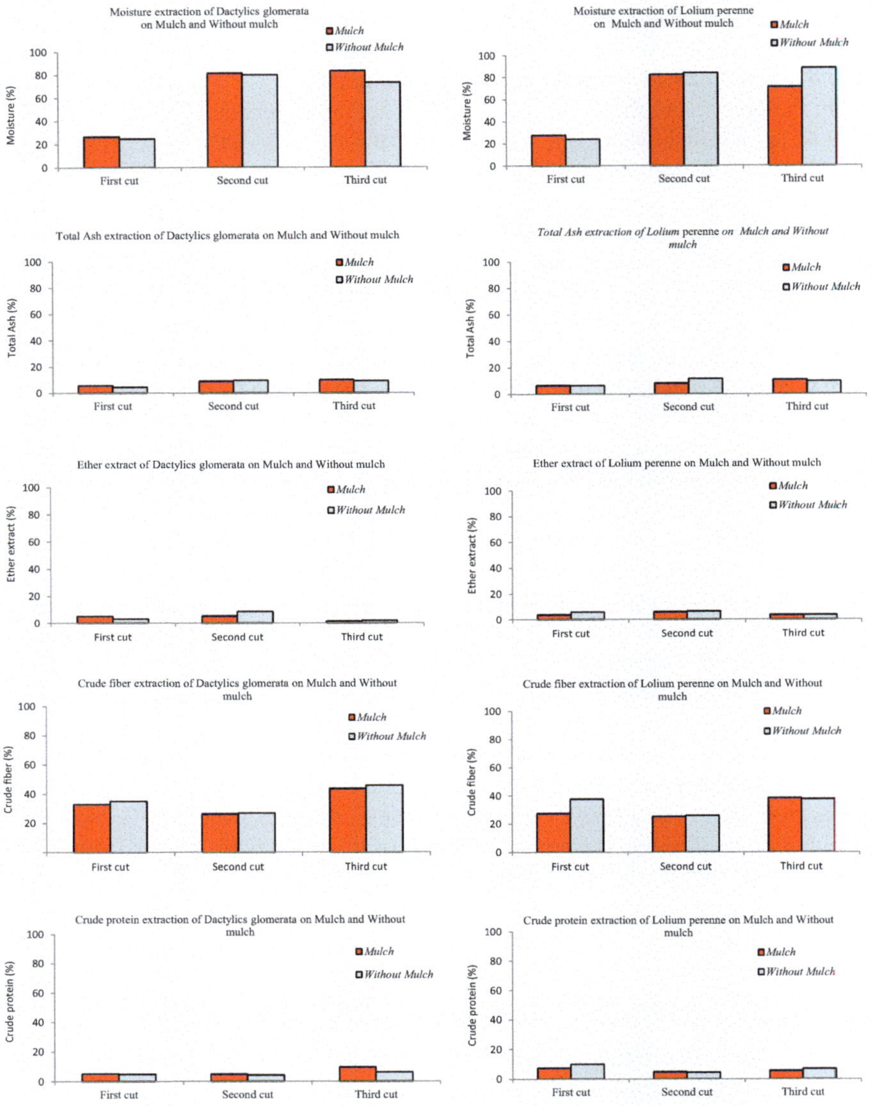

Fig. 5.6 Proximate analysis of fodder crops, dactylics glomerata and lolium perenee

precipitation and temperature data (Fig. 5.7). Particularly in alpine areas, changes in snow patterns and temperatures are already affecting the distribution of some plant species. Alpine ecosystems are known to react sensitively to climate warming since most plant species have altitudinal limits that are set by various climatic parameters and limitation of resources (Gairola et al., 2010). Thus, alpine plant species

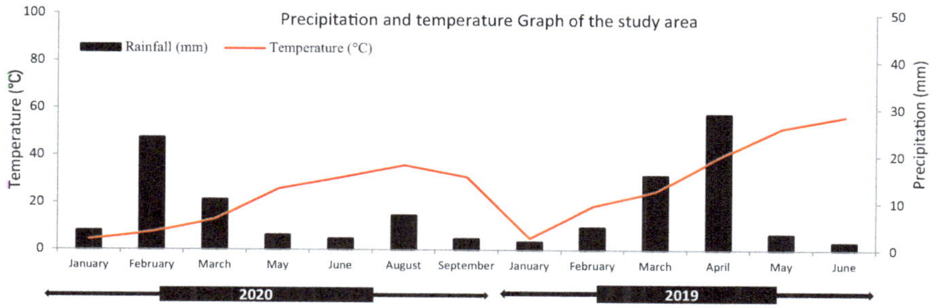

Fig. 5.7 Precipitation and temperature graph of the study area

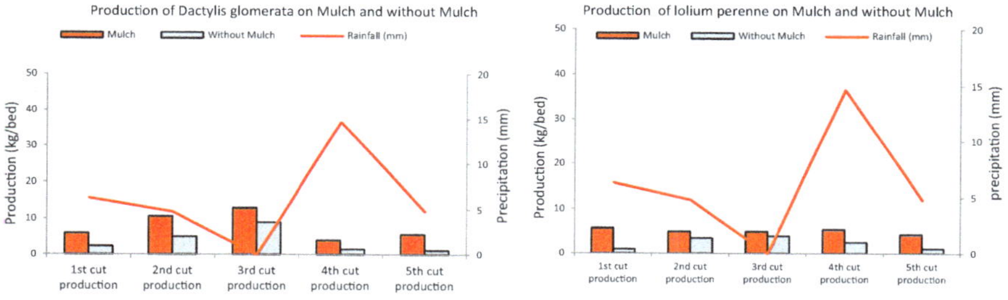

Fig. 5.8 Yield of herbage biomass in different time on mulch and without mulch

have various morphological and physiological means of adaptation against adverse climatic conditions. It is well known that the growth response of plants to climate depends on their life history characteristics and eco-physiology and largely differs between species (Benning et al., 2002).

Yield data were compiled after harvesting biomass in equi-distant time interval and record from the plant species on the bed cover with mulch and without mulch. The production of dactylics *glomerata* and lolium *perenee* ranges from 5 to 11 kg whereas lolium perenee ranges from 4 to 5 kg on mulch on the other hand the yield of dactylics glomerata recorded the ranges from 1 to 9 kg whereas yield of lolium perenee ranges from 1 to 4 kg on without mulch (Fig. 5.8).

5.5 Conclusion

Based on the above research activity in farming techniques at Nyukmadung yak farm, it is feasible and plausible to address the issue of conservation of fodder crops throughout the year, increase the yield by adopting mulching methods. Salix and maize (green forage) practices were promoted extensively for preservation during scarcity days mainly in winter season.

The development of pasture crops at the slope of the lower Himalayas was given more emphasis. Salix and maize (green forage) practises were promoted extensively for preservation during scarcity days, mainly in the winter season. For the first time, *lucernes medicagosativa*, with protein content of 10%, was introduced on the mulching beds. Proximate analysis has helped in understanding the moisture content, ash, crude protein, Fat, and crude fibre. With Proximate analysis, the nutritional value of animal feed can be assessed. Mulching practice in fodder crop is invariably desirable for better yield, good luxuriant and high-density growth, and last but not least, high nutritional value addition to the fodder crop.

Acknowledgements We thank the ICAR-National Research Centre on Yak in Arunachal Pradesh, India, and Dr. Vijay Paul, Principal Scientist, ICAR-NRC, Dr. Narayan Gadagkar, Deputy Director (Retd.) GSI, Hyderabad, Dr. Saidul Islam, and the anonymous reviewers for their invaluable support and assistance during this research.

References

Bahadur, S., Pradhan, S., & Verma, S. Maurya R., Verma, S. K. (2018). Role of plastic mulch in soil health and crop productivity. In B. R. M. Sharma & P. K., Jirli (Eds.), *Climate change and its implications on crop production and food security* (pp. 338–344). Mahima Publications.

Benning, T. L., LaPointe, D., Atkinson, C. T., & Vitousek, P. M. (2002). Interactions of climate change with biological invasions and land use in the Hawaiian Islands: Modeling the fate of endemic birds using a geographic information system. *Proceedings of the National Academy of Sciences, 99*(22), 14246–14249. https://doi.org/10.1073/pnas.162372399

Cochran, R. C., Adams, D. C., Currie, P. O., & Knapp, B. W. (1986). Cubed alfalfa hay or cottonseed meal-barley as supplements for beef cows grazing fall-winter range. *Journal of Range Management, 39*(4), 361–364.

Gairola, S., Shariff, N. M., Bhatt, A., & Kala, C. P. (2010). Influence of climate change on production of secondary chemicals in high altitude medicinal plants: Issues needs immediate attention. *Journal of Medicinal Plants Research, 4*(18), 1825–1829. https://doi.org/10.5897/JMPR10.354

Karatepe, N., & Küçükbayrak, S. (1993). Proximate analysis of some Turkish lignites by thermogravimetry. *Thermochimica Acta, 213*, 147–150. https://doi.org/10.1016/0040-6031(93)80012-Y

Myers, N., Mittermeier, R. A., Mittermeier, C. G., da Fonseca, G. A. B., & Kent, J. (2000). Biodiversity hotspots for conservation priorities. *Nature, 403*(6772), 853–858. https://doi.org/10.1038/35002501

Sommer, A., Vodňanský, M., Petrikovič, P., & Požgaj, R. (2005). Influence of lucerne and meadow hay quality on the digestibility of nutrients in the roe deer. *Czech Journal of Animal Science, 50*(2), 74–80. http://www.agriculturejournals.cz/web/cjas.htm?volume=50&firstPage=74&type=publishedArticle

Wambulwa, M. C., Milne, R., Wu, Z., Spicer, R. A., Provan, J., Luo, Y., Zhu, G., Wang, W., Wang, H., Gao, L., Li, D., & Liu, J. (2021). Spatiotemporal maintenance of flora in the Himalaya biodiversity hotspot: Current knowledge and future perspectives. *Ecology and Evolution, 11*(16), 10794–10812. https://doi.org/10.1002/ece3.7906

Yolcu, H., Tan, M., & Serin, Y. (2006). Effects of early cutting time and stubble height on yield and quality in lucerne. *New Zealand Journal of Agricultural Research, 49*(2), 201–206. https://doi.org/10.1080/00288233.2006.9513710

Chapter 6
Solid Waste as a Sustainable Source of Energy: Current and Future Prospects in India

Prashant Katiyar, Deepshikha Kushwaha, and Amrita Raj

6.1 Introduction

With a population of approximately 1210 million, India has a significant urban population comprising more than 377 million people, accounting for 31.16% of the total population (Census of India, 2011, GOI, 2011). The rapid rate of urbanization has placed additional burdens on the socio-economic and environmental aspects include animal population shifting and the fast depletion of natural resources (Gerdes & Gunsilius, 2010). According to the central pollution control board (CPCB) of India, per capita waste generations have been increase exponentially from 0.26 kg/day to 0.85 kg/day. Unfortunately, approximately 80% to 90% of municipal waste is disposed of in landfills sites or waste incineration without imposing any significant management practices leading to air, water and soil pollution take place (Ahluwalia & Patel, 2018; Joshi & Ahmed, 2016). Although, lack of public awareness regarding the waste management practices and careless attitude dispose unwanted waste at residential areas create significant challenge to the municipal cooperation. Improper waste disposal in landfill sites substantially craft potential threats by emitting unwanted greenhouse gases in environment and affect the environment integrity adversely, as it contaminate groundwater system through the formation of leachates (Ko et al., 2015; Li & Zhao, 2001; Ngwabie et al., 2019). On the other hand, noise and dust pollution during waste transportation cause nuisances for the elderly and newborns as well

P. Katiyar
Department of Biotechnology Engineering, Goel Institute of Technology and Management, Lucknow 226010, UP, India

D. Kushwaha (✉)
Center for Microbiology, Ewing Christian College, Prayagraj 211003, UP, India
e-mail: deep_shikha200@yahoo.com

A. Raj
Department of Applied Science, United University, Prayagraj 211002, UP, India

P. K. Rai and S. Rai (eds.), *Climate, Environment and Agricultural Development*, Advances in Geographical and Environmental Sciences,
https://doi.org/10.1007/978-981-97-8363-2_6

(Yusof et al., 2009). Despite of these facts, the proliferation of single-use plastics has become a global threat on account of their harmful and non-biodegradable nature. An excessive plastic waste generation in cities can lead to blockage of drainage system during monsoons season, resulting in urban flooding. This, in turn, leads to the intermixing of microplastics with water bodies including river and oceans as suggested by Anderson et al. (2016). United Nations, 1992 report shown that microplastics disrupt an aquatic life primarily by affecting the food chain and contribute in global warming. Further, microplastics presence in water system has been identified as a significant cause of species extinction worldwide.

According to a report published by the UNPD, the world produces approximately 300 million tons of plastic waste annually, with only 9% being recycled, around 14% collected for recycling and rest ending up in the ocean (Plastic recycling: an underperforming sector ripe as documented by Remake, (2019). Municipal solid waste management (MSWM) also faces some challenges relevant to the hazardous chemical waste generation from hospitals and industries, leading to respiratory concern problems and resultant into premature death of affected individuals (Joshi & Ahmed, 2016; Mohan, 2019).

India has emerged as a growing market for recycling solid waste in recent years; however, the recycling efforts have not fully met the prescribed standards (Bhattacharya et al., 2018). Improper management of municipal solid wastes (MSWs) at landfill sites often attracts animals, rodents, mosquitoes, vultures and scavengers, posing health risks and even causing harm or death to frontline workers and waste pickers (De Bercegol et al., 2017). Nevertheless, India popular cities including Surat, Alleppy, Bobbili, Panji and Pune had positive intent in developing good solid waste management (SWM) strategies. Further, municipalities primarily concentrate on the solid waste collection, without adopting advanced treatment methodologies. Therefore, in this aspect further upgrading and improvement is needed in order to resolve an issue of solid waste recycling effectively (Annepu, 2012).

Along with this, an effective retrieval of valuable materials impose significant challenges and with the assistance of informal sector, it can be possible to integrate into the mainstream of the municipal solid waste management (MSWM) process. However, the realization of idea can only be achieved when the constant support and funding avail from government agencies, as well as increased public awareness and participation, along with the need to overcome social taboos. Therefore, current book chapter highlight comprehensive review of all possible solid waste management strategies and addressing the current challenges and potential opportunities for future urban cities in India.

6.1.1 Research Significance

Despite numerous attempts to address solid waste management (SWM) issues, a holistic approach to handling solid waste is still lacking. The majority of cities in India continue to collect mixed waste, without enforcing appropriate centralized or decentralized treatment systems. This paper highlights the existing gaps and missing interlinks in addressing the challenges of MSWM in India. The present situation has become critical as many municipal authorities fail to conduct proper assessments of solid waste management services. As a result, unscientific dumpsites lacking proper treatment have now become prevalent on account of absence of good and efficient solid waste management practices.

6.1.2 Available Renewable Energy in India

Now a days, considerable number of areas in India are experiencing insufficient power supply, with regular power cuts lasting for 4–8 h a day. Despite the abundance of renewable energy sources (RES) in the country, the achievement of improved power supply targets has been hindered by a lack of substantial efforts. India's potential for renewable energy varies across different regions due to varying geographical and climatic conditions. Notably, the surrounding sea presents a significant potential source of renewable energy. Furthermore, the eastern region (ER) possesses geothermal energy beneath its surface, while the western, southern, and central regions have ample availability of wind and solar energy. The northern region (NR) of India also holds potential for biomass energy as a renewable source (CEA, 2017). As of May 2017, the utility electricity sector in India had an installed capacity of 330.26 GW. Among the installed capacity, coal sources contributed 59.2%, natural gases contributed 7.6%, diesel energy accounted for 0.25%, nuclear power contributed approximately 2.05%, large hydropower (LHP) plants accounted for nearly 13.5%, and renewable sources contributed around 17.33%. Figure **6.1** provides an updated representation of the energy contribution breakdown. As of 30 November 2021, India's total installed capacity has increased to 392.017 GW, as shown in Fig. 6.1. Fossil fuels account for 60% of the installed capacity, while renewable energy sources contribute 26% to the overall installed capacity. However, in terms of energy generation in 2020–21, renewable energy sources only accounted for 10.7% of the total share, despite their larger installed capacity. In contrast, coal, which makes up 52% of the overall installed capacity, produced nearly 71% of the energy generated in 2020–21.

In July 2021, India achieved highest-ever peak demand of 203,014 MW. Looking at the absolute figures, the installed capacity in July 2021 (386.88 GW) indicate that there should have been sufficient surplus energy to meet the existing demand. However, data released by the Central Electricity Authority (CEA, 2017) explain differently in terms of rising energy demand. The peak demand could not be adequately met and it has been become an ongoing trend. In fact, this trend extends beyond peak demand and affects the overall energy demand as well.

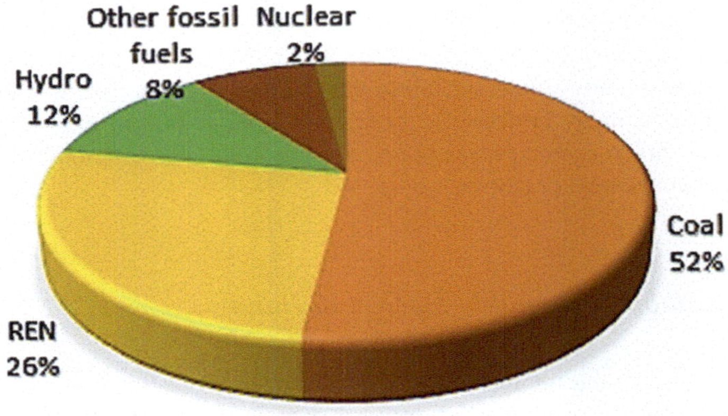

Fig. 6.1 Installed capacity of energy sources in India (as per CEA Report, 2021)

6.2 Waste Characteristics with Its Categories

In broad terms, solid waste refers to waste that is not transported by water and has been deemed unusable for further purposes. This waste category encompasses industrial, mining, municipal and agricultural waste. Solid waste primarily consists of significant amounts of organic matter, ash, fine earth, paper, plastic, glass and metals (Sharholy et al., 2007). However, the composition of solid waste varied as per weather conditions and living standards. Table 6.1 indicate the solid waste classification that is, based on the available origin source and sub-categorized in term of waste types and involved process unit. Further municipal solid waste classified into two sub-categories: putrescible solid waste, referred to as "garbage," and non-putrescible waste, known as "rubbish." Rubbish either include combustible material such as paper and plastic or it may be non-combustible materials including glass and metal. Additionally, other solid waste include construction debris, leaves, street litter, abandoned automobiles and old appliances. The composition of municipal solid waste typically comprises of organic matter (51%), recyclables (17.5%) and other inert materials (31%) (Annepu, 2012). Although solid waste composition variedas per the involve recycling materials and their origin source.

6.3 Solid Waste Generation in India

With the growth of population, increasing urbanization and advancements in technology leading to higher standard of living which significantly raise the quantity and solid waste diversification on account of intense industrial, mining, domestic and agricultural activities. Globally, in 2002, the total amount of waste generated reached to12 billion tonnes, with 11 billion tonnes being industrial waste and rest

Table 6.1 Waste types and source origin

Source of origin	Waste generator	Waste types
Residential	Household activities	Food waste, paper, cardboard, plastics, wood, glass, metals, electronic items etc
Industrial	Manufacturing units, power plants, process industries etc	Housekeeping wastes, hazardous wastes, ashes, special wastes etc
Commercial & Institutional	Hotels, restaurants, markets, office buildings, schools, hospitals, prisons etc	Bio-medical waste, Food waste, glass, metals, plastic, paper, special wastes etc
Construction and Demolition	New construction sites, demolition of existing structures, road repair etc	Wood, steel, concrete, dust etc
Municipal services	Street cleaning, landscaping, parks and other recreational areas, water and wastewater treatment plants	Tree trimmings, general wastes, sludge etc
Agriculture	Crops, orchards, vineyards, dairies, farm etc	Agricultural wastes, hazardous wastes such as pesticides
Mining	Open-cast mining, underground mining	Mainly inert materials such as ash

1.6 billion tonnes of waste treat as municipal solid waste (MSW). Till date, it is projected to 19 billion tonnes of solid waste annually by the year 2025 (Yoshizawa et al., 2004).

Asian countries accounts for the generation of 4.4 billion tonnes of solid waste each year, with MSW comprise of 790 million tonnes. Within Asia, alone India generates around 48 million tonnes (6%) of MSW (CPCB, 2000; Yoshizawa et al., 2004). By the year 2047, MSW generation in India is expected to reach 300 million tonnes, requiring approximately 169.6 square kilometers of land for disposal. However, in 1997, only 20.2 square kilometers were utilized for managing 48 million tonnes of waste (CPCB, 2000). Figure 6.2 presented an overview of the current status of solid waste generation in India, including both non-hazardous and hazardous waste imminent from various sources (Gupta, 1998; Sengupta, 2002). It is noteworthy that in addition to municipal waste, organic waste from agricultural sources alone contributes over 350 million tonnes per year, as depicted in Fig. 6.2.

However, recent reports indicate that agricultural sources alone contribute to the generation of approximately 600 million tonnes (MT) of waste in India (Asokan, 2004). Of which majorly agricultural waste contribute a lot including sugarcane bagasse, paddy, wheat straw, wheat husk, vegetable and food waste including tea waste, oil production residues, jute fibers, groundnut shells, wood mill waste, coconut husks and cotton stalks (Gupta, 1998; Maudgal, 1995; Sengupta, 2002). In this aspect, other non-hazardous industrial inorganic solid waste, notable examples consist of coal combustion residues, bauxite red mud and tailings from primary extraction

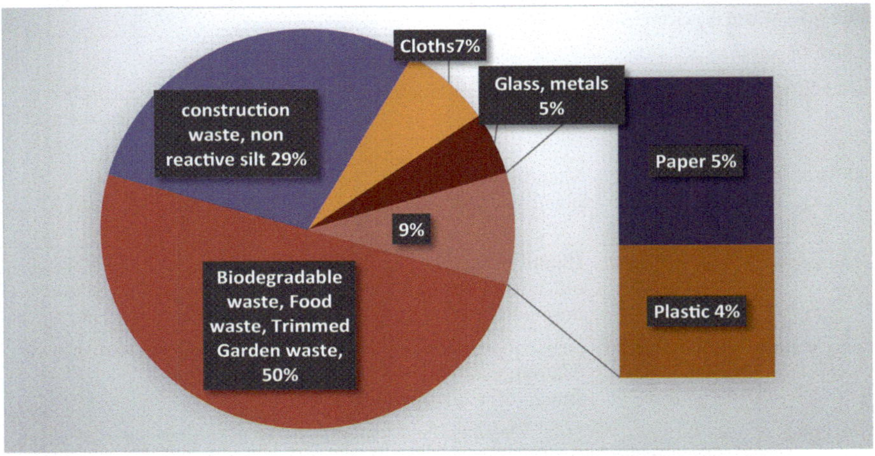

Fig. 6.2 Current scenario of solid waste in Indian cities

processes of aluminum, iron, copper and zinc. The net annual generation of these inorganic industrial wastes in India amounts to 290 MT per year (Gupta, 1998; Ramachandra & Saira, 2004). On the other hand, hazardous wastes, produced from various industrial processes such as electroplating, metal extraction, galvanizing, refining, petrochemicals, pharmaceuticals and pesticide production contribute to an annual generation of approximately 4.5MT of solid waste in India (Asokan, 2004; http://envfor.nic.in, 2004). However, it is important to note that the total volume of solid wastes discharge from different industrial processes in India, is believed to be even higher than the reported data. With the discharge of solid waste in a continuous fashion in an open atmosphere create accumulation of solid waste, which can increase their amount annually and expel toxic gaseous significantly. To address environmental degradation, two major aspects are important including limited use of energy consumption and their financial facts can be control by various organizations of India including, United States Environmental Protection Agency (USEPA), with recommended guidelines in terms of qualitative-wise management of proper-waste generation, transportation, disposal treatment and their recycling (Hammitt, 1989; http://envfor.nic.in, 2004; http://www.epa.gov, 2003; Saxena & Jotshi, 1996).

The safe management of hazardous waste is of utmost importance and maintained the balance between the socio-techno-economic issues and environment concern globally in order to achieve a cleaner and greener environment. Other than this, heterogeneous nature of the vast quantity of available waste adds complexity in recycling and utilization efforts. The physico-chemical properties of hazardous and non-hazardous solid waste and sources in relation to clay and cement are dependent on the raw materials type, mineralogical origins, operating processes and their efficacy. These characteristics categorized wastes indicate their potentiality of recycling and their formulation at production scale for various value-added products generation (Agrawal et al., 2004; Asokan, 2004; Bhattacharyya et al., 2004; Garg et al., 1996; Saxena & Asokan, 2002; Yalcm & Sevinc, 2000).

6.3.1 *Organic Waste Generation with Physiochemical Properties*

Solid waste generated from organic sources encompasses municipal and urban waste, animal waste, farming waste, horticultural waste, domestic refuse and other agro-industrial waste. The availability of agricultural resources, manpower and technological innovations has led to the establishment of various agro-industries in India.

The primary objective of waste management systems is to maximize economic benefits while ensuring environmental protection. Urban waste predominantly consists of organic matter (46%), paper (6%), glass (0.7%), rags (3.2%), plastic (1%), with the remaining portion being moisture (Maudgal, 1995). Animal waste primarily consists of organic matter and moisture. The decomposition of both animal and urban organic waste can be achieved through aerobic or anaerobic digestion. Considering the substantial annual production of these organic wastes in India, there is significant potential for CH_4 production, which not only helps in reducing greenhouse gas emissions but also contributes to mitigating global warming.

India boasts abundant agricultural resources, and agricultural waste is generated as a byproduct of various agricultural activities such as crop production, crop harvest, saw milling and agro-industrial processing, among others. For instance, the sugar industry alone produces around 90 million tonnes (MT) of sugarcane bagasse annually, which is utilized in the manufacturing of insulation boards, wall panels, printing paper and corrugating medium (BMTPC, 2005; Sengupta, 2002). There is a growing concern regarding the burning of agricultural waste, as it significantly contributes to global warming. Efforts have been made to explore and utilize organic waste materials such as peanut husk, mahau and linseed residues, coconut coir dust, rubber seed pods and spent cashew nut shells for various commercial applications (Khan et al., 2022).

6.4 Current Waste Management State in India

Current estimates indicate that the global annual production of municipal solid waste (MSW) amounts to 2.01 billion tons. Unfortunately, approximately 33% of this generated MSW is not being properly handled, leading to a widespread issue, particularly in developing countries (Shah et al., 2021). In the case of India, for example, around 52.9 million tons of MSW were generated in 2018, followed by 53.2 million tons in 2019 (Shah et al., 2021). Although solid waste management (SWM) in India has improved over the years, it is still an evolving process, and the percentage of waste processed annually remains relatively low compared to developed nations, currently standing at approximately 60% (Kumar et al., 2020; Ramaiahet al., 2017; Ren et al., 2021; Varjani & Upasani, 2021).

The increasing population in India further exacerbates the issue as more waste is generated, resulting in a higher proportion of unmanaged MSW (Kumar et al., 2020; Ramaiah et al., 2017; Ren et al., 2021; Varjani & Upasani, 2021). Unmanaged MSW poses several challenges for SWM in India, particularly in the areas of waste collection, storage and transportation, which require immediate attention (Gallardo et al., 2015; Gupta et al., 2015; Ramachandra et al., 2018).

The SWMinfluence especially in developing countries is often flawed on account of various factors. Firstly, the diverse composition of MSW necessitates costly and time-consuming handling procedures (Das & Bhattacharyya,2015; Soni et al., 2016). Secondly, developing countries typically lack doorstep MSW collection facilities and struggle with lower recycling rates (Negi et al., 2019). As a result, only a small portion of MSW is processed, while the majority is unscientifically disposed of in landfills (Negi et al., 2019). Thirdly, developing countries face a shortage of facilities to adequately treat the growing volumes of MSW, leading to unregulated dumping that contaminates water sources (such as groundwater) and the air, posing risks to humans, animals, and plants (Datta & Kumar, 2017). Consequently, significant efforts are required to implement robust solid waste management techniques and minimize the amount of unmanaged MSW (Dutta & Jinsart, 2020; Rajmohan et al., 2021).

6.5 Solid Wastes Composition in Indian Cities

The composition of waste plays a crucial role in determining waste management practices. Higher-income groups tend to use more packaged products, resulting in increased volume of paper, glass, metals, plastics and textiles compared to lower-income groups (Sridevi et al., 2012). Municipal solid waste (MSW) also contain hazardous materials such as paints, used medicines, pesticides, e-waste and batteries. The informal sector's economy relies heavily on the composition of the generated waste. Previous studies have shown that approximately 40% to 50% of solid waste composition is organic, around 30% consists of inert and construction/demolition waste and the remaining portion is recyclable waste. The average calorific value of solid waste in India ranging from 1500 to 2200 kcal/kg, which is lower compared to developed countries due to the lesser use of paper and plastic waste (de Souza et al., 2017). The moisture content in solid waste is higher in India compared to developed countries. Different Indian states exhibit regional differences in solid waste generation, including in composition, moisture content and calorific value, within cities and across states, union territories and regional zones of India.

6.6 Waste-to-Energy (WtE) Related Environmental Issues

6.6.1 *Environment and Health Issues*

The degradation of the environment is closely linked to human health concerns (Shukla et al., 2000). Workers, both in formal and informal sectors, are at direct risk to their health when proper protective gear, such as gloves, uniforms and safety equipment, is not used. Individuals living near disposal sites exposes to a high risk of gastrointestinal parasitic worm infections (Giusti, 2009). Open dump yards release methane and other gases during the anaerobic decomposition of biodegradable waste, leading to potential fires, explosions and significantly contributing in global warming (Slagstad & Bratteb, 2013). Odor issues and the migration of leachates into water bodies and soil are associated with these problems (Dasgupta et al., 2013; Muhammad et al., 2020; Unnikrishnan et al., 2006). Rubber discarded at dumpsites wherein; it collect water, establish breeding grounds for mosquitoes and increasing the risk of diseases like malaria, dengue and West Nile fever. Uncontrolled burning of waste at dumpsites release fine particles and smog, which are major causes of respiratory diseases (Annepu, 2012; Ghosh, 2016a; Sridevi et al., 2012). Harmful gases such as CO, CO_2, PM_2, mercury, polycyclic aromatic hydrocarbons (PAHs) and plastic pyrolysis, as well as direct release of arsenic in water, have severe impacts on newborns and adults, leading to cancer and death (Ahamed et al., 2020; Azar & Azar, 2016; Ghosh, 2016a). Microplastics discharge in the ocean can disrupt the marine ecosystem and food chain (Ivar Do Sul & Costa, 2014). ZnOnanosheets (ZnO-nS) are used to remove total H_2S and CO from syngas (Da Oh et al., 2018).

6.6.2 *Infrastructure and Finance Management*

According to the expert committee of the Ministry of Urban Development in India, it has been estimated that the investment and maintenance costs in Class 1A cities amount to Rs. 900 and Rs. 269 per capita/year, respectively (Mani & Singh, 2015). However, municipalities and urban local bodies (ULBs) face challenges in the waste sector due to insufficient financial resources and management systems, resulting in the absence of comprehensive strategies (Ghosh, 2016b). Consequently, unsegregated waste is not economically viable and often left unaddressed. In order to promote Solid Waste Management Systems (SWMS) across the country, the Government of India has sanctioned grants and funds under initiatives such as the 12th and 13th Finance Commission, JnNURM, UIDSSMT, and the Swachh Bharat Mission, with an allocation of Rs. 20,000 crores (Mani & Singh, 2015).

6.6.3 *Improper Implementation of Government Policies*

Effective management of solid waste generated within a country should be on priority basis when formulating national policies is imposed (Abas & Wee, 2014). This is particularly crucial in urban areas as a result of economic growth, urban migration, unplanned land use and the lack of appropriate legislation on solid waste management. The institutional framework for managing solid waste is still under development stage. In this regard, the Swachh Bharat Mission (SBM) initiative is started by the Ministry of Urban Development (MoUD) under the government of India has played a significant role in this aspect. As a nationwide campaign, SBM focuses on making such smart cities, where in; open defecation-free and emphasis on the social priorities of solid waste management systems, involving local municipal bodies (ULBs) (Mani & Singh, 2015).

6.7 Regulatory Hurdles

The regulatory landscape surrounding waste-to-energy technologies can be highly intricate and challenging to navigate. On one hand, certain methods, such as incineration, may be bannedas it disturb air quality index and produce byproducts like ash, which may be classified as hazardous materials. On the other hand, although advancements in the power industry have created opportunities for small producers in order to compete with established utilities, additional regulations, which govern the electrical grid, posing barriers for aspiring waste-to-energy producers.

6.7.1 *High Capital Costs*

The installation of waste-to-energy systems can be a costly endeavor. While these install systems offer potential financial advantages and surpass waste reduction and energy losses, securing the necessary financing packages significantly, especially for emerging technologies in respect to waste disposal are also exist and show a strong presence in the market.

6.8 Waste-to-Energy (WtE) Challenges and Prospectus

The increasing global concern over municipal solid waste (MSW) is shoot out on account of various factors such as rising waste volume, environmental pollution, social inclusiveness, economic sustainability and flexible environmental regulations related to MSW disposal (Zhu et al., 2008). A significant issue revolves around the high organic content of MSW and their proper disposal. Integrated technologies hold promise for large-scale production of biomass, biopolymers, and energy from waste in the future. Waste-to-Energy (WtE) systems, when integrated with other technologies, can be an effective means of treating MSW to generate energy in the form of steam and electricity (Singh, 2020).

To achieve sustainability, it is crucial to develop waste biorefineries that utilize recent advancements in resource recovery from MSW.

WtE scenarios offer viable solutions for MSW disposal and present significant economic and environmental benefits. A WtE supply chain provides a practical approach to energy generation and offers an integrated solution, including waste reduction and greenhouse gas (GHG) emissions mitigation, leading to economic and environmentally friendly outcomes (Pan et al., 2015). Furthermore, WtE helps minimize the amount of MSW deposited in landfills, typically reducing volume by about 90% and mass by 80% (Young, 2010). However, a report by an NGO stated that WtE does not contribute to the circular economy due to the absence of residues after incineration (Zero Waste European Commission, 2017).

6.9 Waste to Energy Technologies in India

Energy generation from waste is a topic that demands significant attention, although it is not a new concept. Several technologies exist to process solid waste into energy, and the process selection depend on the physicochemical properties of the waste, the type and amount of waste feedstock, and the desired energy form. The conversion of solid waste into energy typically involves three primary process technologies: thermochemical, biochemical, and mechanical extraction (McKendry, 2002).

6.9.1 Biochemical Conversion

Biochemical conversion processes involve microbial enzymes to break down feed biomass. This method of conversion offers an eco-friendly approach to obtaining energy fuel from MSW. Anaerobic digestion and fermentation employed microbial agent in this conversion process.

1. Anaerobic digestion is the direct conversion of organic matter into biogas, released gas with a calorific value ranging from 20 to 25 MJ/Nm^3 containing methane (45% to 75%) and carbon dioxide (CO_2), along with hydrogen gas.
2. Fermentation is extensively used in various countries for large-scale production of ethanol from sugar-crops. This process yielded diluted alcohols, which then require distillation. However, fermentation tends to have lower overall efficacy rate and higher plant costs.

6.9.2 Thermochemical Conversion

Thermal conversion plays a significant role in various integrated waste management strategies.

It involves four main technologies such as: combustion, gasification, pyrolysis and liquefaction. Each serving as different forms of energy carriers and to treat dried and solid waste as well.

1. Combustion is biomass burning in air. It involve widely in commercial and industrial combustion plant outputs to convert the chemical energy stored in solid waste into heat or electricity. Using boilers and turbines, are employed in this energy conversion phenomenon. Combustion of biomass with a moisture content below 50% is typically suitable, otherwise; pre-drying or it may be required as a feed biomass.
2. Gasification involve treating a carbon-based material with oxygen or steam to produce a gaseous fuel. The resulting gas can be further cleaned and burned in a gas engine or chemically transformed into methanol, a useful synthetic compound.
3. Pyrolysis entails heating biomass in the absence of oxygen, resulting in the production of liquid (referred to as bio-oil or bio-crude material), solid and gaseous fractions. The yield of each fraction varied as per heating rate, temperature, particle size, and retention time.
4. Liquefaction is the low temperature cracking of biomass molecules at high pressure, leading to the formation of a liquid-diluted fuel. This method operates at relatively low temperature (around 200 °C to 400 °C), as it compensate lower yield and require sophisticated equipment wherein; upto (50 to 200 bars) pressure level is needed

6.9.3 Mechanical Extraction

Solid waste can serve as a valuable source for producing oil, particularly from the seeds it contains. For instance, rapeseed oil can be extracted from the seeds found in solid waste. This oil can then undergo further processing through a chemical reaction called esterification, where it is combined with alcohol to produce biodiesel.

6.10 Biogas Market Options

The kind of energy derived from biogas is determined and is based on the specific requirements of the user. A form of energy obtained from biogas: electricity, heat and steam generation and transportation fuel (Munster & Lund, 2009).

6.10.1 Electricity Generation

The most common form of energy produced in modern facilities is through combined heat and power (CHP) generation. CHP, also known as cogeneration, is an efficient and reliable method of generating both power and thermal energy from solid waste. Implementing a CHP system tailored to meet a thermal and electrical needs, which can significantly enhance operational efficacy value by reducing energy cost and greenhouse gas emissions in surrounding space.

Fuel cell technology: Another promising technology is a fuel cell technology, which convert biogas into electricity with increased efficiency. While some biogas installations currently employ molten carbonate fuel cell technology, solid oxide fuel cell technology and are consider as the most promising one for future applications on account of higher power density and scalability nature.

Biogas Engines: Biogas based engines is another good option for generating electricity as it utilize biogas energy as a fuel source. The biogas based engines typically have a capacity convert 18% to 25% of the biogas into electricity, depending on the engine design and load factor.

Micro-gas turbines: The micro-gas turbines are small gas turbines specifically designed to utilize biogas energy. This technology show fabulous advantage as it lower nitrogen oxide (NOx) emissions in atmosphere at reduced maintenance cost. However, it should be noted that micro-gas turbines are slightly less energy-efficient machines compared to internal combustion (IC) engines and come with higher upfront costs.Overall, these various technologies provide diverse options for harnessing energy from biogas, catering needs and offering potential benefits in terms of efficacy, emissions reduction and cost-effectiveness.

6.10.2 Heat and Steam Generation

Producing and selling heat and steam requires the existence of available industrial consumers and matching the supply with their needs. It is also possible to use steam at institutional or domestic complexes.

6.10.3 Transportation Fuel

Biogas is utilized as transportation fuel in number of countries. It can be upgraded to natural gas quality in order to be use in natural gas based normal vehicles.

6.11 Recommendations

The following recommendations are put forth in this chapter:

1. Solid waste can be utilized as an energy source in India. However, it is crucial for waste-to-energy (WTE) and to adhere to follow strict guidelines that prioritize environmental concern and precautionary principle.
2. The implementation of landfill disposal techniques should be promoted to harness the biogas potential.
3. Waste-to-energy and the valorization of solid waste represent an emerging areas of research that require further development and exploration.

References

Abas, M. A., & Wee, S. T. (2014). The issues of policy implementation on solid waste management in Malaysia. *International Journal of Concept Management and Social Sciences, 2*(3), 12–17.

Agrawal, A., Sahu, K. K., & Pandey, B. D. (2004). Solid waste management in nonferrous industries in India resources. *Conservation and Recycling, 42*, 99–120.

Ahamed, A., Veksha, A., Yin, K.,Weerachanchai, P., Giannis, A., & Lisak, G. (2020). Environmental impact assessment of converting flexible packaging plastic waste to pyrolysis oil and multi-walled carbon nanotubes. *Journal of Hazardous Materials, 390*, 121449.

Ahluwalia, I. J., & Patel, U. (2018). *Solid waste management in India An assessment of resource recovery and environmental impact* (Working paper No. 356). Isher Judge Ahluwalia. Indian council for research on international economic relations (356). https://icrier.org/pdf/Working_Paper_356.pdf

Anderson, J. C., Park, B. J., & Palace, V. P. (2016). Microplastics in aquatic environments: Implications for Canadian ecosystems. *Environmental Pollution, 218*, 269–280.

Annepu, R. K. (2012). *Sustainable solid waste management in India* (MS Dissertation) pp. 1–189. https://doi.org/10.1007/978-981-4451-73-4

Asokan, P. (2004). *Application of coal combustion residues for hazardous waste management* (Third Annual PhD, progress report). Indian Institute of Technology, Bombay, India.

Azar, S. K., & Azar, S. S. (2016). Waste related pollutions and their potential effect on cancer incidences in Lebanon. *Journal of Environmental Protection, 07*(06), 778–783.

Bhattacharyya, J. K., Shekdar, A. V., & Gaikwad, S. A. (2004). Recyclability of some major industrial solid wastes. *Journal of Indian Association for Environmental Management, 31*, 71–75.

Bhattacharya, R. R. N. S., Chandrasekhar, K., Deepthi, M. V., Roy, P., Khan, A. (2018). Challenges and opportunities: Plastic waste management in India. *The Energy and Resources Institute, 24*(2). https://www.teriin.org/sites/default/files/2018-06/plasticwaste-management 0.pdf

BMTPC. (2005, February). Home page. http://www.bmtpc.org/fibre.pdf

CEA Report. (2017). *All India installed capacity of power station.* GOI. http://www.cea.nic.in/reports/monthly/installedcapacity/2017/installedcapacity-01.pdf

Census of India, GOI. (2011). Census of India 2011 provisional population totals. https://censusindia.gov.in/2011-prov-results/paper2/data files/india2/1 data highlight.pdf

Central Pollution Control Board. (CPCB). (2000). Report on management of municipal solid wastes.

Das, S., Bhattacharyya, B. K. (2015). Optimization of municipal solid waste collection and transportation routes. *Waste Management, 43*, 9–18.

Dasgupta, B., Yadav, V. L., & Mondal, M. K. (2013). Seasonal characterization and present status of municipal solid waste (MSW) management in Varanasi India. *Advances in Environmental Research, 2*(1), 51–60.
Datta, M., & Kumar, A. (2017). Assessment of subsurface contamination potential of municipal solid waste (MSW) dumps. *Indian Geotechnical Journal, 47*, 410–420.
Da Oh, W., Lei, J., Veksha, A., Giannis, A., Lisak, G., Chang, V. W. C., Hu, X., & Lim, T. T. (2018b). Influence of surface morphology on the performance of nanostructured ZnO-loaded ceramic honeycomb for syngas desulfurization. *Fuel, 211*, 591–599.
De Bercegol, R., Cave, J., & Nguyen Thai Huyen, A. (2017). Waste municipal service and informal recycling sector in fast-growing Asian cities: Co-existence, opposition or integration. *Resources, 6*(4), 70. https://doi.org/10.3390/resources6040070
Do Sul, J. A. I., & Costa, M. F. (2014). The present and future of microplastic pollution in the marine environment. *Environmental Pollution, 185*, 352–364.
Dutta, A., & Jinsart, W. (2020). Waste generation and management status in the fast-expanding Indian cities: A review. *Journal of the Air & Waste Management Association, 70*, 491–503.
U.S. Environmental Production Agency Web site. (August, 2003). Federal register environmental documents. Hazardous waste management system; identification and listing of hazardous waste; proposed exclusion. http://www.epa.gov/fedrgstr/EPA-WASTE/2002/February/Day-26/f4530.htm
Gallardo, A., Carlos, M., Peris, M., & Colomer, F. J. (2015). Methodology to design a municipal solid waste pre-collection system. A case study. *Waste Management, 36*, 1–11.
Garg, M., Singh, M., & Kumar, R. (1996). Some aspects of durability of a phospogypsum-lime-fly ash binder. *Construction and Building Materials, 101*(4), 173–279.
Gerdes, P., Gunsilius, E. (2010). *The waste experts: Enabling conditions for informal sector integration in solid waste management*. In Management. www.gtz.de/recyclingpartnerships
Ghosh, S. K. (2016a). Sustainable SWM in developing countries focusing on faster growing economies, India and China. *Procedia Environmental Science, 35*, 176–184.
Ghosh, S. K. (2016b). Swachhaa Bharat Mission (SBM)—A paradigm shift in waste management and cleanliness in India. *Procedia Environmental Sciences, 35*, 15–27. https://doi.org/10.1016/j.proenv.2016.07.002
Giusti, L. (2009). A review of waste management practices and their impact on human health. *Waste Management*, 2227–2239.
Gupta, T. N. (1998). *Building materials in India: 50 years, a commemorative volume*. Building materials technology promotion council. Government of India.
Gupta, N., Yadav, K. K., & Kumar, V. (2015). A review on current status of municipal solid waste management in India. *Journal of Environmental Sciences, 37*, 206–217.
Hammitt, J. K. (1989). Illegal hazardous-waste disposal and enforcement in the United States: A preliminary assessment. *Journal of Hazardous Materials, 22*, 101–119.
Joshi, R., & Ahmed, S. (2016). Status and challenges of municipal solid waste management in India: A review. *Cogent Environmental Science, 2*(1), 1–18.
Khan, A. H., Lopez-Maldonado, E. A., Khan, N. A., Villarreal-Gomez, L. J., Munshi, F. M., Alsabhan, A. H., et al. (2022). Current solid waste management strategies and energy recovery in developing countries—State of art review. *Chemosphere, 291*, 133088.
Ko, J. H., Xu, Q., & Jang, Y. C. (2015). Emissions and control of hydrogen sulfide at landfills: A review. *Critical Reviews in Environment Science and Technology, 45*(19), 2043–2083.
Kumar, G., Reddy, K. R., McDougall, J. (2020). Numerical modeling of coupled biochemical and thermal behavior of municipal solid waste in landfills. *Computers and Geotechnics, 128*(8), 103836.
Li, X. Z., & Zhao, Q. L. (2001). Efficiency of biological treatment affected by high strength of ammonium-nitrogen in leachate and chemical precipitation of ammonium-nitrogen as pretreatment. *Chemosphere, 44*(1), 37–43.

Mani, S., & Singh, S. (2015). Sustainable municipal solid waste management in India: A policy agenda. *International society of waste management, air and water* (pp. 206–213). https://smartnet.niua.org/sites/default/files/webform/SWM-APolicyAgenda.pdf

Maudgal, S. C. (1995). Waste management in India. *Journal IAEM*, 22, 203–208.

McKendry, P. (2002). Energy production from biomass (part-2): Conversion technologies. *Bioresource Technology, 83*, 47–54.

Ministry of Environment & Forests Web site. Hazardous Wastes (Management and Handling) Amendment Rules, 2003-Notification. http://envfor.nic.in/, 2004. Hazardous wastes resource center web site. http://www.etc.org/technologicalandenvironmentalissues/treatmenttechnologies/landfills/index.cfm, August 2004.

Mohan, V. (2019). 91% premature deaths due to air pollution in low and middle-income countries: UNEP. *The Economics Times*. https://economictimes.indiatimes.com/news/environment/pollution/91-premature-deaths-due-to-air-pollutionin-low-and-middle-income-countriesunep/articleshow/69264417.cms

Muhammad, N., Nafees, M., Khan, M. H., Ge, L., & Lisak, G. (2020). Effect of biochars on bioaccumulation and human health risks of potentially toxic elements in wheat (*Triticumaestivum L.*) cultivated on industrially contaminated soil. *Environmental Pollution, 260*, 113887.

Munster, M., & Lund, H. (2009). Use of waste for heat, electricity and transport-challenges when performing energy system analysis. *Energy, 34*, 636–644.

Negi, H., Agrawal, R., Verma, A., & Goel, R. (2019). Municipal solid waste to bioenergy: Current status, opportunities, and challenges in Indian context. In *New and future developments in microbial biotechnology and bioengineering* (pp. 191–203). Elsevier.

Ngwabie, N. M., Wirlen, Y. L., Yinda, G. S., & VanderZaag, A. C. (2019). Quantifying greenhouse gas emissions from municipal solid waste dumpsites in Cameroon. *Waste Management, 87*, 947–953.

Pan, S. Y., Du, M. A., Huang, I. T., Liu, I. H., Chang, E. E., & Chiang, P. C. (2015). Strategies on implementation of waste-to-energy (WTE) supply chain for circular economy system: A review. *Journal of Cleaner Production, 108*, 409–421.

Plastic recycling: An underperforming sector ripe for a remake, 2019. United Nation environment programme. https://www.unenvironment.org/news-and-stories/story/plasticrecycling-underperforming-sector-ripe-remake

Rajmohan, K., Ramya, C., & Varjani, S. (2021). Trends and advances in bioenergy production and sustainable solid waste management. *Energy & Environment, 32*, 1059–1085.

Ramachandra, T. V., Bharath, H. A., Kulkarni, G., & Han, S. S. (2018). Municipal solid waste: Generation, composition and GHG emissions in Bangalore, India. *Renewable and Sustainable Energy Reviews, 82*, 1122–1136.

Ramachandra, T. V., & Saira, V. (2004). Exploring possibilities of achieving sustainability in solid waste management. *Indian Journal of Environmental Health, 45*(4), 255–264.

Ramaiah, B. J., Ramana, G. V., & Datta, M. (2017). Mechanical characterization of municipal solid waste from two waste dumps at Delhi, India. *Waste Management, 68*(9), 275–91.

Ren, X., Liu, T., Awasthi, M. K., Varjani, S., Pandey, A., & Zhang, Z. (2021). Municipal solid waste biorefineries: A case study in China. In *Waste Biorefinery* (pp. 439–57). Elsevier.

Saxena, S. C., & Jotshi, C. K. (1996). Management and combustion of hazardous wastes. *Progress in Energy and Combustion Science, 22*, 401–425.

Saxena, M., & Asokan, P. (2002). Timber substitute products from industrial solid wastes. In R. Srivastava, (Ed). *Proceedings of the 18th national convention of environmental engineers and national seminaron solid waste management* (pp. 192–200).

Shah, A. V., Srivastava, V. K., Mohanty, S. S., & Varjani, S. (2021). Municipal solid waste as a sustainable resource for energy production: State-of-the-art review. *Journal of Environmental Chemical Engineering, 9*, 105717.

Sharholy, M., Ahmad, K., Vaishya, R. C., & Gupta, R. D. (2007). Municipal solid waste characteristics and management in Allahabad, India. *Waste management, 27*(4), 490–496.

Shukla, S. R., Akolkar, A. B., Bhide, A. D., Dhussa, A. K., Varshney, A. K., Acharya, D. B., Datta, M. M., Dutta, M., Mazumdar, N. B., Asnani, P. U., Ramanathan, R., Ramaprasad, V. B., Uppal, S. B. B., & Dhinadhayalan, M. (2000). *Manual on municipal solid waste management, 663*. The Ministry of Urban Development, the Government of India. http://scholar.google.com/scholar

Singh, S. (2020). Decentralized solid waste management in India: A perspective on technological options. *National Institute of Urban Affairs*, 290–304.

Sengupta, J. (2002). Recycling of agro-industrial wastes for manufacturing of building materials and components in India. An overview. *Civil Engineering Construction Review, 15*(2), 23–33.

Slagstad, H., & Bratteb, H. (2013). Influence of assumptions about household waste composition in waste management LCAs. *Waste Management, 33*(1), 212–219.

Soni, A., Patil, D., & Argade, K. (2016). Municipal solid waste management. *Procedia Environmental Sciences, 35*, 119–126.

de Souza, V., Melare, A., Montenegro Gonzalez, S., Faceli, K., & Casadei, V. (2017). Technologies and decision support systems to aid solid-waste management: A systematic review. *Waste Management, 59*, 567–584.

Sridevi, V., Modi, M., Lakshmi, M. V. V. C., & Kesavarao, L. (2012). A review on integrated solid waste management. *International Journal of Advances in Engineering Science Technology, 2*(5), 1491–1499.

United Nations. (1992). Agenda 21. Report of the United Nations conference on environment and development, Rio de Janeiro 1992 United Nations publication (pp. 1–486). 92-1-100498–5; In preparation.

Unnikrishnan, H., Gowrav, B., Jathanha, S. (2006). *Sustainable decentralized model for solid waste management in urban India* (pp. 1–10).

Varjani, S., & Upasani, V. N. (2021). Bioaugmentation of *Pseudomonas aeruginosa* NCIM 5514 - A novel oily waste degrader for treatment of petroleum hydrocarbons. *Bioresource Technology, 319*, 124240.

Yalcm, N., Sevinc, V. (2000). Utilisation of Bauxite waste in ceramic glazes. *Ceramics International, 26*(5), 485–493.

Yoshizawa, S., Tanaka, M., & Shekdar, A. V. (2004). Global trends in waste generation. In I. Gaballah, B. Mishar, R. Solozabal, & M. Tanaka (Eds.), *Recycling, waste treatment and clean technology* (1541–52 [II]). TMS Mineral, Metals and Materials Publishers.

Young, G. C. (2010). *Municipal solid waste to energy conversion processes: Economic, technical, and renewable comparisons* (1st ed.). John Wiley.

Yusof, N., Haraguchi, A., Hassan, M. A., Othman, M. R., Wakisaka, M., & Shirai, Y. (2009). Measuring organic carbon, nutrients and heavy metals in rivers receiving leachate from controlled and uncontrolled municipal solid waste (MSW) landfills. *Waste Management, 29*(10), 2666–2680. https://doi.org/10.1016/j.wasman.2009.05.022

Zhu, D., Asani, P. U., Zurbrugg, C., Anapolsky, S., Mani, S. (2008). *Improving municipal solid waste management in India: A sourcebook for policy makers and practitioners*. The World Bank https://doi.org/10.1596/978-0-8213-7361-3

Zero Waste Europe. (2017). *Harmful subsidies to waste-to-energy incineration*. Europe Commission.

Chapter 7
Morphometric Analysis and Flood Hazard Assessment of Damodar River Basin of India Through Multi-criteria Decision Analysis

Dipjyoti Gogoi and G. Bhaskaran

7.1 Introduction

With increasing anthropogenic activity or a combination of anthropogenic and natural factors around the world (Lyu et al., 2016), flooding has become a major issue in developing countries like India. Natural flooding occurs when water levels rise to the point that they breach natural or man-made levees, submerging typically dry areas (Luino, 2016). In general flood has been categorised into three types i.e., Fluvial flood, Pluvial flood (Kundzewicz & Pińskwar, 2022), and Coastal flood (Turkington et al., 2016). The flood may be coastal, surface, or riverine, which depends on hydrological and topographical conditions (Singh et al., 2020). The frequency and magnitude of hydrological hazards (Costache et al., 2020) directly or indirectly affect land use/ land cover, and land management. Now a days, most of the researchers are working on predicting the spatial distribution of flood susceptibility and providing solutions to the various natural and physical conditions (Akay, 2021).

A GIS approach to flood hazard mapping may be used effectively at the state or regional scale or at the basin level to provide interim flood hazard information prior to detailed hydrologic studies on a local basis (Baker, 1975). Watershed characteristics are obtained spatially by extracting available data using a geographic information system (Akay, 2021) to provide a level of expectation for spatial hazard consequences. At present, remote sensing also represents useful tools for assessing land-use changes as well as flood potential (Costache et al., 2020) at an advanced level using different models. Flood risk management procedures include hazard evaluation, execution, and

D. Gogoi (✉) · G. Bhaskaran
Centre for Water Resources Management, University of Madras, Chennai, India
e-mail: dipgogoi.iitr7@gmail.com

P. K. Rai and S. Rai (eds.), *Climate, Environment and Agricultural Development*, Advances in Geographical and Environmental Sciences,
https://doi.org/10.1007/978-981-97-8363-2_7

exploration of alternatives to mitigate flooding damage (Singh et al., 2020; Timbadiya et al., 2014).

India is a monsoon-dominated country, receiving massive amounts of rainfall and being severely affected by floods (Ghosh & Guchhait, 2016), and millions of people suffer throughout the year. The study area experiences several flood situations due to seasonal rains during the South-West and North-East Monsoon every year (Chandra, 2003), and it varies depending on the intensity of the rain. Historically, the Damodar river basin suffered major and moderate floods in the following years: 1990, 1991, 1993, 1994, 1995, 1996, 1998, 1999, 2000, 2006, 2007, 2009, 2011, 2013, 2015, and 2017 (Singh et al., 2020). So the DRB needs to be studied properly for further prevention of flood occurrences.

7.2 Study Area

The Damodar River Basin (DRB) (Fig. 7.1), located between 23° 38′ 39.90″ N and 84° 39′ 37.01″ E to 22° 16′ 35.17″ N and 88° 04′ 58.15″ E, shares two states, namely Jharkhand and West Bengal. The origin of the Damodar River is in the Palamu Hills of Jharkhand, and it joins the Barakar River just down-stream of the Panchet and Maithon dams (Singh et al., 2020). The basin shares its boundary in two districts i.e., Jharkhand and West Bengal respectively.

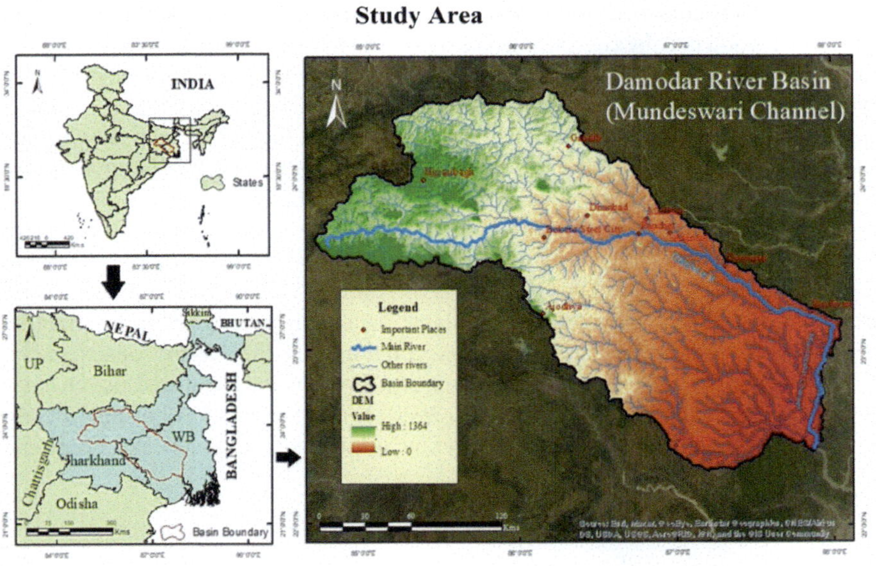

Fig. 7.1 Study area

The total area of the Damodar river basin is nearly 37,677.17 km^2. The Damodar River flows through six districts in Jharkhand, such as Bokaro, Chatra, Dhanbad, Giridih, Hazaribag, Jamtra, Kodarma, Letehar, Purba Singhbhum, Ranchi, and six districts in West Bengal, such as Bankura, Barddhaman, Howrah, Hugly, West Midnapore, and Purulia.

7.3 Materials and Methods

7.3.1 Materials

The primary and secondary data set have been collected from various sources to perform the multi-criteria and analytical hierarchy process (Costache et al., 2020) techniques in the present study. Further the following spatio-temporal remote sensing data has been used to prepare various thematic layers to support the multi-criteria analysis (Table 7.1). The rainfall data from Indian Meteorological Department, geomorphology and structure layers from Bhukosh (Geological survey of India) and Soil maps from NBSS&LUP have been collected and compiled.

Table 7.1 Collected data sets and its sources

Data set	Path/Row	Resolution	Date of acquisition	Data source
LANDSAT-8 Satellite image	Path-139	30 m	23–11-2020	USGS Earth Explorer
	Row-44			
	Path-140		30–11-2020	
	Row-43			
	Path-140		30–11-2020	
	Row-44			
	Path-141		07–12-2020	
	Row-43			
	Path-141		07–12-2020	
	Row-44			
SRTM DEM		30 m	23–09-2014	USGS Earth Explorer
Rainfall				IMD (India Meteorological Department)
Geomorphology				Bhukosh (Geological survey of India)
Soil Texture data				NBSS & LUP

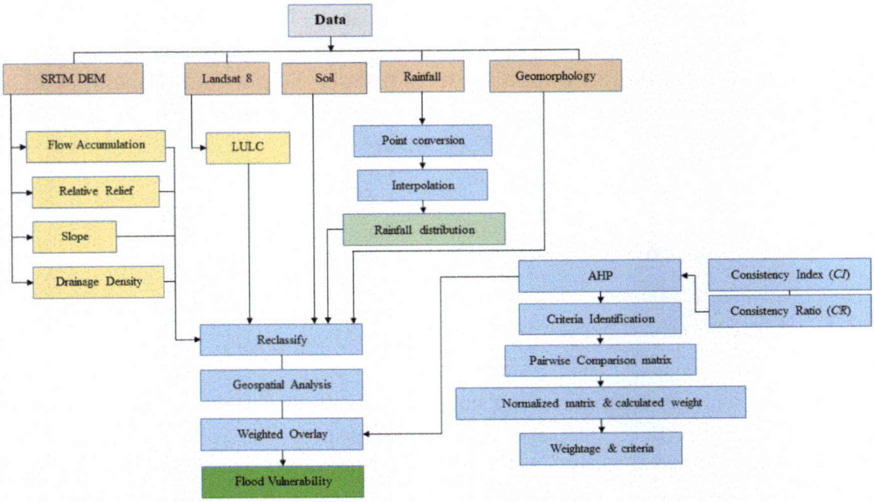

Fig. 7.2 Methodological flow chart

7.3.2 *Methods*

The GIS-based spatial data has been used to identify the flood hazard zones through Multi-Criteria Decision Analysis techniques. The GIS based spatial data sets like land use/land cover, rainfall intensity, drainage density, flow accumulation, slope, relative relief, soil texture, and geomorphology have been prepared according to the NRSC guidelines. Then the relative weights of the criteria were calculated through the AHP method (Saaty, 1980) by analysing the relationships between selected criteria in a structured way, considering their influence on flood generation potential. The detailed methodological flow chart (Fig. 7.2) followed in this study has been shown below.

7.3.3 *Analytical Hierarchy Process (AHP)*

The Analytical Hierarchy Process (AHP) is a method for organising and analysing complex decisions using math and psychology that was developed by Thomas L. Saaty in the year 1980. The weight of each parameter is defined following the AHP (Dipjyoti et al., 2019), a structured technique used for analysing complex problems where a large number of interrelated objectives or criteria are involved. It is a very popular approach to multi-criteria decision-making (MCDM) that involves qualitative data (Alonso & Lamata 2006), and the weights of the criteria are defined after they are ranked according to their relative importance. Thus, once all criteria are sorted in a hierarchical manner, a pairwise comparison matrix for each criterion is created to enable a significance comparison. The relative significance between the criteria is evaluated from 1 to 9, indicating less important to much more important criteria, respectively (Kazakis et al., 2015). The MCDA method permits the assessment of a region based on multiple criteria and supports decision-making in the flood hazard assessment process (Abdelkarim et al., 2020).The principal eigenvalue (λ_{max}) was calculated using eigenvector techniques (Machiwal et al., 2011).

$$\lambda_{max} = \frac{1}{n}\sum_{W_i=1}^{n} \frac{(AW)_i}{W_i} \tag{7.1}$$

where,

W = corresponding eigenvector of λ_{max},
$Wi\ (i = 1, 2\ldots n)$ = weight value for ranking,
AW = consistency vector (Tiwari et al., 2019)
n = number of classes.
The CI i.e., Consistency Index can be calculated as:

$$CI = \frac{(\lambda_{max} - n)}{n - 1} \tag{7.2}$$

where,

λ_{max} = the principal eigenvalue of matrix (Eq. 7.1),
n = number of components or factors in the matrix

$$CR = \frac{CI}{RI} \tag{7.3}$$

where,

CR = consistency ratio
CI = consistency index
RI = random index

The order of the matrix (n) and the equivalent Random Index (RI) can be derived from the Random Index table (Saaty, 1980) (Table 7.2).

The validation of the consistency ratio shows that the higher the CR, the less consistent it is, and vice versa. If the CR ≤ 0.1, the pairwise comparison is relatively consistent and no further correction is needed (Saaty, 1980) in the pairwise matrix, and it is highly acceptable (Table 7.3).

Table 7.2 Random consistency Index (*RI*) (Saaty, 1980)

Matrix size (n)	1	2	3	4	5	6	7	8	9	10
RI	0	0	0.52	0.89	1.11	1.25	1.3	1.4	1.45	1.49

Table 7.3 Relative importance of criteria (Saaty, 1980)

Scale of importance	Definition	Explanation
1	Equal importance	Two activities contribute equally to the objective
2	Weak or slight	
3	Moderate importance	Experience and judgement slightly favour one activity over another
4	Moderate plus	
5	High importance	Experience and judgement strongly favour one activity over another
6	Strong plus	
7	Very strong or demonstrated importance	An activity is favoured very strongly over another; its dominance demonstrated in practice
8	Very, very strong	
9	Extremely importance	The evidence favouring one activity over another is of the highest possible order of affirmation

7.4 Result and Discussion

7.4.1 Land Use/Land Cover

The use of accurate classification of land use/land cover maps can improve the results of flood risk management (Mousavi & Rostamzadeh, 2019), so supervised classification is one of the common methods (Gogoi et al., 2023) used to extract the land uses/land covers (Costache et al., 2020). The level II classification has been carried out based on the standard classes prescribed by the NRSC, 2014 (Fig. 7.3). The study area is being classified into ten classes: water bodies, deciduous forest, riverine sandy area, evergreen forest, scrub forest, barren land, built-up land, crop land, canal, fallow land, etc.

Supervised classification, one of the most known methods is modified k-nearest neighbour (Costache et al., 2020) and shows that the crop land covers nearly 14,290.86 km^2 which is a maximum area of 38.46% (Fig. 7.4) entire basin area followed by evergreen forest area of nearly 5644.41 km^2, deciduous forest of nearly 4764.22 km^2 which found in southern, western, north & north-western portion of the study area with mostly the hilly regions. The built-up area covers nearly 3964.86 sq. km, which is seen along with the river course; the scrub forest area covers 2805.98 sq. km; and the other features like barren land, fallow land, and canal cover the smallest portion of area at 2421.36 sq. km, 2150.06 sq. km, and 153.84 sq. km, respectively.

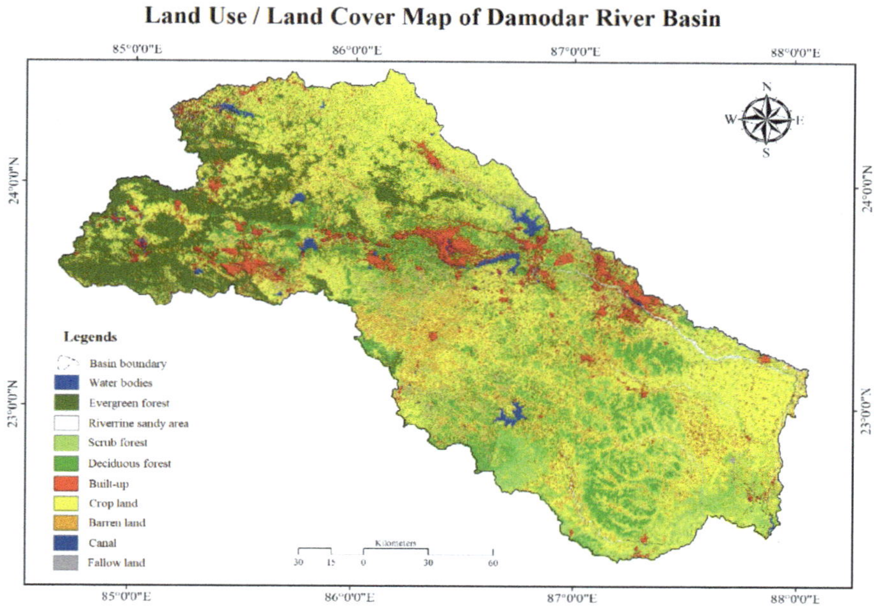

Fig. 7.3 Land use/land cover map of DRB

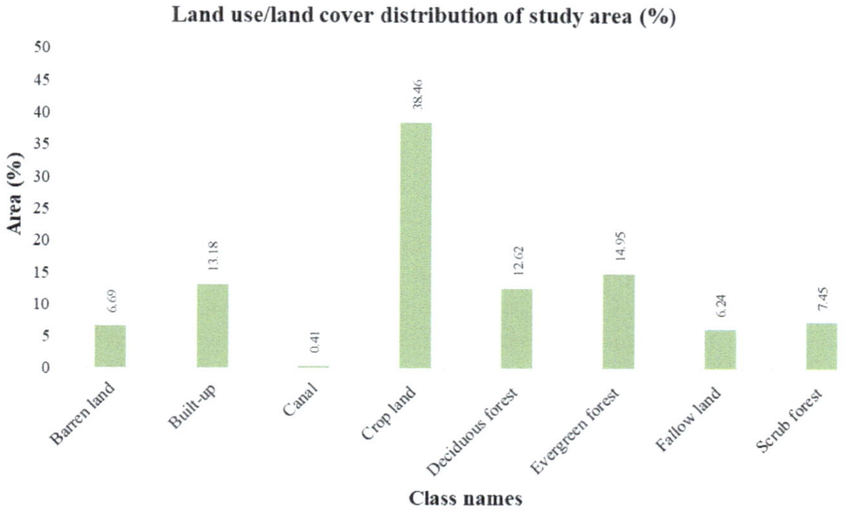

Fig. 7.4 Land use/land cover distribution (%) of DRB

7.4.2 Drainage Density (Dd)

The concentration of drainage network has a direct effect on flood events, specifically in the monsoonal seasons. The higher the drainage density, the greater the risk of inundation, and vice versa. Drainage density less than 2 indicates very coarse drainage texture, 2–4 indicates coarse drainage texture, 4–6 indicates moderate drainage texture, and 6–8 and greater than 8 indicate fine and very fine drainage texture (Ghosh et al., 2021). From the analysis carried out for the drainage density (Dd) of the Damodar river basin, a very coarse texture ranged from 0 to 1.8 (Fig. 5a). Further, for the process of AHP, it is divided into five classes, from low density to high.

7.4.3 Flow Accumulation

The flow accumulation is an important indicator in defining the flood hazard. At first, the algorithm has been generated the flow direction and then accumulated the flow sums of the water flowing downslope into cells of the output raster (Kazakis et al., 2015). After classified into the five classes (Fig. 5b), higher value represented as areas of concentrated flow and consequently higher the probability of flood hazard occurrence. The highest value observed in the main river flow, which is the Damodar River, and lower values are observed in the streams of lower orders.

7.4.4 Relative Relief

The region is situated between the higher relief orders to the lower plain's plateau regions. Higher flood hazard is found at lower elevations (Feloni et al., 2020). According to the relative height calculation of the study area (Fig. 5c), the North-western part is comprises with the Mountainous region and a small part in the northern side with dissected plateau. Towards the southern side there are scattered dissected plateaus over the region. The overall relief variation in the basin ranges from 4.74 to 1055.57 m. Categorically, relative relief values are classified into five classes, such as 4.74–74.79 (very low), 74.79–140.73 (low), 140.73–235.51 (moderate), 235.51–408.59 (high), and 408.59–1055.57 (very high). The maximum higher dissected plateaus are located in the north-northwest, portions and a few are located in the western and eastern parts of the drainage basin, and the southern part of the drainage basin is essentially a low relief area.

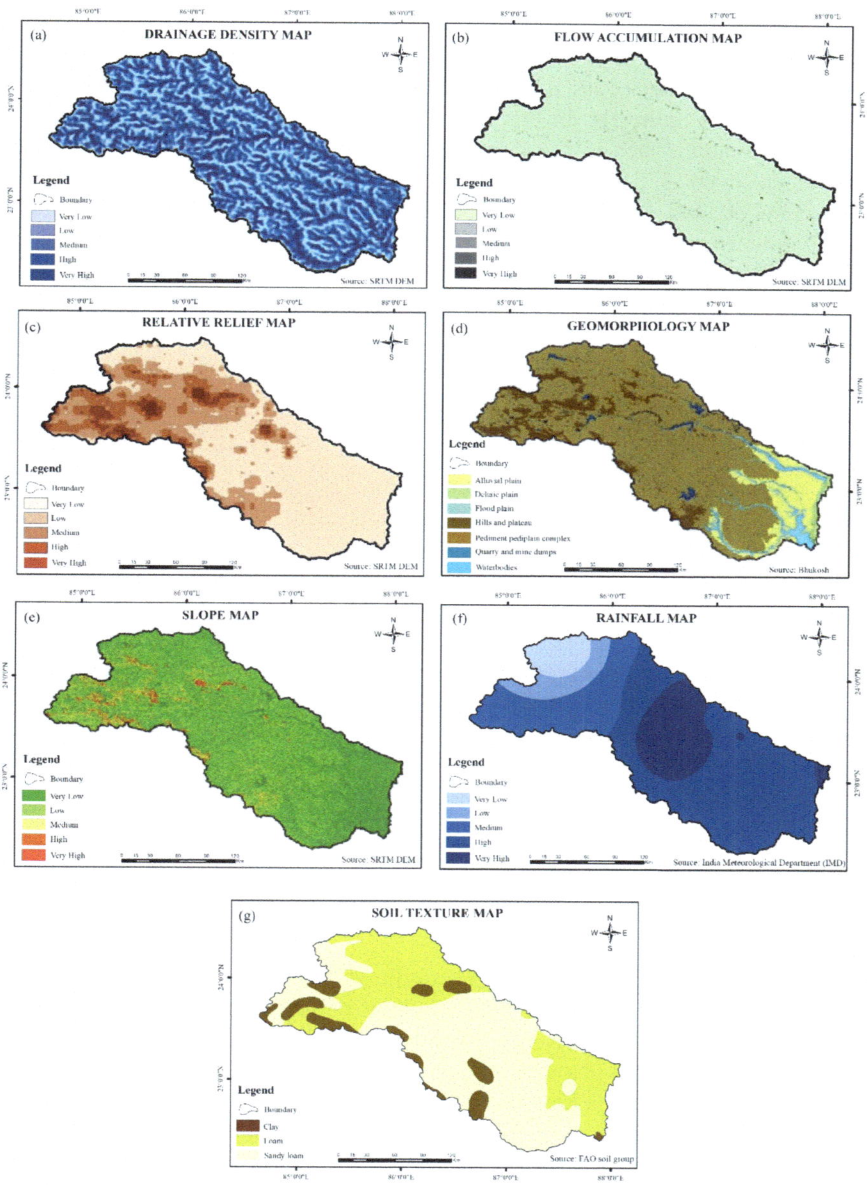

Fig. 7.5 Physical characteristics of the DRB, **a** drainage density, **b** flow accumulation **c** relative relief, **d** geomorphology, **e** slope, **f** rainfall intensity, **g** soil texture

7.4.5 Geomorphology

Geomorphology is one of the most important indicators that controls flood hazard, either directly or indirectly. In this study area it is documented that there are six geomorphic features such as (Fig. 5d), the southern part of the basin showing alluvial plain which covers, 3701.78 km^2, flood plain covers 2,011.24 km^2, hills and plateaus covers 3384.60 km^2, water bodies including rivers, dams, pond, lakes covers nearly 1139.99 km^2 area, the maximum portion of the study area cover by pediment, pediplain complex is nearly 27,241.69 km^2, and lastly quarry and mine dump covers 197.86 km^2 area.

7.4.6 Slope

High surface slope facilitates quick runoff, thus lowering the risk of inundation by overland flow or rainfall (Mousavi & Rostamzadeh, 2019; Feloni et al., 2020), whereas low surface slope facilitates inundation, thus aggravating the risk of floods. In the study area, which is mostly alluvial in nature, the class of slopes ranges from 0° to 2.16° (very low), 2.16° to 5.40° (low), 5.4° to 10.27° (moderate), 10.27° to 19.5° (high), and 19.27° to 68.9° (very high) (Fig. 5e). The slopes range from 10.27° to 68.9° and are mostly observed in the highly elevated terrain of Jharkhand and the Western Bengal district of Puruliya. Slopes are observed almost analogous to the location of the residual hills such as Parashnath, Tundipahar, Panchet, and Ajodhya Hills, among others. Since low slopes favour surface water retention by the soil, flood risk increases in accordance with the low slope of most of the study area.

7.4.7 Rainfall

Rainfall pattern is an important impacts on flood hydrologic response (Zhu et al., 2018) according to the impact of other hydrologic units such as soil moisture texture, formation, etc. High rainfall, coupled with high runoff, can be particularly effective for generating flash floods in a short amount of time. From the analysis, it is shown that most of the area receives a high amount of rainfall, ranging from 1337–1418 mm (very high), 1296–1337 mm (high), 1244–1295 mm (moderate), 1187–1244 mm (low), and 1116–1187 mm (very low) amount of net annual rainfall (Fig. 5f). The majority of these are monsoonal rains that fall between June and September.

7.4.8 Soil

Soil texture denotes the relative proportion of sand, silt, or clay particles in the soil. Soil texture plays an important role in the flood hazard potential of an area (Arya & Singh, 2021; Azizat & Wan Omar, 2018; Cao et al., 2016). Most of the soil in the study area (Fig. 5g) is of loamy texture and has been designated as loam. Sandy Loamy soil predominates in the south eastern, central, and north-eastern parts of the basin area, followed by loamy soil that is mostly present in the north-eastern and eastern parts of the basin. Clayey soil is observed in scattered patches, which coincide with areas of higher slopes and thus indicate more runoff than the other soil types in the area.

7.4.9 Flood Hazard Assessment

The Hazard map was created and displayed the various flood-occurring areas throughout the study area (Tempa, 2022) after assigning weights to each category using the AHP method (Fig. 7.6). The weightage has been assigned in the pairwise comparison matrix (Table 7.4) given to each parameter (Tiwari et al., 2019) with potential impact on the study area according to the physiographic conditions. The weight has been assigned according to the scale of importance given by Saaty (1980), with a scale of 1–9 (Table 7.3). Before checking the consistency level, the weightage has been normalised (Table 7.5), and λ max (Eq. 7.1) is calculated as 8.54 and CI (Eq. 7.2) is calculated as 0.07. The accuracy level of the assigned weightage, the CR (Eq. 7.2), can be derived from the CI and RI. The RI for this study is 1.4 as the total number of parameters is 8 (Table 7.3). The Consistency Ratio (CR) was observed to be 0.05, which can be acceptable for further analysis.

The flood hazard map has been created through overlay analysis and categorised into very low, low, moderate, high, and very high. The western part of the basin has a very low flood hazard zone that covers 5672.05 kms^2., mostly hilly areas with steep slopes, followed by a low flood hazard zone that covers nearly 8457.9 square kilometres in the southern and south-eastern parts of the basin. The moderate flood hazard zone covers an area of 8413.6 kms^2. and is mostly observed along the north-western part of the area, around some industrial cities like Hazaribagh, Bokaro, Dhanbad, Asansol, and Durgapur. The high flood risk zones and very high-risk zones are more vulnerable, which are mostly observed in the flat terrain near the central and south-eastern regions with 11,659.3 and 3467.16 km^2., respectively.

Large portions of both the states of West Bengal and Jharkhand fall under the "highly vulnerable" category. The prepared flood hazard zone map has been validated through previous incidents and articles. Among the affected regions the Ghatal is one of the worst affected during 2009 situated in east Medinipur district, the Burnpur Airport, alongside the Damodar River, also experienced flood inundation in the year 2015, in the month of September. The areas just east of the Kangsabati dam also

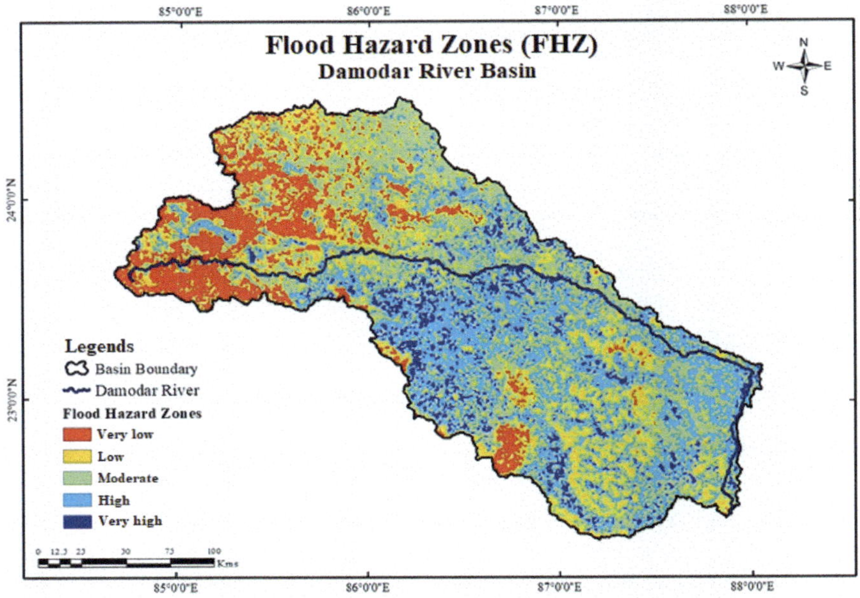

Fig. 7.6 Flood Hazard map of Damodar river basin

face inundation during 2018. The flood hazard map based on the spatial units will be useful for planners to prepare countermeasures (Dandapat & Panda, 2017).

7.5 Conclusion

The above study provides an idea for the hazard assessment for the different physical characteristics, and for DRB, it is shown as a highly flood-prone area. This has been a well-known fact for a long time; as such, the river was once called the "Sorrow of Bengal". The most flood vulnerable zone accounts for 40% of the total basin area, which is to be concentrated by the policymakers to mitigate the floods in the study area. As a result, the high-flood vulnerable areas are falls on the intensive agricultural areas in the plains, which causes heavy damages to the regional economy of the region (Fig. 7.5) and records more loss of life. Towards the southern side there are several dissected plateaus to seen with irregular shape, this also decrease the runoff during the rainy period. The present work identified the frequent flooding areas in the DRBs and addressed the magnitude of the issue for suggesting alternative preventive measures to safeguarding the community. This is a preliminary validation of the prepared flood hazard map performed by observing post-flood incidents for some selected localities in the DRB, which further validates the nature of the area's high flood risk. Hence,

Table 7.4 Pairwise comparison matrix for the selected physical features of the study area

	Drainage density	Flow accumulation	Relative relief	Slope	Rainfall	Soil	Geomorphology	LULC
Drainage density	1.00	0.25	0.17	0.33	0.50	0.33	2.00	0.33
Flow accumulation	4.00	1.00	0.50	3.00	4.00	3.00	6.00	3.00
Relative Relief	6.00	2.00	1.00	2.00	6.00	4.00	3.00	5.00
Slope	3.00	0.33	0.50	1.00	3.00	2.00	5.00	2.00
Rainfall	2.00	0.25	0.17	0.33	1.00	0.50	3.00	0.50
Soil	3.00	0.33	0.25	0.50	2.00	1.00	5.00	2.00
Geomorphology	0.50	0.17	0.33	0.20	0.33	0.20	1.00	0.25
LULC	3.00	0.33	0.20	0.50	2.00	0.50	4.00	1.00

Table 7.5 Normalized pairwise comparison matrix for the selected physical features of the study area

	Drainage density	Flow accumul-ation	Relative relief	Slope	Rainfall	Soil	Geomorphology	LULC	Weightage	λ
Drainage density	0.04	0.05	0.05	0.04	0.03	0.03	0.07	0.02	0.04	8.44
Flow accumulation	0.18	0.21	0.16	0.38	0.21	0.26	0.21	0.21	0.23	8.77
Relative Relief	0.27	0.43	0.32	0.25	0.32	0.35	0.10	0.36	0.30	8.74
Slope	0.13	0.07	0.16	0.13	0.16	0.17	0.17	0.14	0.14	8.67
Rainfall	0.09	0.05	0.05	0.04	0.05	0.04	0.10	0.04	0.06	8.44
Soil	0.13	0.07	0.08	0.06	0.11	0.09	0.17	0.14	0.11	8.62
Geomorphology	0.02	0.04	0.11	0.03	0.02	0.02	0.03	0.02	0.03	8.22
LULC	0.13	0.07	0.06	0.06	0.11	0.04	0.14	0.07	0.09	8.48

this output will help to boost the need for further detailed research to observe minute things in the area and prepare mitigation strategies for a better livelihood.

Acknowledgements I would like to thank the Centre for Water Resources Management (CWRM), University of Madras, Chennai for providing the requirements for the research. I thank ICSSR (Indian Council for Social Sciences Research) for funding this research work, which has helped me to carry out my work in an effective way.

Conflict of Interest The authors declare no conflict of interest.

References

Abdelkarim, A., Al-Alola, S. S., Alogayell, H. M., Mohamed, S. A., Alkadi, I. I., & Ismail, I. Y. (2020). Integration of GIS-based multicriteria decision analysis and analytic hierarchy process to assess flood hazard on the Al-Shamal train pathway in Al-Qurayyat region, kingdom of Saudi Arabia. *Water, 12*(6), 1702. https://doi.org/10.3390/w12061702

Alonso, J. A., & Lamata, M. T. (2006). Consistency in the analytic hierarchy process: A new approach. *International Journal of Uncertainty, Fuzziness and Knowledge-Based Systems, 14*(04), 445–459.

Akay, H. (2021). Flood hazards susceptibility mapping using statistical, fuzzy logic, and MCDM methods. *Soft Computing, 25*(14), 9325–9346.

Arya, A. K., & Singh, A. P. (2021). Multi criteria analysis for flood hazard mapping using GIS techniques: A case study of Ghaghara river basin in Uttar Pradesh, India. *Arabian Journal of Geosciences, 14*(8). https://doi.org/10.1007/s12517-021-06971-1

Azizat, N., & Wan Omar, W. M. S. (2018). Assessment of three flood hazard mapping methods: A case study of Perlis. *E3S Web of Conferences, 34*, 02028. https://doi.org/10.1051/e3sconf/20183402028

Baker, V. R. (1975). Flood hazards along the Balcones escarpment in central Texas; *Alternative approaches to their recognition, mapping, and management*. Virtual Landscapes of Texas

Cao, C., Xu, P., Wang, Y., Chen, J., Zheng, L., & Niu, C. (2016). Flash flood hazard susceptibility mapping using frequency ratio and statistical index methods in coalmine subsidence areas. *Sustainability, 8*(9), 948. https://doi.org/10.3390/su8090948

Chandra, S. (2003). India: Flood management-Damodar river basin. *World Meteorological Organization and the Associated Programme on Flood Management. Integrated Flood Management*. Case Study.

Costache, R., Bao Pham, Q., Corodescu-Roşca, E., Cîmpianu, C., Hong, H., Thi Thuy Linh, N., ... & Thai Pham, B. (2020). Using GIS, remote sensing, and machine learning to highlight the correlation between the land-use/land-cover changes and flash-flood potential. *Remote Sensing, 12*(9), 1422.

Dandapat, K., & Panda, G. K. (2017). Flood vulnerability analysis and risk assessment using analytical hierarchy process. *Modeling Earth Systems and Environment, 3*(4), 1627–1646.

Dipjyoti, G., Dhanusree, M., & Bhaskaran, G. (2019). *Mapping of flood hazard zones of Tuthapuzha sub-basin, Kerala* (p. 245). Governing Council of the Indian Geographical Society.

Feloni, E., Mousadis, I., & Baltas, E. (2020). Flood vulnerability assessment using a GIS-based multi-criteria approach—The case of Attica region. *Journal of Flood Risk Management, 13*, e12563.

Ghosh, S., & Guchhait, S. K. (2016). Dam-induced changes in flood hydrology and flood frequency of tropical river: A study in Damodar river of West Bengal India. *Arabian Journal of Geosciences, 9*(2), 1–26. https://doi.org/10.1007/s12517-015-2046-6

Ghosh, P. K., Mukhopadhyay, R., & Jana, N. C. (2021). Quantitative analysis of drainage basin parameters towards better management of Damodar river, Eastern India. *Journal of the Geological Society of India, 97*(7), 711–734.

Gogoi, D., Bhaskaran, G., & Gogoi, A. (2023). An analysis of land dynamics in relation to urban sprawl in the Guwahati city of Assam India. *Ecocycles, 9*(1), 49–60.

Kazakis, N., Kougias, I., & Patsialis, T. (2015). Assessment of flood hazard areas at a regional scale using an index-based approach and analytical hierarchy process: Application in Rhodope-Evros region, Greece. *Science of the Total Environment, 538*, 555–563. https://doi.org/10.1016/j.scitotenv.2015.08.055

Kundzewicz, Z. W., & Pińskwar, I. (2022). Are pluvial and fluvial floods on the rise? *Water, 14*(17), 2612.

Luino, F. (2016). Floods. *Encyclopedia of engineering geology*. Encyclopedia of Earth sciences series (pp. 506–507). Springer. https://doi.org/10.1007/978-3-319-12127-7_126-1

Lyu, H. M., Wang, G. F., Shen, J. S., Lu, L. H., & Wang, G. Q. (2016). Analysis and GIS mapping of flooding hazards on 10 May 2016, Guangzhou China. *Water, 8*(10), 447.

Machiwal, D., Jha, M. K., & Mal, B. C. (2011). Assessment of groundwater potential in a semi-arid region of India using remote sensing, GIS and MCDM techniques. *Water Resources Management, 25*(5), 1359–1386.

Mousavi, S. M., & Rostamzadeh, H. (2019). Estimation of flood land use/land cover mapping by regional modelling of flood hazard at sub-basin level case study: Marand basin. *Geomatics, Natural Hazards and Risk*.

NRSC. (2014). Land use/Land cover database on 1:50,000 scale, Natural resources census project, LUCMD, LRUMG, RSAA, National Remote Sensing Centre, ISRO.

Saaty, T. L. (1980). *The analytic hierarchy process*. McGraw-Hill.

Singh, R. K., Villuri, V. G. K., Pasupuleti, S., & Nune, R. (2020). Hydrodynamic modeling for identifying flood vulnerability zones in lower Damodar river of eastern India. *Ain Shams Engineering Journal, 11*(4), 1035–1046.

Tempa, K. (2022). District flood vulnerability assessment using Analytic Hierarchy Process (AHP) with historical flood events in Bhutan. *PLoS ONE, 17*(6), e0270467.

Timbadiya, P. V., Patel, P. L., & Porey, P. D. (2014). One-dimensional hydrodynamic modelling of flooding and stage hydrographs in the lower Tapi river in India. *Current Science, 106*(5), 708–716. http://www.jstor.org/stable/24099969

Tiwari, A., Ahuja, A., Vishwakarma, B. D., & Jain, K. (2019). Groundwater Potential Zone (GWPZ) for urban development site suitability analysis in Bhopal, India. *Journal of the Indian Society of Remote Sensing, 47*(11), 1793–1815.

Turkington, T., Breinl, K., Ettema, J., Alkema, D., & Jetten, V. (2016). A new flood type classification method for use in climate change impact studies. *Weather and Climate Extremes, 14*, 1–16.

Zhu, Z., Wright, D. B., & Yu, G. (2018). The impact of rainfall space-time structure in flood frequency analysis. *Water Resources Research, 54*(11), 8983–8998.

Chapter 8
Climate Change and Sustainable Agriculture in India: Navigating the Current Scenario

Kaushiki Singh and Sanjeev Kumar Singh

8.1 Introduction

Agriculture is critical to India's food, nutrition, and livelihood security, employing about two-thirds of the workforce. A country's total economy benefits from the development of agriculture because of its close linkages to other economic sectors. Even though Indian agriculture has made significant strides recently, there are still several challenges facing it today. The main issues facing Indian agriculture include decreasing per capita land availability, yield plateauing, deteriorating soil quality, net sown area stagnation, and the adverse consequences of climate change (Schipper, 2020). The vast and varied agricultural sector in India is being significantly impacted by climate change (Malhi et al., 2021). The current situation necessitates a thorough grasp of how climate change affects agriculture and how sustainable methods might prevent its negative impacts (Das, 2015). A defining force in the complicated web of global issues, climate change has an impact on many different sectors globally.

Among these, agriculture is a fragile linchpin, susceptible to subtle fluctuations in climate patterns (Gupta & Pathak, 2016). With its large and diverse environment, India is grappling with the fundamental ramifications of climate change, which pose severe risks to agricultural output and food security (Naresh Kumar et al., 2020). Ecosystems are drastically shifted by land use changes brought about by rapid urbanisation, changed river courses, evolving agricultural techniques, erosion, and desertification, according to experts (Chakraborty, 2009; Hussain et al., 2024;

K. Singh (✉)
Department of Commerce, Dr. Shakuntala Misra National Rehabilitation University, Lucknow, India
e-mail: kaushikinirupamsingh@gmail.com

S. K. Singh
School of Business Management, Chhatrapati Shahu Ji Maharaj University, Kanpur, India
e-mail: drsanjeevksingh@csjmu.ac.in

P. K. Rai and S. Rai (eds.), *Climate, Environment and Agricultural Development*, Advances in Geographical and Environmental Sciences,
https://doi.org/10.1007/978-981-97-8363-2_8

Sreenivasulu & Bhaskar, 2010). India, a swiftly advancing nation, has witnessed a concerning warming trend over the years, giving rise to extreme weather events and an increased demand for cooling during the hot summer months, driven by both population growth and the impacts of climate change.

8.2 Climate Change Affects Agriculture

Climate change, as seen by rising temperatures and irregular rainfall, has important implications for Indian agriculture. Crop yields are at risk as conventional growing seasons become less predictable. Traditional farming methods have been impacted by changing climate patterns, which are marked by erratic rainfall, rising temperatures, and an increase in the frequency of catastrophic weather events (Pathak et al., 2014). These climate changes present significant difficulties to crop yields, soil health, and overall agricultural output (Rai et al., 2022). Water scarcity caused by unpredictable precipitation, as well as increased flood and drought hazards, have far-reaching consequences for rural communities' socioeconomic fabric. The chapter presents a thorough examination of how climate change functions as a compounding factor, necessitating adaptive techniques and governmental interventions to strengthen Indian agriculture against the varied challenges posed by the changing climate scenario.

Degraded ecosystems should be restored and existing green spaces conserved because they both support greater biodiversity, ecological balance, and the development of more resilient urban environments (Hussain et al., 2024). Planning for mitigation and adaptation is made more difficult by the extreme unpredictability of climate change and its effects. This entails creating climate-resilient technology using an interdisciplinary strategy unique to the area (Malhi et al., 2021). Climate change diminishes food production and lowers the nutritional value of produce. Droughts and other extreme events have an impact on food and nutrient intake, as well as on farmers. The Indian government has developed schemes/plans to strengthen agriculture's resilience to climate change. A key initiative within the National Action Plan on Climate Change (NAPCC), the National Mission for Sustainable Agriculture (NMSA) seeks to develop and apply methodologies to bolster the resilience of Indian agriculture against the impacts of climate change (pib.gov.in).

8.3 Sustainable Agriculture Practices

To address these difficulties, sustainable farming techniques are emerging as a critical answer. Precision farming, which focuses on optimizing resource usage, and organic agriculture, which promotes soil health, is gaining popularity. Agroforestry, which involves incorporating trees into farming systems, not only helps with carbon sequestration but also gives farmers with extra income opportunities. Amidst the present scenario of climate change influencing agriculture in India, implementing sustainable agriculture practices is critical as a transformative approach to mitigating environmental consequences and strengthening the agrarian sector's resilience. Sustainable agriculture requires a paradigm change that aligns agricultural practices with ecological balance, social equality, and financial sustainability. The adoption of climate-resilient crop cultivars is fundamental to sustainable agricultural practices (Rai & Prasad, 2023). As the environment grows more unpredictable, cultivating crops and its microbiome that are resistant to extreme weather and adaptive to shifting conditions becomes increasingly important (Rai et al., 2020).

The chapter emphasizes the need of using advances in agricultural science to produce and promote such varieties, guaranteeing that farmers can maintain productivity despite the uncertainties associated with climate change. As a crucial component in the arsenal of sustainable agriculture, precision farming optimizes the utilization of resources such as pesticides, fertilizers, and water. The use of technology, such as sensor-based systems and data analytics, allows farmers to make more informed decisions, increasing efficiency while lowering environmental impact.

Furthermore, agroecological, microbial and biotechnological techniques dominate the conversation about sustainable agriculture (Rai et al., 2023). Agroforestry, conservation tillage, and cover cropping are emphasized as solutions that not only improve sustainability but also help with carbon sequestration and overall environmental health. The chapter envisions a future in which Indian agriculture not only withstands the challenges of climate change, but also becomes a beacon of sustainability, ensuring food security and environmental stewardship for future generations.

8.4 Water Management in Agriculture

Water scarcity, intensified by climate change, poses a severe threat to Indian agriculture. Innovations in water management, such as drip irrigation and rainwater harvesting, play a pivotal role in adapting to changing precipitation patterns. Assessing the impact of these technologies on water use efficiency and crop productivity is imperative for devising resilient agricultural systems. Navigating through unpredictable rainfall, heightened occurrences of droughts, and the imminent threat of water scarcity, India stands at a pivotal juncture. In the current scenario of climate change impacting agriculture in India, water management has emerged as a critical domain in the pursuit of sustainable practices.

The changing climate patterns, erratic rainfall, and heightened instances of drought have necessitated a re-evaluation of water use in agriculture. The adoption of precision irrigation techniques, emphasis on water-use efficiency, and the integration of innovative water harvesting and storage solutions are pivotal aspects of navigating the current water-related challenges in agriculture. Initiatives led by the government, like the Pradhan Mantri Krishi Sinchayee Yojana (PMKSY), hold a key position in advancing practices for sustainable water management.

8.5 Government Policies and Initiatives

As part of the XII Five Year Plan under the National Action Plan on Climate Change (NAPCC), the Indian Government has introduced eight National Missions. Among these, the National Mission for Sustainable Agriculture (NMSA) and the Mission on Strategic Knowledge on Climate Change (NMSKCC) are specifically designed to attain a 4% annual growth in agriculture, simultaneously equipping the nation to confront the challenges posed by a shifting climate. These missions represent a concerted effort to harmonize agricultural development with climate resilience objectives outlined in the NAPCC. The government's involvement in supporting sustainable agriculture is also examined, with an emphasis on regulations and initiatives that reward environmentally beneficial practices. The National Mission for Sustainable Agriculture (NMSA) and the Paramparagat Krishi Vikas Yojana (PKVY) are being evaluated for their contributions to sustainable agriculture. These initiatives offer farmers financial assistance, knowledge transfer, and incentives to adopt ecologically benign and climate-resilient techniques (Bahinipati et al., 2021). Sustainable agriculture is more than a set of methods; it is a fundamental change toward a more peaceful relationship with nature.

Assessing the effectiveness of these policies and pinpointing areas for enhancement contributes to fortifying a more resilient framework for sustainable agriculture. Amid the current scenario, the connection between government policies and sustainable agriculture becomes pivotal in bolstering India's agrarian resilience against climate change. Acknowledging the pressing need to tackle climate-related challenges, the government has instituted numerous initiatives dedicated to fostering sustainability and adaptability in agriculture. An exemplary instance is the National Mission for Sustainable Agriculture (NMSA), encompassing varied programs and strategies with the goal of enhancing resource-use efficiency, promoting climate-resilient crop practices, and equipping farmers with the capacity to adapt to the challenges presented by climate change. Furthermore, the Pradhan Mantri Krishi Sinchayee Yojana (PMKSY) is directed towards enhancing water use efficiency by implementing effective water management practices.

This is particularly critical in alleviating the consequences of changing precipitation patterns and the escalating challenge of water scarcity. The PMKSY underscores the significance of judicious water utilization as a proactive approach to mitigate the impacts of evolving climatic conditions on agriculture. The Soil Health Card Scheme is another pivotal initiative providing farmers with information on soil health, enabling them to make informed decisions about nutrient management and fostering sustainable soil practices. Furthermore, the Paramparagat Krishi Vikas Yojana (PKVY) encourages the adoption of organic farming methods, promoting ecological balance and reducing the environmental footprint of agriculture. These initiatives collectively underscore the government's commitment to steering Indian agriculture towards sustainability, resilience, and climate adaptation in the current era of climate change challenges.

The country has initiated the National Livestock Mission (NLM) to boost livestock production while simultaneously ensuring environmental conservation, preservation of animal biodiversity, bio-security, and livelihood support for farmers, where initiatives such as creating hay/silage, establishing fodder banks, densification, and implementing feed enrichment processing units, coupled with the establishment of feed processing units, are expected to mitigate the production of greenhouse gases. The NLM, therefore, represents a comprehensive effort to promote sustainable livestock practices that align with environmental preservation and contribute to the welfare of farmers.

8.6 Challenges and Opportunities

Crucial for ensuring food, nutrition, and livelihood security in India, agriculture has made significant strides in the past but currently faces various challenges (Pathak, 2023). Issues like a stagnant net sown area, plateauing yield levels, soil quality degradation, reduced per capita land availability, and the adverse effects of climate change collectively present significant hurdles for Indian agriculture. In the current context, addressing these challenges becomes imperative to sustain and enhance the productivity and resilience of the agricultural sector.

Economic constraints, limited awareness among farmers, and the need for extensive capacity-building efforts pose hurdles. In the current landscape where climate change exerts an ever-intensifying influence on agriculture in India, the intricate interplay between challenges and opportunities is brought to the forefront. The challenges are stark and multifaceted, stemming from the increasingly unpredictable weather patterns, water scarcity issues, and the vulnerability of farmers, especially those with smallholdings, to the repercussions of a shifting climate. The reliance on traditional farming practices faces disruption, exacerbating the risks to crop yields and livelihoods.

However, amidst these challenges lie profound opportunities for transformative change. Opportunities emerge for the adoption of climate-resilient crop varieties, efficient water management, and the integration of agroecological approaches. This nuanced exploration doesn't merely acknowledge challenges; it reframes them as catalysts for innovation and adaptation.Beyond the agronomic sphere, the socio-economic dimensions of these challenges and opportunities become evident. Rural communities face the risk of livelihood disruptions, yet the same challenges propel opportunities for community-driven resilience initiatives and inclusive policies. It envisions a future where the current challenges act as pivotal moments for change, catalyzing a shift towards a more resilient, adaptive, and sustainable agricultural paradigm in India.

To tackle the dual challenges of food security and climate change, the nation needs to reorient its land use and agricultural practices by incorporating state-of-the-art technologies and policy initiatives. A particular focus should be directed towards understanding the gender-specific impacts of climate change. Evaluating the costs linked to adaptation and considering economic aspects is crucial for both future climate scenarios under business-as-usual conditions and altered management approaches aimed at expanding adaptation measures across larger regions. A coherent policy framework is imperative for the effective implementation of adaptation and mitigation options, safeguarding farmers from the adverse consequences of climate change.

8.7 Role of Technology

In the contemporary landscape, the integration of these technological advancements is instrumental in fostering a more sustainable and resilient agricultural sector (Prasad et al., 2015). These innovations not only empower farmers with real-time insights but also contribute to the overall efficiency and environmental sustainability of farming practices (Chandra et al., 2017). Evaluating the impact and scalability of these technologies contributes to a comprehensive understanding of their role in sustainable agriculture. Precision farming serves as a guiding light, utilizing technological progress like sensors, drones, and data analytics to enhance the efficient utilization of resources (Ouya et al., 2023). This approach not only boosts agricultural productivity but also mitigates the environmental impact by minimizing inputs like water, fertilizers, and pesticides.

The integration of climate-smart technologies becomes paramount, offering farmers real-time weather information and tailored solutions to navigate unpredictable climatic patterns (Hellin et al., 2023). Furthermore, embracing genetically modified and climate-resistant crop varieties, cultivated through biotechnology, is crucial in upholding food security amidst evolving environmental conditions. Digital platforms and mobile applications empower farmers with access to market information, agricultural best practices, and financial services. This democratization of information enhances the decision-making capacity of farmers, particularly smallholders, contributing to their overall resilience in the face of climate-induced uncertainties.

Government initiatives supporting agritech and technological innovation, such as the Digital India campaign and the National Mission on Sustainable Agriculture, are also scrutinized in the chapter. These initiatives serve as catalysts for fostering a tech-enabled agricultural ecosystem, ensuring that farmers across diverse landscapes can harness the benefits of cutting-edge technologies. By embracing technology as a cornerstone, the sector not only adapts to climate change but also positions itself as a dynamic and sustainable force in feeding the nation amid the evolving challenges of a changing climate.

8.8 Community Involvement and Awareness

Engaging local communities is paramount for the success of sustainable agriculture initiatives. Building awareness among farmers about climate-resilient practices and involving them in decision-making processes are critical components. Community-driven approaches ensure the relevance and acceptance of sustainable practices at the grassroots level. By promoting awareness campaigns, workshops, and knowledge-sharing platforms, the chapter envisions a grassroots movement where farmers are equipped with the information and tools to adapt to changing climatic conditions. Recognizing the interconnectedness of communities and their reliance on agriculture, the chapter advocates for collaborative efforts that empower farmers to become stewards of sustainable practices.

8.9 Economic Viability and Market Access

The widespread adoption of sustainable agriculture practices hinges on their economic viability, necessitating a thorough analysis of the associated benefits such as cost savings and potential income diversification. Furthermore, facilitating market access for sustainably produced goods serves as an encouragement for farmers to invest in climate-smart practices. As the agrarian landscape grapples with climate-induced challenges, ensuring the economic sustainability of farming practices becomes imperative.

The importance of aligning sustainable agricultural initiatives with market dynamics, fostering a system where climate-resilient produce not only meets environmental goals but also addresses economic imperatives. It advocates for policies that enhance market access for farmers practicing sustainable agriculture, creating avenues for them to reap the economic benefits of environmentally friendly practices. By integrating economic viability with sustainable agriculture, the chapter envisions a future wherein farmers not only adapt to climate change but also flourish within a market-oriented agricultural landscape, securing the long-term economic resilience of India's agricultural sector.

8.10 Global Collaboration for Knowledge Exchange

Acknowledging the worldwide scope of climate change, promoting international collaboration becomes essential, with knowledge exchange, the sharing of best practices, and joint research initiatives contributing to a united endeavor in tackling the impacts of climate change on agriculture. Collaborative platforms facilitate the transfer of technology and expertise to enhance sustainability. Recognizing that climate change transcends national boundaries, the chapter advocates for a collaborative approach that harnesses global expertise, research, and technology. It underscores the importance of knowledge exchange platforms, international partnerships, and collaborative research initiatives to share best practices and innovative solutions. Through nurturing global collaboration, India has the opportunity to leverage a wealth of diverse insights and experiences, thereby strengthening its ability to tackle the intricate challenges presented by climate change in agriculture.

8.11 Conclusion

Conclusively, addressing the complex interplay between climate change and sustainable agriculture in India requires a comprehensive and adaptable approach. Balancing technological innovations, government policies, community involvement, and economic considerations is essential for building resilient agricultural systems. As India strives to ensure food security and environmental sustainability, ongoing research, and collaborative efforts will be instrumental in shaping the future of agriculture in the face of a changing climate. The journey towards sustainable agriculture is a collective endeavor, requiring concerted efforts from farmers, policymakers, and the broader community.

The anticipated impacts of climate change on agriculture are expected to be far-reaching, affecting both direct and indirect aspects. It becomes imperative to preserve global plant health not only for mitigating climate change but also for conserving biodiversity and sustaining ecosystem services amid global transformations. Upholding ecosystem health and services in uncertain conditions requires fostering resilience, decentralization, participatory research, and collaborative breeding networks, underscoring the necessity for increased involvement of diverse stakeholders and scientists beyond the realm of plant pathology to address trade-offs with other objectives.

Promotion of the Integrated Farming System (IFS) involves introducing supplementary livelihood support activities beyond traditional crops to enhance the resilience and sustenance of farmers during extreme climatic events. The Government of India has instituted a National Agro-forestry Policy to encourage agro-forestry as a comprehensive farming system. Recognizing the necessity for agricultural diversification in the face of adverse conditions and extreme climate variability, there is a shift towards relief-based management approaches. This necessitates the exploration and investments in alternative sustainable agricultural methods tailored to local and agro-climatic conditions, aiming to generate economic benefits for local communities, optimize natural resource use, and concurrently improve health and nutrition. These approaches underscore the minimization of inputs, prioritize farmers' roles, address climate change challenges, reverse ecological deterioration, and enhance farmers' resilience and incomes.

References

Bahinipati, C. S., Kumar, V., & Viswanathan, P. K., (2021). An evidence-based systematic review on farmers' adaptation strategies in India. *Food Security*, 1–20. https://doi.org/10.1007/s12571-020-01139-3

Chakraborty, K. (2009). Vegetation change detection in Barak Basin. *Current Science 96*, 1236–1242. http://www.jstor.org/stable/24105415

Chandra, A., McNamara, K. E., & Dargusch, P. (2017). The relevance of political ecology perspectives for smallholder climate-smart agriculture: A review. *J Polit Ecol*. https://doi.org/10.2458/v24i1.20969

Das, D. (2015). Changing climate and its impacts on Assam, Northeast India. *Bandung Journal of the Global South, 2*, 26. https://doi.org/10.1186/s40728-015-0028-4

Gupta, A., & Pathak, H. (2016). *Climate change and agriculture in India: A thematic report of National Mission on Strategic Knowledge for Climate Change (NMSKCC) under National Action Plan on Climate Change* (p. 68). Department of Science and Technology, Ministry of Science & Technology, Government of India.

Hellin, J., Fisher, E., Taylor, M., et al. (2023). Transformative adaptation: From climate-smart to climate-resilient agriculture. *CABI Agric Biosci, 4*, 30. https://doi.org/10.1186/s43170-023-00172-4

Hussain, S, Hussain, E, Saxena, P, Sharma, A, Thathola P & Sonwani, S. (2024). Navigating the impact of climate change in India: A perspective on climate action (SDG13) and sustainable cities and communities (SDG11).

Malhi, G. S., Kaur, M., & Kaushik, P. (2021). Impact of climate change on agriculture and its mitigation strategies: A review. *Sustainability, 13*, 1318. 10.3390/su13031318

Naresh Kumar, S., Chander, S., Sinha, P., & Padaria, R. N. (2020). Climate change impact, vulnerability and adaptation: Agriculture, a report for India's third national communication to UNFCCC, ICAR-Indian Agricultural Research institute, New Delhi, p. 21.

Ouya, F. O., Murage, A. W., Pittchar, J. O., et al. (2023). Impacts of climate-resilient push–pull technology on farmers' income in selected counties in Kenya and Tanzania: Propensity score matching approach. *Agric & Food Secur, 12*, 15. https://doi.org/10.1186/s40066-023-00418-4

Pathak, H., Pramanik, P., Khanna, M., & Kumar, A. (2014). Climate change and water availability in Indian agriculture: Impacts and adaptation. *Indian Journal of Agricultural Sciences, 84*(6), 671–679.

Pathak, H. (2023). Impact, adaptation, and mitigation of climate change in Indian agriculture. *Environmental Monitoring and Assessment, 195*, 52. https://doi.org/10.1007/s10661-022-10537-3

Prasad, Y. G., Srinivasa Rao, C. H., Prasad, J. V. N. S., Rao, K. V., Ramana, D. B. V., Gopinath, K. A., Srinivas, I., Reddy, B. S., Adake, R., Rao, V. U. M., Maheswari, M., Singh, A. K., & Sikka, A. K. (2015). *Technology demonstrations: Enhancing resilience and adaptive capacity of farmers to climate variability* (p. 109). National Innovations in Climate Resilient Agriculture (NICRA) project, ICAR-Central Research Institute for Dryland Agriculture.

Rai, S., Omar, A. F., Rehan, M., Al-Turki, A., Sagar, A., Ilyas, N., Sayyed, R. Z., & Mirza, H. (2022). Crop microbiome: Their role and advances in molecular and omic techniques for the sustenance of agriculture. *Planta, 257*, 27. https://doi.org/10.1007/s00425-022-04052-5

Rai, S., & Prasad, R., (2023). Trichoderma as a toolbox: Biotic and climate resilient agriculture. In: S. Rai & R. Prasad (Eds), *New and future developments in microbial biotechnology and bioengineering* (pp. 173–202), Elsevier.

Rai, S., Solanki, M. K., Solanki, A.C., & Samal, S., (2023). *Microbial endophytes as probiotics for the plant health: An overview* (pp. 269-281). In: Solanki, M.K., Yadav, M.K., Singh, B.P., Gupta, V.K. Academic Press.

Rai, S., Solanki, A. C., & Anal, A. K. D. (2020). Modern biotechnological tools: An opportunity to discover complex phytobiomes of horticulture crops. In M. K. Solanki, P. L. Kashyap, R.

A. Ansari, & B. Kumari (Eds.), *Microbiomes and plant health panoply and their applications* (pp. 85–124). Academic Press.
Schipper, E. L. F. (2020). Maladaptation: When adaptation to climate change goes very wrong. *One Earth, 3*, 409–414.
Sreenivasulu, V., & Bhaskar, P. U. (2010). Change detection in landuse and landcover using remote sensing and GIS techniques. *International Journal of Engineering, Science and Technology, 2*, 7758–7762.
https://pib.gov.in/PressReleaseIframePage.aspx?PRID=1909206#:~:text=Climate%20change%20reduces%20crop%20yields,more%20resilient%20to%20climate%20change.
https://www.epa.gov/climateimpacts/climate-change-impacts-agriculture-and-food-supply#topc
https://dst.gov.in/sites/default/files/Report_DST_CC_Agriculture.pdf
https://www.vasudha-foundation.org/wp-content/uploads/Full-Action-Plan-Rajkot.pdf
https://www.ceew.in/sites/default/files/CEEW-FOLU-Sustainable-Agriculture-in-India-2021-20Apr21.pdf

Chapter 9
Estimation of Water Use Efficiency for Winter Cereals Through Evapotranspiration Using Remote Sensing Data

Chanev Milen, Filchev Lachezar, and Ilina Kamenova

9.1 Introduction

Among environmental factors, water deficit during the reproductive phase of development is the most important limiting factor for winter cereals (Blum, 2009). Figure 9.1 shows the crop calendar of main cereals grown in Bulgaria, France and India, these calendars show us the need for water in different months for the main cereal crops. Regarding the moisture requirement of wheat, it refers to crops with medium demand, to obtain a yield of 500–600 kg/da, about 500–550 mm of precipitation is needed during the growing season, distributed in a way to cover the critical periods. At the beginning of the budding phase, the need for moisture must be met by the fall precipitation during the fall season. When mass budding, spindling and growth begins in the spring, the need for water increases sharply. In terms of soil moisture content, the spindle and heading period is critical for wheat crops. During the period of flowering, fertilization and pouring of the grain, the presence of moisture helps to set a larger number of grains and for their normal pouring. Drought in this period has a negative effect, the number of grains in the class and their mass are reduced. Moisture is one of the main factors limiting wheat yield (Chanev & Filchev, 2021; Delchev & Zhelyazkova, 2016). This shows that in order to have an optimal yield in cereal crops,

C. Milen · F. Lachezar (✉) · I. Kamenova
Space Research and Technology Institute at Bulgarian Academy of Sciences, Sofia, Bulgaria
e-mail: lachezarhf@space.bas.bg

C. Milen
e-mail: mchanev@space.bas.bg

I. Kamenova
e-mail: ilina.kamenova@space.bas.bg

P. K. Rai and S. Rai (eds.), *Climate, Environment and Agricultural Development*,
Advances in Geographical and Environmental Sciences,
https://doi.org/10.1007/978-981-97-8363-2_9

it is necessary for the plants in them to maintain their water regime during the generative phases, and the accumulated amount of biomass during the vegetation period is not of decisive importance (Kato et al., 2008). It has been found, that with efficient use of water from crops in the appropriate phenological phases, their yield also increases (Varga et al., 2013). In countries such as India, rice is the cereal crop that requires the largest amounts of water, Davis et al. (2018) suggest replacing it with alternative crops such as maize, millet, pearl millet or sorghum, thereby reducing the water footprint. This will reduce the demand for irrigation water by 33%. This gives us various benefits in terms of both reducing fresh water use and improving nutrient production (Davis et al., 2018).

Utilizing remote sensing imagery offers a dependable method for tracking the physiological progress of crops and has the potential to gauge both crop evapotranspiration and the necessary irrigation water quantities (Ihuoma et al., 2021). Determining the water needs of agricultural crops at the field level and the different conditions within the field allows farmers to better manage crops in terms of water use. With the development of technology and the ability to process large arrays of satellite data in cloud arrays, it allows us to map the earth's surface and quickly identify problem areas. This is of great importance for regions of the world where there is drought and food security is threatened (Abbasi et al., 2023). Another advantage of satellite data is that it can compensate for the lack of meteorological data (Talebi et al., 2023). The so-called Drought Index (DI), see **Eq. 9.1**, is used to determine to what extent the rainfall during the growing season is sufficient. DI is determined

* Corresponding Author

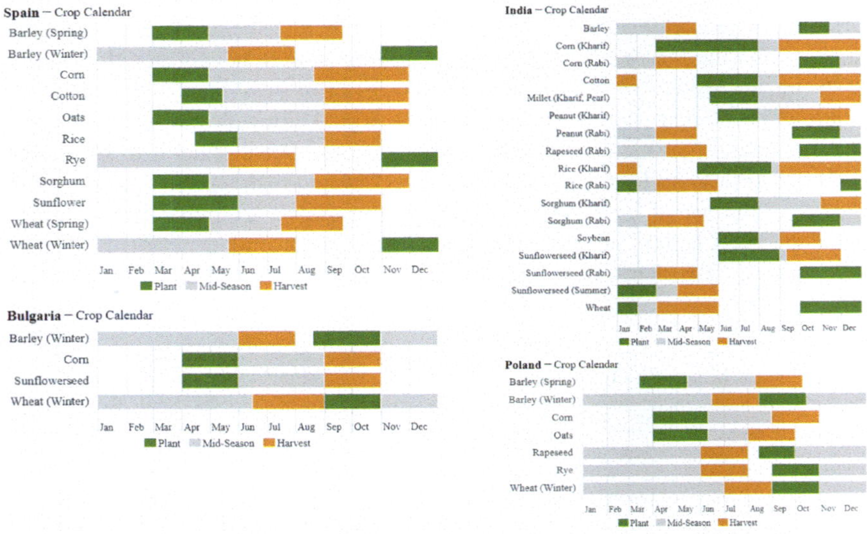

Fig. 9.1 Crop calendars of India, Bulgaria, Poland, and Spain. image source https://ipad.fas.usda.gov/ogamaps/cropcalendar.aspx

using the subsequent formula:

$$DI = (A_r - M)/\sigma \quad (9.1)$$

where Ar is the annual rainfall (the sum of fallow rainfall and seasonal rainfall), M is the average annual rainfall, and s is the standard deviation for annual rainfall. DI is used to distinguish between the wet (DI > 0.35), normal ($-0.35 \leq$ DI ≤ 0.35), and dry (DI < $-$ 0.35) growing seasons (Guo et al., 2012). Similarly, the DIs for fallow rainfall and seasonal rainfall were also calculated to assess variations and status in rainfall among different growing seasons (Wang et al., 2018).

Vegetation indices generated by different satellite platforms can be used to calculate evapotranspiration, respectively to model water stress and cereal yields (Abbasi, 2023; Conrad et al., 2013; García-León et al., 2019; Kogan, 1995; 1997; Li et al., 2008). Normalized Difference Vegetation Index (NDVI) and soil moisture index (SMI) were used to assess drought in spring cereal crops in central-western Poland. **Figure 9.1** shows the calendar of the most common cereal crops in Poland.

In-situ data collected during this study are meteorological, phenological and cereal yield data. With the collected in-situ data and satellite data obtained and the use of the FAO CROPWAT Smith (1992) model, the water requirement during different phenological phases (sowing, emergence, tillering, shooting, heading, milk maturity and wax maturity) was determined. Crop evapotranspiration was determined using Etcrop, see **Eq. 9.2** (Ekwue et al., 2015; Martyniak et al., 2007). Evapotranspiration was calculated from the FAO CROPWAT model (Martyniak et al., 2007; Smith, 1992). It calculates the water balance in the soil, where the income is precipitation and the expenditure is evaporation.

$$ET_{crop} = k_c ET_0 \quad (9.2)$$

where ETcrop—crop evapotranspiration under optimal agronomic, no water stress [mm d $-$ 1]; kc—crop coefficient, which refers to the growth and development of crops, no water stress; ETo—climatic parameter computed from weather data; this is used as "a hypothetical reference crop with an assumed height of 0.12 m, a fixed surface resistance of 70 sm^{-1} and an albedo of 0.23" (Allen et al., 1998; Martyniak et al., 2007).

In examining the relationships between the vegetation index NDVI and evapotranspiration of cereal crops, Martyniak et al. (2007) found a significant correlation (r = 0.59), but not satisfactory, while for individual cereal crops, for example, for barley, they found a very high correlation (r = 0.82). It has been found that generating NDVI and SMI vegetation indices from the NOAA satellite with meteorological data increases the possibility of modeling and assessing drought in agricultural areas, as it increases the possibility of identifying drought zones.

The use of satellite data enables us to assess how crops over large areas are absorbing rainfall, whether they are experiencing water stress, and to predict yields. By employing various drought indices extracted from satellite data, we are able to compute the drought parameter and crop yield within statistical models that have been adjusted to the local context (García-León et al., 2019). Farm-level yield data can be obtained from the relevant agricultural agency for each country. For example, for the European Union, this type of data is available on the site (http://ec.europa.eu/agriculture/rica/), which describes the productivity of agricultural holdings across the European Union by country (García-León et al., 2019). To establish the level of water stress in cereal crops in Spain for a twelve-year period from 2003 to 2015, García-León et al. (2019) uses the vegetation indices (VI) Vegetation Condition Index (VCI) serves us to measure the humidity conditions, the Temperature Condition Index (TCI) is more closely related to the heat stress of the vegetation, and the average of the two, the Vegetation Health Index (VHI), see **Eq. 9.3** (Kogan, 1995, 1997). All three VIs are widely used to determine the level of drought impact on crops (Bokusheva et al., 2016; García-León et al., 2019; WMO and GWP, 2016). The most suitable for determining the level of crop stress is the VI VHI, as it is calculated as the average of the VCI and TCI, this allows us to calculate the combined effect of the two stressors. and thus represents a good general indicator of vegetation health (García-León et al., 2019). The Temperature Condition Index (TCI) is more closely related to the heat stress of the vegetation, and the average of the two, the Vegetation Health Index (VHI) (Kogan,). All three vegetation indices (VIs) are extensively utilized for assessing the degree of drought's influence on crops (Bokusheva et al., 2016; García-León et al., 2019; WMO and GWP, 2016).

$$
\begin{aligned}
VCI_t &= 100 * \frac{NDVI_t - NDVI_{min}}{NDVI_{max} - NDVI_{min}} \\
TCI_t &= 100 * \frac{LST_{max} - LST_t}{LST_{max} - LST_{min}} \\
VHI_t &= \frac{VCI_t - TCI_t}{2}
\end{aligned} \tag{9.3}
$$

The Crop Water Stress Index (CWSI) is a widely used indicator for assessing the water stress level experienced by crops. It is calculated based on the temperature difference between the actual canopy temperature and the theoretical canopy temperature under non-stressed conditions, normalized by a reference temperature range. The formula for calculating CWSI is as follows, see **Eq. 9.4**:

$$
CWSI = \frac{Tc - Tcr}{Tc - Tca} \tag{9.4}
$$

where: *CWSI* is the Crop Water Stress Index, *Tc* is the actual canopy temperature, *Tcr* is the reference canopy temperature under non-stressed conditions, and *Tca* is the air temperature.

The reference canopy temperature *Tcr* is estimated using the following formula, see **Eq. 9.5**:

$$
Tcr = Ta + \frac{(Tmax - Ta) * Kc}{Krs} \tag{9.5}
$$

where: *Ta* is the air temperature, *Tmax* is the maximum air temperature, *Kc* is a crop coefficient representing the ratio of actual to potential crop evapotranspiration, *Krs* is the crop resistance to water stress.

The CWSI ranges from 0 to 1, where values closer to 0 indicate no water stress (optimal conditions), and values closer to 1 indicate severe water stress.

It's important to note that the specific calculation of CWSI may vary slightly depending on the study or model being used, as different researchers may incorporate additional factors or modify the formula to suit their specific needs or environmental conditions.

9.2 Satellite Remote Sensing Data for Water Use Efficiency for Winter Cereals

To ascertain crop distribution and their water needs, multispectral satellite data from RapidEye and Landsat, obtained prior to the main summer growth stages, can be integrated with the CropWat model developed by the Food and Agriculture Organization (FAO). This model aids in classifying crops within a specific region. Enhancements to the model can be implemented, and the FAO's ClimWat database is utilized for computing crop water requirements (CWR). These requirements are determined based on the reference evapotranspiration for the respective crops (ETc, Allen et al., 1998). Depending on the needs and using the CropWat model, different levels of detail can be obtained using different crop coefficients based on agro-meteorological data, soil and plant information. The selection of images needs to comply with the crop calendar for the region of the studied territory. In Fig. 9.2 presents a calendar of the main crops in Kazakhstan Central Asia As the classification itself is divided into two stages: the first stage includes the separation of agricultural grasslands from other types of land cover, and the second represents the determination of the different types of crops. The 'random forest' algorithm according to Breiman (2001) was used for field crop classification (Conrad et al., 2013). The first stage involves the separation of agricultural grasslands from other land cover types, and the second represents the determination of the different types of crops.

An area where crops experience significant water stress is The North China Plain (NCP). Growing winter wheat there consumes more than half of the region's total water use. Satellite data is also used to achieve better management of water resources in agriculture in this area. The obtained satellite data are used to estimate evapotranspiration and water productivity in wheat. To achieve this, the Surface Energy Balance Algorithm for Land (SEBAL) model is used, which uses satellite Earth observation, a method developed by WaterWatch. Daily evapotranspiration of wheat is estimated using SEBAL. The model combines three types of satellite data, respectively normalized difference vegetation Index (NDVI), surface albedo, and surface temperature. Satellite data from NOAA were used to feed the model, requiring a minimum of one cloud-free image for each of the months during the crop's growing season. The other

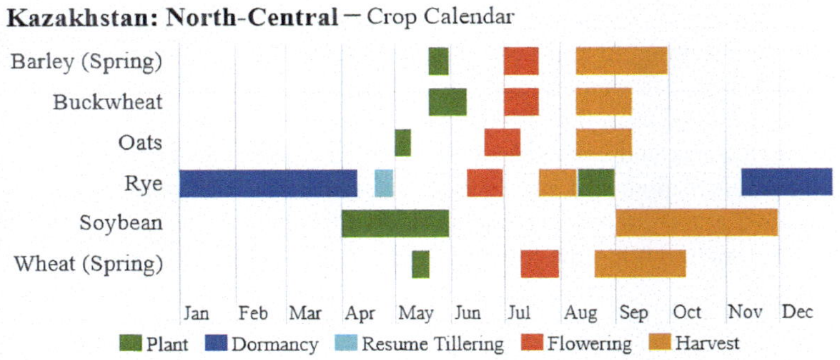

Fig. 9.2 Crop calendar of North-Central Kazakhstan

type of data that is needed to feed the model is meteorological data that is collected from meteorological stations located in the study area. For the model to work, it is necessary to perform a spatial interpolation of the meteorological data to achieve the same resolution as with the satellite data. Yield data can be obtained either from the farmers or from the relevant agricultural offices that also collect this type of data. The calculation of evapotranspiration is done by means of the energy balance, see **Eq. 9.6**, at the moment when the satellite orbits over the studied territory (Li et al., 2008). With a minimum of one cloud-free image required for each of the months during the growing season of the crop. The other type of data that is needed to feed the model is meteorological data that is collected from meteorological stations located in the study area. For the model to work, it is necessary to perform a spatial interpolation of the meteorological data to achieve the same resolution as with the satellite data. Yield data can be obtained either from the farmers or from the relevant agricultural offices that also collect this type of data.

$$\lambda ET = R_n - G - H \tag{9.6}$$

where λET is the latent heat flux (W m − 2), *Rn* the net radiation (W m − 2), *G* the soil heat flux (W m − 2), and *H* is the sensible heat flux (W m − 2).

The estimation of surface albedo is given by the following **Eq. 9.7**:

$$\alpha = 0.545\alpha_1 + 0.320\alpha_2 + 0.035 \tag{9.7}$$

where $\alpha 1$, $\alpha 2$ are the reflectance rates of the first and second channel of NOAA satellite data.

The land surface temperature is calculated using, see **Eqs. 9.8** (Coll et al., 1994; Li et al., 2008).

$$\begin{aligned} T &= T_4 + [1.29 + 0.28(T_4 - T_5)](T_4 - T_5) + 45(1 - \varepsilon_4) - 40\Delta\varepsilon \\ \varepsilon_4 &= 0.9897 + 0.029\ln(NDVI) \\ \Delta_\varepsilon &= 0.01019 + 0.01344\ln(NDVI) \end{aligned} \tag{9.8}$$

where T4, T5 are the brightness temperature of the 4th and 5th band of NOAA data (K), $\varepsilon 4$ the surface emissivity of the 4th band of NOAA data, and $\Delta\varepsilon$ is the differences in surface emissivity between the 4th and 5th band of NOAA data (Li et al., 2008).

In recent decades, the possibilities of using satellite data for monitoring in agriculture have developed enormously, this also gives an opportunity for an adequate assessment of the evapotranspiration of cereal crops on a regional scale as well as on a field level. The SEBAL model can be very successfully applied to this type of monitoring in agriculture, while at the same time it is very easy to use because the calculations are not complicated and it does not require much in-situ data (Li et al., 2008).

Satellite data are also used in Africa for studies related to climate change, drought, and management of cereal crops in the continent's arid regions. As such, an area experiencing difficulties in monitoring cereal crops in Africa is Ethiopia. Data from the Moderate Resolution Imaging Spectroradiometer (MODIS) satellite, developed by NASA, are used to improve the management of cereal crops in the region. MODIS data were used to construct an evapotranspiration estimation method, the vegetation index NDVI, as well as a satellite-derived estimate of precipitation and ground-based yield data for teff, sorghum, corn, barley, and wheat. The research was carried out with the aim of assisting the decision-makers for the management of yields and increasing food security in the country, as well as to evaluate the feasibility of using such type of data for mining management. The results of this study showed high correlations ($r = 0.99$) between evapotranspiration and rainfall for all crops grown throughout Ethiopia. Regarding the correlation dependences between the seasonal evapotranspiration and the vegetation index NDVI, a smaller dependence ($r = 0.7$) was found, which varies in different regions of the country. A similar trend of correlation between the seasonal values of evapotranspiration and the yield of cereal crops in the country is noticed. Ultimately, it was concluded that the method is suitable for monitoring cereal yields, and can also be used as an early warning method by identifying drought zones (Tadesse et al., 2015).

Another aspect of using satellite data to assess water productivity and thus cereal yields is that these data can be used not only for earlier yield predictions and farm management decisions, but where natural, infrastructural and socio-economic conditions do not allow action to be taken to reduce drought damage, earlier action to secure the necessary food from imports will allow. Such a country is Morocco, and for this reason a study was made on the possibility of predicting the yields there of the main cereal crops in the country (soft wheat, barley and durum wheat). To achieve the goal of the research, various data were used, meteorological data (precipitation and air temperature), regional climate indices (particularly the North Atlantic Oscillation) and drought indices derived from satellite observations at different wavelengths. The vegetation indices used are the Vegetation Condition Index (VCI), the Temperature Condition Index (TCI) Kogan (1995) and the Soil Moisture Condition Index (SMCI) Zhang & Jia (2013). These indices were chosen because a study by Bouras et al. (2020) found good correlations between them and cereal yields in Morocco and were used to assess crop drought in the country in previous studies. The vegetation index VCI very well shows the density of the crops and the activity of the vegetation in them, attack also in the months of March and April shows a good correlation with the yield of cereals, the TCI well describes the thermal stress in the crops and shows a high correlation between the yields of cereal crops in the months January and February, and SMCI is well correlated with soil moisture and shows a very good correlation between it and cereal yields in the early phases of crop development (Bouras et al., 2021). Thus, these three indices can be used in the different phases of crop development to monitor yields. As a result of the research done, it is proven that the use of remote sensing data by accurately generating vegetation indices from them in combination with meteorological data to be used as predictors for the yield of cereal crops is a more suitable method for forecasting the yields than using only of in-situ data (Bouras et al., 2021).

Research conducted in Spain utilizing satellite models demonstrated high accuracy, both in predicting crop yields and identifying anomalies in cereal yields. The models for yield levels successfully accounted for approximately 70% of the variability in winter wheat and barley yields, with comparable results observed for oats and rye (Ceglar et al., 2017; García-León et al., 2019; Lobell et al., 2014; Quiring & Papakryiaokou, 2003).

India is the second most populous country in the world, therefore the issue of food security is very important there. At the same time, there are also areas that are affected by drought and management of water productivity and cereal yields is essential. For better management of water productivity and yields respectively, Singh et al. (2022) suggest using a combination of GIS data, hydrological model output and open access satellite data. They use Budyko's (1974) water balance hypothesis where water consumption is said to be an integral part of the water balance components and propose that water consumption be estimated through satellite image analysis. The proposed model is based on eight layers and maps which are: (1) Resource base; (2) Evapotranspiration; (3) Agricultural services; (4) Stream used; (5) surface water; (6) underground water; (7) ecosystem services; (8) Sustainability. The study was conducted for the Subarnarekha Basin, India for the period 2003 to 2014 to assess both water productivity and land productivity. It has been found that the method can be used to model and analyze water consumption, land productivity and rainfed crops. The advantage of the method is the easy analysis of the obtained data, their visibility in the form of maps and tables. They also allow easy identification of farms with low water productivity. This will allow efforts to be focused on these farms and this should lead to an increase in food security and poverty in the area (Singh et al., 2022).

In addition, to all the presented ways of managing water resources in agriculture, there is a developed scale for evaluating water productivity based on satellite data for the main cereal crops. Different satellite platforms provide the ability to measure water productivity in cereal crops by using pixel-by-pixel values, which provides the ability to determine local, regional and global water productivity values for comparison. The idea is to create a single scale on a global level, such as there are for example for earthquakes (Richter), as well as for wind force (Beaufort). With the help of this scale to identify in time the areas where a food crisis caused by drought may occur. Areas where different cereal crops are grown are used to establish local production potential. The scale also takes into account the presence of irrigation systems in addition to the general physical-graphical conditions. Accordingly, the base values for the climatic normalized values of water productivity for three main cereal crops on a global scale were established to be 2.45, 2.3, and 4.9 kg/m^3 for wheat, rice and corn, respectively. The scale can be used both at the farm level to make the best decisions to increase yields, and at the government level to set national water and food policies and use it as a reporting tool against the standards of the Sustainable Development Goals (SDGs) development (Bastiaanssen & Steduto, 2017).

Currently, the SentinelHub platform provides the ability to calculate evapotranspiration based on data from the Sentinel-2 satellite. This occurs using a script that composition composite between channels (B11, B09, B02). This provides an opportunity to establish évapotranspiration rates for a region to be detected easily and with great precision. Thus, vegetation is visible in 6 tons, they indicate the degree of eupotranspiration, at the time of filming with the Sentinel-2 satellite. The interpretation of the data is based on the level of absorption of water vapor and humidity, the higher it is the lower the level of évapotranspiration. In this case, the colors will be darker, as évapotranspiration increases, the reflectivity of the vegetation therefore the colors will be lighter. Figure 9.3 presents the evapotranspiration of part of Parvomay municipality, Plovdiv region, in central southern Bulgaria for the months of June and August. When interpreting, it should be borne in mind that the water is observed in dark blue to black tones, the buildings are in shades of gray, ranging from light gray to dark gray. This makes it possible to identify areas that are without vegetation and areas that are highly urbanized.

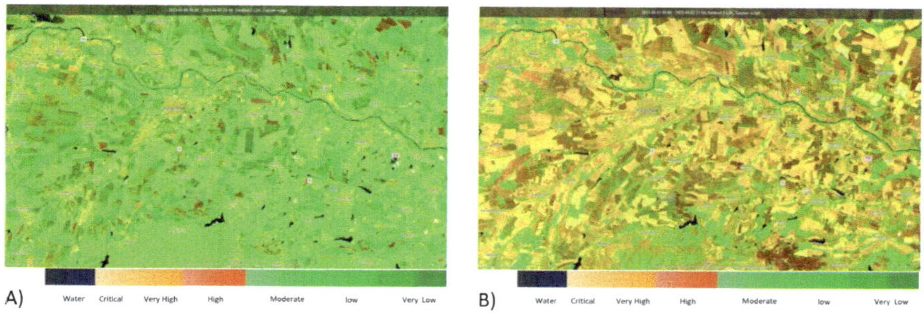

Fig. 9.3 Evapotranspiration of part of the municipality of Pârvomay, Plovdiv region, in central southern Bulgaria, as of (A) June 9, 2023 and (B) August 3, 2023. (image credit Sentinelhub/Detection of Evapotranspiration Levels Composite—https://custom-scripts.sentinel-hub.com/custom-scripts/sentinel-2/evapotranspiration_levels/)

9.3 UAV Data Water Use Efficiency for Winter Cereals

Unmanned aerial vehicle (UAV) technology has also developed in the last decade and it has allowed the use of very high resolution sensors to be used in agriculture to fine-tune its operations. This technology provides spectral, spatial and temporal resolution, plant height data can be obtained in crops, thus assessing damage caused by abiotic stress such as strong winds, torrential rain or hail (Gracia-Romero et al., 2019; Maes & Steppe, 2019). Using sensors in the red–green–blue (RGB), multispectral, and thermal range mounted on UAVs can be used for genetic and phenotyping, and from there to determine yields in agriculture. In fact, this technology is very suitable for determining crop stress and yield. At the same time, it is an extremely cheap technology for the level of precision it provides and has the other advantage that it can be used at a time when it is needed by the farmer and is not affected by cloud cover. Interestingly, even the use of RGB images is shown to be a very cheap and efficient method for assessing the condition and yields of crops. The fact that the height of surveying agricultural areas does not give particular differences in accuracy when determining the yields of agricultural crops is another factor supporting the use of UAV technology in agriculture (Gracia-Romero et al., 2019) that even the use of RGB images is shown to be a very cheap and efficient method for assessing crop condition and yields.

In order to establish the water stress in the crops, the strength of the plant growth and predict the yield from the crops through the UAV technology, it is necessary to combine the data received by the UAV with multispectral and thermal. These data should be applied as integrated monitoring in crop growth and development models. And the data obtained should be used to refine the operations in the individual fields and their individual parts. The UAV data can be used for the precise irrigation of crops, which is of great importance, because in solving the problems related to the lack of water (Fig. 9.4). At the same time, this type of data will also help to better manage soils, their protection from salinization, which will lead to preservation or even increase of natural fertility in soils (Ding, 2019; Maes & Steppe 2019). At present some off-the-shelf commercial products of ICI 8640 P-series (Infrared Cameras Inc., USA), FLIR Vue Pro R 640 (FLIR Systems, USA), thermoMap (senseFly, Switzerland) are used frequently in the research community. The main application areas are Forest Monitoring: These cameras were tested in St. Louis, Missouri, for forest monitoring; Vegetation Stress Detection: In Columbia, Missouri, they were used to detect plant stress and phenotyping; High Throughput Phenotyping: The Maricopa, Arizona site focused on high throughput phenotyping.

Fig. 9.4 UAV Wingtra and image mosaic of a field near Byala Reka village, municipality of Pârvomay, Plovdiv region, in South-Central Bulgaria. Image credit Geovara Ltd. (Bulgaria)

9.4 Thermal Imaging

Thermal imaging is well suited for establishing the water stress experienced by crops, as transpiration is a high-energy physiological process that linearly reduces the surface temperature of leaves and vegetation, for this reason thermal imaging is well suited for early detection of drought stress (Ding, 2019; Gago et al., 2015; Maes & Steppe, 2012; Maes & Steppe, 2019).

Thermal imaging has emerged as a valuable tool for assessing water stress experienced by winter cereals, offering non-destructive and efficient means of monitoring plant health and physiological responses. By capturing the infrared radiation emitted by plants, thermal imaging enables the visualization of temperature variations associated with water stress, providing insights into plant water status and vigor. Studies have demonstrated the efficacy of thermal imaging in detecting subtle differences in canopy temperature that correspond to varying levels of water availability, thereby facilitating early detection of water stress in winter cereals (). Moreover, thermal imaging allows for the quantification of crop water stress indices, such as Crop Water Stress Index (CWSI), which can serve as reliable indicators of plant water status and inform irrigation management decisions (Moriondo et al., 2013). Overall, thermal imaging presents a promising approach for assessing water stress in winter cereals, offering practical applications in precision agriculture and resource-efficient crop management.

It has been found that in cereal crops such as barley and wheat, thermal imaging can be used to detect diseases in them even before the first visible signs are visible to the human eye. The principle of action is that the rate of transpiration serves as an immediate indicator of the physiological processes and health of the plant. Thus, by using thermal and RGB images obtained from the UAV, the total energy balance in the crop can be calculated. This can be achieved by using the Surface Energy Balance System (SEBS) and making a daily model of evapotranspiration in the crop

(Woodgate et al., 2020). Taking daily measurements of evapotranspiration in wheat and barley crops using UAV-mounted thermal cameras and calculating evapotranspiration for the fields from the Landsat satellite showed a very high similarity in measurements. Evapotranspiration was a better discriminator of invisible infections in May than the NDVI vegetation index. This shows a promising transformation from satellite to UAV imagery for evapotranspiration estimation. However, further research is needed in this direction so that this type of data can be used more often by farmers (Ishaq et al., 2021).

Thermal proximal imaging has emerged as a valuable technique for estimating water-use efficiency (WUE) in winter cereals, offering high-resolution insights into plant water status and physiological responses. By capturing detailed thermal infrared images at close range, thermal proximal imaging enables the assessment of canopy temperature variations, which are indicative of plant water stress levels. Several studies have highlighted the effectiveness of thermal proximal imaging in estimating WUE in winter cereals, demonstrating strong correlations between thermal-derived metrics and traditional field-based measurements of WUE (Rosenthal et al., 2014). These studies have emphasized the advantages of thermal proximal imaging in capturing spatial and temporal variations in crop water dynamics, thereby facilitating precise irrigation management decisions and enhancing agricultural productivity. Furthermore, the integration of thermal proximal imaging with other proximal sensing techniques, such as spectral reflectance measurements, further enhances the accuracy of WUE estimation models, providing comprehensive insights into crop water management strategies (Woodgate et al., 2020). In **Fig. 9.5** two thermal images of two different crops of the barley culture are presented.

The high spatial resolution images presented clearly show the temperature differences on the surface and interior of crops and how they respond to heat stress. since transpiration is a high-energy physiological process that linearly reduces the surface temperature of leaves and vegetation, thermal imaging is therefore well suited for early detection of drought stress (Ding, 2019; Gago et al., 2015; Maes & Steppe, 2012; Maes & Steppe, 2019).

Satellite thermal data has been increasingly utilized for estimating water-use efficiency (WUE) in winter cereals, offering a remote sensing approach to monitor

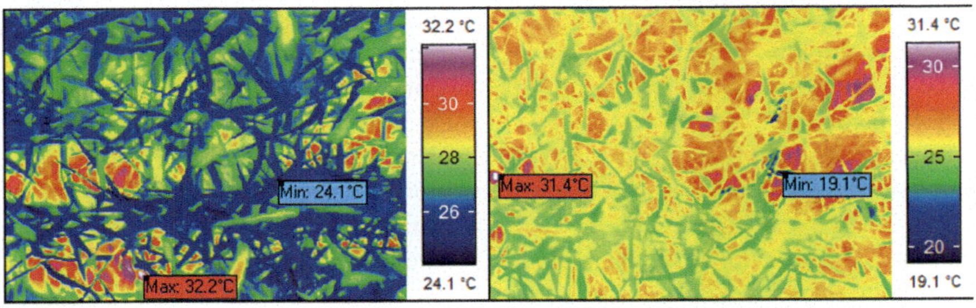

Fig. 9.5 Thermal images of a wheat crop acquired with handheld thermal camera

crop water dynamics and optimize irrigation management strategies. By capturing the thermal infrared radiation emitted by plants, satellite thermal data enable the calculation of crop canopy temperature, which serves as a proxy for plant water status. Studies have demonstrated the utility of satellite thermal data, particularly derived from platforms such as Landsat and MODIS, in estimating WUE in winter cereals across different agricultural landscapes (Jackson et al., 1981). These studies have shown strong correlations between thermal-derived metrics and field-based measurements of WUE, indicating the potential of satellite thermal data for large-scale monitoring of crop water use efficiency. Moreover, the integration of thermal data with other remote sensing variables, such as vegetation indices and meteorological data enhances the accuracy of WUE estimation models, providing valuable insights for sustainable water management in agriculture (Vuolo et al., 2015).

Acknowledgements This study was carried out within the Dragon 5 project ID. 57160: Monitoring water productivity in crop production areas from food security perspectives. This work was supported/(partially supported) by the Bulgarian Ministry of Education and Science under the National Research Programme "Young scientists and postdoctoral students—2" approved by DCM 206 / 07.04.2022 to which Milen Chanev, PhD is beneficiary. We express our gratitude to the company Geovara Ltd. (Bulgaria), who performed the aerial imaging with the WingtraOne UAV during the flowering and milk ripeness stages.

Funding: This work was supported under contract № 4000137730/22/I-NB with the European Space Agency as part of the program "Support to European Senior & Young Scientists within the Dragon 5 Cooperation ", Dragon-5 project (ID. 57160).

Conflict of Interest Authors declare no conflict of interest. The funding entities did not contribute to the study's design, data collection, analysis, interpretation, manuscript drafting, or the decision to publish the findings.

References

Abbasi, N., Nouri, H., Didan, K., Barreto-Muñoz, A., Chavoshi Borujeni, S., Opp, C., ... & Siebert, S. (2023). Mapping vegetation index-derived actual evapotranspiration across croplands using the google earth engine platform. *Remote Sensing, 15*(4), 1017. https://doi.org/10.3390/rs1504 1017

Allen, R. G., Pereira, L. S., Raes, D., & Smith, M. (1998). Crop evapotranspiration-guidelines for computing crop water requirements-FAO irrigation and drainage paper 56. *FAO, Rome, 300*(9), D05109.

Bastiaanssen, W. G., & Steduto, P. (2017). The water productivity score (WPS) at global and regional level: methodology and first results from remote sensing measurements of wheat, rice and maize. *Science of the Total Environment, 575*, 595–611.

Budyko, M. I. (1974). *Climate and life*. Academic press.

Bouras, E. H., Jarlan, L., Er-Raki, S., Albergel, C., Richard, B., Balaghi, R., & Khabba, S. (2020). Linkages between rainfed cereal production and agricultural drought through remote sensing indices and a land data assimilation system: A case study in Morocco. *Remote Sensing, 12*(24), 4018.

Bouras, E. H., Jarlan, L., Er-Raki, S., Balaghi, R., Amazirh, A., Richard, B., & Khabba, S. (2021). Cereal yield forecasting with satellite drought-based indices, weather data and regional climate indices using machine learning in Morocco. *Remote Sensing, 13*(16), 3101.

Blum, A. (2009). Effective use of water (EUW) and not water-use efficiency (WUE) is the target of crop yield improvement under drought stress. *Field Crops Research, 112*(2–3), 119–123.

Breiman, L. (2001). *Random Forests. Machine Learning, 45*, 5–32.

Bokusheva, R., Kogan, F., Vitkovskaya, I., Conradt, S., & Batyrbayeva, M. (2016). Satellite-based vegetation health indices as a criteria for insuring against drought-related yield losses. *Agricultural and Forest Meteorology, 220*, 200–206.

Ceglar, A., Turco, M., Toreti, A., & Doblas-Reyes, F. J. (2017). Linking crop yield anomalies to large-scale atmospheric circulation in Europe. *Agricultural and Forest Meteorology, 240*, 35–45.

Chanev, M., & Filchev, L. (2021). Satellite Monitoring of Wheat Yields in Bulgaria. *Journal of Mountain Agriculture on the Balkans., 24*(5), 373–395.

Coll, C., Caselles, V., & Schmugge, T. J. (1994). Imation of land surface emissivity differences in the split-window channels of AVHRR. *Remote Sensing of Environment, 48*(2), 127–134.

Conrad, C., Rahmann, M., Machwitz, M., Stulina, G., Paeth, H., & Dech, S. (2013). Satellite based calculation of spatially distributed crop water requirements for cotton and wheat cultivation in Fergana Valley, Uzbekistan. *Global and Planetary Change, 110*, 88–98.

Davis, K. F., Chiarelli, D. D., Rulli, M. C., Chhatre, A., Richter, B., Singh, D., & DeFries, R. (2018). Alternative cereals can improve water use and nutrient supply in India. *Science Advances, 4*(7), eaao1108.

Ding, K. (2019). State of knowledge of irrigation techniques and practicalities within given socio-economic settings. *Irrigation and Drainage, 68*(1), 31–45.

Delchev, G. & Zhelyazkova, T. S. (2016). *Horticulture, scipress*, ISBN 978-954-92614-5-5 (in Bulgarian).

Ekwue, E. I., Rebekah, C. C., & Birch, R. (2015). Simulation of irrigation water requirements of some crops in trinidad using the CROPWAT irrigation software. *Notes, 136*(3.77), 148.

Gago, J., Douthe, C., Coopman, R. E., Gallego, P. P., Ribas-Carbo, M., Flexas, J., ... & Medrano, H. (2015). UAVs challenge to assess water stress for sustainable agriculture. *Agricultural water management, 153*, 9–19.

García-León, D., Contreras, S., & Hunink, J. (2019). Comparison of meteorological and satellite-based drought indices as yield predictors of Spanish cereals. *Agricultural Water Management, 213*, 388–396.

Gracia-Romero, A., Kefauver, S. C., Fernandez-Gallego, J. A., Vergara-Díaz, O., Nieto-Taladriz, M. T., & Araus, J. L. (2019). UAV and ground image-based phenotyping: A proof of concept with durum wheat. *Remote Sensing, 11*(10), 1244.

Guo, S., Zhu, H., Dang, T., Wu, J., Liu, W., Hao, M., ... & Syers, J. K. (2012). Winter wheat grain yield associated with precipitation distribution under long-term nitrogen fertilization in the semiarid Loess Plateau in China. *Geoderma, 189*, 442–450.

Ihuoma, S. O., Madramootoo, C. A., & Kalacska, M. (2021). Integration of satellite imagery and in situ soil moisture data for estimating irrigation water requirements. *International Journal of Applied Earth Observation and Geoinformation, 102*, 102396.

Ishaq, R. A. F., Shah, S. R. A., & Shabbir, Y. (2021, December). UAV assessment of crop evapotranspiration dynamics in winter wheat and barley under varying pressures of fungal diseases. In *2021 Seventh International Conference on Aerospace Science and Engineering (ICASE)* (pp. 1–8). IEEE.

Jackson, R. D., Reginato, R. J., Idso, S. B., & Pinter, P. J. (1981). Canopy temperature as a crop water stress indicator. *Water Resources Research, 17*(4), 1133–1138. https://doi.org/10.1029/WR017i004p01133

Kato, Y., Kamoshita, A., & Yamagishi, J. (2008). Preflowering abortion reduces spikelet number in upland rice (Oryza Sativa L.) under water stress. *Crop Science, 48*(6), 2389–2395.

Kogan, F. N. (1995). Application of vegetation index and brightness temperature for drought detection. *Advances in Space Research, 15*(11), 91–100.

Kogan, F. N. (1997). Global drought watch from space. *Bulletin of the American Meteorological Society, 78*(4), 621–636.

Li, H., Zheng, L., Lei, Y., Li, C., Liu, Z., & Zhang, S. (2008). Estimation of water consumption and crop water productivity of winter wheat in North China plain using remote sensing technology. *Agricultural Water Management, 95*(11), 1271–1278.

Lobell, D. B., Roberts, M. J., Schlenker, W., Braun, N., Little, B. B., Rejesus, R. M., & Hammer, G. L. (2014). Greater sensitivity to drought accompanies maize yield increase in the US Midwest. *Science, 344*(6183), 516–519.

Maes, W. H., & Steppe, K. (2012). Estimating evapotranspiration and drought stress with ground-based thermal remote sensing in agriculture: A review. *Journal of Experimental Botany, 63*(13), 4671–4712.

Maes, W. H., & Steppe, K. (2019). Perspectives for remote sensing with unmanned aerial vehicles in precision agriculture. *Trends in Plant Science, 24*(2), 152–164.

Martyniak, L., Dabrowska-Zielinska, K., Szymczyk, R., & Gruszczynska, M. (2007). Validation of satellite-derived soil-vegetation indices for prognosis of spring cereals yield reduction under drought conditions–case study from central-western Poland. *Advances in Space Research, 39*(1), 67–72.

Moriondo, M., Orlandini, S., Giannakopoulos, C., & Bindi, M. (2013). Climate change impact assessment: The role of climate extremes in crop yield simulation. *Climatic Change, 117*(1–2), 29–41. https://doi.org/10.1007/s10584-012-0565-8

Quiring, S. M., & Papakryiakou, T. N. (2003). An evaluation of agricultural drought indices for the Canadian prairies. *Agricultural and Forest Meteorology, 118*(1–2), 49–62.

Rosenthal, D. M., Ruiz-Vera, U. M., Siebers, M. H., Gray, S. B., Bernacchi, C. J., & Ort, D. R. (2014). Biochemical acclimation, stomatal limitation and precipitation patterns underlie decreases in photosynthetic stimulation of soybean (Glycine max) at elevated [CO2] and temperatures under fully open-air field conditions. *Plant Science, 226*, 136–146. https://doi.org/10.1016/j.plantsci.2014.06.013

Singh, P. K., Jain, S. K., Mishra, P. K., & Goel, M. K. (2022). An assessment of water consumption patterns and land productivity and water productivity using WA+ framework and satellite data inputs. *Physics and Chemistry of the Earth, Parts a/b/c, 126*, 103053.

Smith, M., (1992). *CROPWAT A computer program for irrigation planning and management.* Irrigation and Drainage, Paper 46, Rome, FAO, pp. 1–132.

Tadesse, T., Senay, G. B., Berhan, G., Regassa, T., & Beyene, S. (2015). Evaluating a satellite-based seasonal evapotranspiration product and identifying its relationship with other satellite-derived products and crop yield: A case study for Ethiopia. *International Journal of Applied Earth Observation and Geoinformation, 40*, 39–54.

Talebi, H., Samadianfard, S., & Kamran, K. V. (2023). Investigating the roles of different extracted parameters from satellite images in improving the accuracy of daily reference evapotranspiration estimation. *Applied Water Science, 13*(2), 59.

Varga, B., Varga-László, E., Bencze, S., Balla, K., & Veisz, O. (2013). Water use of winter cereals under well-watered and drought-stressed conditions. *Plant, Soil and Environment, 59*(4), 150–155.

Vuolo, F., D'Urso, G., De Michele, C., Bianchi, B., & Cutting, M. (2015). Satellite-based irrigation advisory services: A common tool for different experiences from Europe to Australia. *Agricultural Water Management, 147*, 82–95. https://doi.org/10.1016/j.agwat.2014.08.004

Wang, L., Palta, J. A., Chen, W., Chen, Y., & Deng, X. (2018). Nitrogen fertilization improved water-use efficiency of winter wheat through increasing water use during vegetative rather than grain filling. *Agricultural Water Management, 197*, 41–53.

Woodgate, W., van Gorsel, E., Hughes, D., Suarez, L., Jimenez-Berni, J., & Held, A. (2020). THEMS: An automated thermal and hyperspectral proximal sensing system for canopy reflectance, radiance, and temperature. *Plant Methods, 16*, 1051.

WMO, GWP (2016). *Handbook of drought indicators and indices*. Integrated drought management tools and guidelines. World Meteorological Organization and Global Water Partnership.

Zhang, A., & Jia, G. (2013). Monitoring meteorological drought in semiarid regions using multi-sensor microwave remote sensing data. *Remote Sensing of Environment, 134*, 12–23.

Chapter 10
Flood Inundation Mapping Over Kerala Region During 2018 Flood—Based on Cloud Computing and Automatic Threshold Detection Approach

Jishnu V, Mohit Arora, Harpinder Singh, Pritam Chand, Raj Setia, and Brijendra Pateriya

10.1 Introduction

Floods stand out as the most prevalent and impactful natural calamity, characterized by their frequent occurrence and significant scale all over the world. Flooding represents a transient state where sudden water accumulates, either from inland sources, tidal influences, or rapid runoff resulting from high frequency and intensity rainfall, leading to complete water inundation. Such flood disaster triggered by diverse weather conditions, have affected numerous countries, resulting in the loss of countless lives, substantial property damage, and disruption of overall economic activities (Brivio et al., 2002). Between 1995 and 2020, weather-related disasters accounted for 47% of all incidents, with flooding being responsible for a considerable portion, affecting around 2.3 billion people over the world, primarily concentrated in Asia (Wahlstrom & Guha-Sapir, 2015). Bangladesh as the most flood-affected country, followed by India. In India, out of 28 states, 23 states are susceptible to flooding, covering approximately 40 million hectares (ha), which accounts for around one-eighth of the nation's total geographical area (approx. 329 million ha) which accounts for approximately 12.16% of the entire area. India alone experiences almost one-fifth of the global death rate resulting from floods. The flooding-prone area in India spans around four Lakh square km, constituting about one-eighth of its total topographical

J. V · P. Chand
Department of Geography, School of Environment and Earth Sciences, Central University of Punjab, Bathinda, Punjab, India

M. Arora (✉) · H. Singh · R. Setia · B. Pateriya
Punjab Remote Sensing Centre, Punjab Agricultural University Campus, Ludhiana, Punjab, India
e-mail: mohitarorakuk12@gmail.com

P. K. Rai and S. Rai (eds.), *Climate, Environment and Agricultural Development*, Advances in Geographical and Environmental Sciences,
https://doi.org/10.1007/978-981-97-8363-2_10

territory (Sengupta & Kakran, 2011). Populated regions typically located in tropical areas are susceptible to floods, which are classified as hazardous natural calamities. Excessive rainfall leads to the rise of water level in catchment areas, causing water to overflow and breach its usual boundaries, resulting floods. Central Water Commission (CWC) reported that heavy rains and floods caused a total of 107,535 casualties and led to damages worth 53,425.50 million USD (equivalent to INR 378,247.047 crores) to public utilities, houses, and crops throughout the country over a span of 64 years, from 1953 to 2017 (Nabangchang et al., 2015; Nace, 2017). The severity and timing of floods are varied across different locations, influenced by various factors. The effective planning and management of flood mitigation necessitate understanding of land usage and precisely identifying and mapping areas prone to flooding. Timely and accurate estimation of inundated regions during floods is crucial for monitoring and assessing risks promptly. This precise and valuable information aids policymakers and decision-makers in implementing sustainable management practices and preparing for the future (Tramblay et al., 2012).

Approximately 80% of the annual rainfall in India occurs during the monsoon season, as reported by Parthasarathy et al. (1994). In recent years, many area of India have experienced severe flooding events. For instance, on 26 July 2005, Mumbai experienced its worst flooding in recorded history, receiving 942 mm of rainfall in a single day (Prasad & Singh, 2005). Similarly, on 17 June 2013, the state of Uttarakhand received more than 340 mm of rainfall, resulting in catastrophic floods and landslides that caused extensive damage to life and property (Dube et al., 2014; Martha et al., 2015). Another significant event occurred during the November 2015 Chennai floods, which resulted in over 500 deaths as Chennai received three times its usual rainfall (Ray et al., 2019). Annually, flooding in India from extreme rainfall leads to economic losses of approximately $3 billion, representing about 10% of global economic losses (Roxy et al., 2017).

In July 1924, Kerala state received a huge torrential rainfall, over 3000 mm, for a short period of time resulted in overflowing rivers, dams, and swamp lands. The Mullape riyar dam, one of the largest reservoirs of water, overflowed which caused in the overflowing of rivers in the mainland's located downstream (George et al., 2012). The hydrological model-based approach faced challenges due to the need for up-to-date and precise DEM (Digital Elevation Model), AWS (Automatic Weather Station) data, and sufficient computing infrastructure, which were often unattainable. Instead, many researchers chosen for a different method, utilizing SAR data (Sentinel-1 & ALOS-2 PALSAR) (Lal et al., 2020; SK, 2020; Vishnu et al., 2019) and combined this data with a combination of manual threshold-based, conventional, and machine learning-based classification techniques to generate flood extent maps and evaluate the damage caused by the flood (Bernhofen et al., 2018; Comer et al., 2017; Cook et al., 2018).

Many researchers have used different optical and microwave remote sensing data like MODIS, Landsat, Sentinel-1 & 2, TerraSAR-X, Radarsat-1 & 2, ALOS PALSAR, COSMO-SkyMed SAR etc. for flood monitoring and management. This study established the potential of high-resolution SAR data in capturing detailed flood patterns and dynamics (Bhaduri & Ramachandran, 2013; Chen & Lin, 2015; Clement et al., 2018; Khan et al., 2018; Mudi & Das, 2022). Flood inundation mapping using SAR data has become an increasingly significant topic in the field of remote sensing and disaster management. SAR technology provides all-weather and day-and-night capabilities, making it highly suitable for monitoring and mapping flood events and also provide valued remote sensing data that is vital for flood monitoring and management. The microwave signal backscatter is responsive to alterations in the target's dielectric properties and structure, resulting in a unique response compared to water-covered regions. This separate ability allows exact determination of the extent of water pixels using SAR data (Gaume et al., 2016; Hinkel et al., 2014). Though, challenges persist, including issues related to data processing, calibration, validation, and the integration of multi-source data. As technology continues to evolve, further advancements in SAR-based flood mapping are anticipated, contributing to improved flood management strategies in India. Much research explores the potential of multi-frequency and multi-polarization SAR data for flood monitoring, particularly focusing on flooded vegetation areas and examine the sensitivity of different SAR parameters to inundated vegetation and provides visions into their applicability for flood mapping (Rözer & Surminski, 2020). This makes it an optimal choice for expediting flood map generation and facilitating swift responses to flood-related emergencies (Kirkpatrick & Olbert, 2020).

The objective of the study to evaluate the potential of automatic threshold detection techniques using microwave SAR images for rapid and precise mapping of flood inundated areas over Kerala region. This study also estimate statistics of the flood affected land use land cover. However, an alternative solution lies in using a cloud-based image processing platform with automatic threshold methods.

10.2 Study Area

Indian state of Kerala region situated in the southwestern part of India, is often referred to as "God's Own Country." Its geographical coordinate range lies between 8°18' and 12°48' N latitudes and 74°52' and 77°22' E longitudes with covering an area of 38,863 square kilometre. The state boasts a coastline spanning 590 kilometres, with its width varying from 11 to 121 kilometres. This tropical haven is flanked by the Arabian Sea on its western side and the Western Ghats on its eastern side. Approximately 48% of its terrain is occupied by the Western Ghats, also known as Sahyadri. Kerala can be divided into three main physiographic regions: highlands (above 75.0 meters), midlands (7.5–75.0 meters), and coastal plains (below 7.0 meters). These areas exhibit diverse geomorphic features, encompassing notable landmarks such as the Anaimalai mountain peak (2695 meters), westward-flowing small rivers, and a coastal plain adorned with lagoons and barriers.

Kerala's elevation ranges from the coastal plains along the Arabian Sea to the towering Western Ghats Mountain range that runs parallel to its eastern border. Its rich geological history has played a significant role in shaping its diverse landscape, culture, and natural resources. From its coastal plains and backwaters to the Western

Ghats Mountain range. These hills are a result of ancient tectonic collisions and volcanic activities, creating a complex terrain of valleys, plateaus, and peaks. The Ghats not only contribute to Kerala's stunning topography but also play a crucial role in its climatic patterns, influencing the region's monsoons and bi8odiversity. The state's extensive network of backwaters, lagoons, and estuaries showcase the interplay between the sea and the rivers, creating a unique ecosystem that supports a variety of aquatic life and human activities. Kerala's soil diversity can be attributed to its varying topography and climatic conditions. The state can be broadly divided into three agro-ecological zones: the coastal lowlands, the midlands, and the highlands. Each of these zones presents distinct soil characteristics and challenges. The coastal lowlands, influenced by the proximity to the sea and the presence of backwaters, often have alluvial soils that are rich in organic matter. These soils are ideal for cultivating crops like rice, coconut, and various vegetables. However, saltwater intrusion and soil erosion can pose significant threats to these fertile lands. Moving to the midlands, the soil type transitions to lateritic soils. These soils are derived from weathered rock and tend to be well-draining but relatively low in fertility. Proper soil management practices are crucial in this zone to enhance agricultural productivity and prevent erosion. In the highlands, which include the Western Ghats Mountain range, the soil types vary widely depending on altitude and slope. Here, you can find soils ranging from clay to loam, often layered and interspersed with rocky terrain. Tea, coffee, spices, and other cash crops flourish in these areas, with terrace farming being a common sight. Kerala experiences a tropical climate with distinct wet and dry seasons. The temperature in Kerala varies throughout the year, with relatively consistent temperatures due to its proximity to the sea. The state enjoys a warm and humid climate, which can be categorized into three main seasons: summer, monsoon, and winter. During the summer months, from March to May, Kerala witnesses warmer temperatures, with coastal areas experiencing average highs of around 30 to 35 degrees Celsius (86 to 95 degrees Fahrenheit). Inland areas, especially those closer to the Western Ghats, tend to be slightly cooler. The monsoon season, which lasts from June to September, is a defining period for Kerala. The state receives heavy rainfall during these months, with the southwest monsoon winds bringing moisture-laden clouds from the Arabian Sea. This contributes to the lush greenery and vibrant landscapes that Kerala is known for. The temperature during the monsoon months remains relatively cooler, with averages ranging from 25 to 30 degrees Celsius (77 to 86 degrees Fahrenheit). As the monsoon recedes, Kerala enters the winter season from October to February. During this time, temperatures become milder and more comfortable, with average highs ranging from 28 to 32 degrees Celsius (82 to 90 degrees Fahrenheit) (Fig. 10.1).

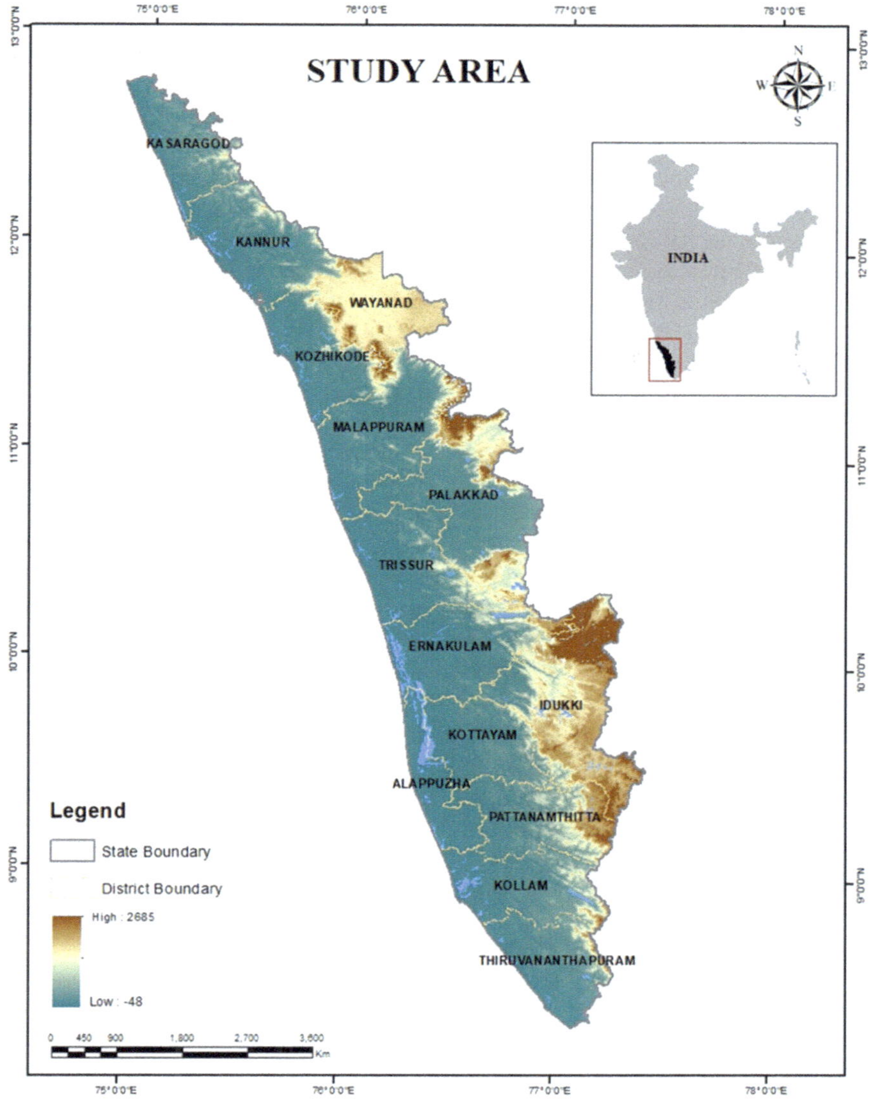

Fig. 10.1 Study area showing the Kerala state overlaid on DEM

10.3 Materials and Methodology

This study incorporates various space-based satellite data including Sentinel-1 SAR images, GPM (Global Precipitation Measurement) daily rainfall data and Land use land cover (LULC) derived from sentinel-2 data. Additionally, ancillary data includes a 30-m resolution shuttle radar topography mission (SRTM) digital elevation models (DEM) is also used in this study to see the topographic variations over the study area. DEM resolution and accuracy have a significant impact on water surface elevations and flood extent (Abrams et al., 2020). SAR is an active remote sensing technology that emits a RADAR pulse and measures returning backscatter. The strength of the backscattering coefficient ($\sigma 0$) is a function of sensor specification (like wavelength, polarisation, incidence angle) as well as environmental factors at the surface (like roughness, geometry and water content) (Hofste et al., 2021). The Sentinel-1part of the European Space Agency (ESA) operates within C-band (5.407 GHz) frequencies in one of four acquisition modes: Strip map (SM), Interferometric Wide swath (IW), Extra-Wide swath (EW) and Wave (WV). Sentinel-1 mission consists of two systematically acquiring satellite sensors (S-1A/1B) which operate at an altitude of 693 km with 12 days of repeat cycle of each sensor. Sentinel-1A was launched on April 3, 2014, while Sentinel-1B was launched on April 25, 2016. These two satellites follow the same orbital path and work together to capture images every six days, regardless of weather conditions or time of day. This enables them to cover the same area every 12 days (Gorelick et al., 2017). The data specifications of Sentinel-1 satellites are mentioned in (Table 10.1). The Sentinel-1 C-band SAR instruments consist of a transmission chain and two parallel reception chains that handle horizontal (H) and vertical (V) polarizations. The sensors operate in either single polarization (HH or VV) or dual polarization (HH + HV or VV + VH) modes. Data acquired in SM, IW, and EW modes can be obtained in either single or dual polarization, Flood extents are generally derived by a histogram based, backscatter approach. Identification of flooding is very challenging task from SAR compared to optical systems, but frequent revisits of SAR make it ideal for flood monitoring. The Sentinel-2 datasets are used to demarcate two land use/land cover (LULC) classes of the Kerala state. The Sentinel-2 image is classified into two major classes namely: agricultural land, built-up. The lands which are under cropping practices were considered as agricultural lands. The build-up class comprised of constructed areas such as residential, commercial, industrial etc. The flow chart of the detailed methodology adopted in the present study is shown in Fig. 10.2.

10.3.1 SAR Data Processing

The Sentinel-1 Ground Range Detected (GRD) level-1 processing data are collected over the study region during 21 August 2018 for analysing and find the flood inundation area. Sentinel-1 image was initially pre-processed to mitigate the SAR-typical speckle noise signatures from the images for flood classification because SAR images suffers from speckle due to the variations in the radar return within the pixel caused by multiple scattering sources (Cozzolino et al., 2013; Lee et al., 2008). Speckle noise removal from SAR image was carried out by using median filter with 3×3 kernel sizes. Radiometric calibration of the SAR image was carried out to generate the backscatter values. The GRD products did not have phase information but only

Table 10.1 Data specifications for the sentinel-1 satellites

Data Specifications	
Frequency/wavelength	C-band (5.407 GHz)
Orbit	693 km (sun-synchronous)
Interval	12 days
Acquisition Dates	Before flood (21–07-2018, 28–07-2018) During/After flood (14–08-2018, 21–08-2018)
Mode	Interferometric Wide swath (Default mode over land)
Swath width	250 km
Product	Level-1 Ground Range Detection (GRD)
Spatial Resolution	10 m
Polarisation	VV & VH
% Cloud Cover	-

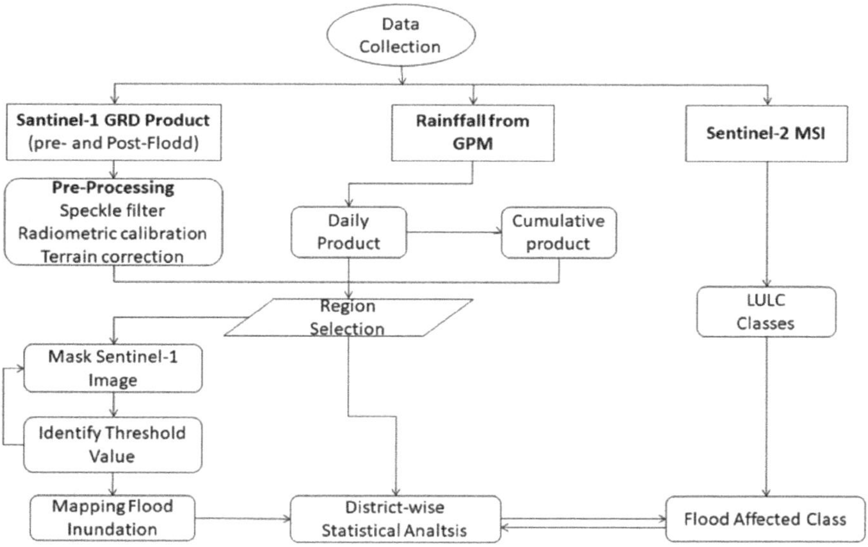

Fig. 10.2 Flow chart for remote sensing data analysis

frequency information. The Multi-looked level-1 GRD products were projected to UTM 1984 projection. The SAR image has been processed through ESA's Sentinel Application Platform: SNAP (version 8.0) and followed a step-by-step procedure of calibration, filtering, masking, and rectification. he resulting images have pixels with approximately square size (10 m × 10 m). Two sets of SAR data (before flood and during or after flood) have been used for detecting the extent of flood pixels and mapping of flood inundated area. The water pixels have a very low backscatter

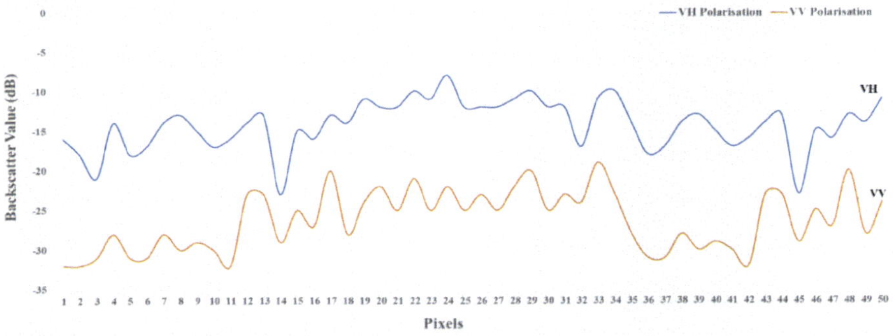

Fig. 10.3 Signature profiles of backscatter in VH and VV polarizations during flood over Kerala

coefficient and appear darker due to the specular nature of water against microwave signals. A total of more than 50 sampling point during the flooding period were taken to identify threshold value over VH and VV polarization image and such optimum backscatter threshold values were used in the binarization process to separate flood pixels (Fig. 10.3). A fixed range of dB values is applied to classify and map the flood inundated areas. Suitable threshold gives more accurate value which could be used further in other operations (Mullissa et al., 2020; Singh & Kumar, 2017). This study used Google Earth Engine (GEE) approach to monitor flood inundated area. GEE has become the most popular and successful virtual globe tool and can effectively address global environmental change in a way that was systematic and integrated (Torres et al., 2012). The utilization of Google Earth Engine (GEE) facilitates cloud-based computation and serves as a powerful instrument for conducting extensive analyses on large-scale global geospatial data. Google Earth Engine (GEE) is a cloud computing platform that was launched by Google in 2010 (Moskolaï et al. 2020). GEE offers publicly accessible Earth observation data on a massive scale, along with sophisticated algorithms for processing large geographic datasets and an interactive coding environment.

10.3.2 Global Precipitation Measurement (GPM) Rainfall Data

The Global Precipitation Measurement (GPM) mission's Integrated Multi-Satellite Retrievals for GPM provides valuable rainfall data on a global scale. The GPM mission is a collaborative effort between NASA and the Japan Aerospace Exploration Agency (JAXA), along with other international partners, launched on 27th February 2014 and is a follow-on, expanded mission to TRMM (Tropical Rainfall Measuring Mission) which provides high spatial resolution (10 × 10 km) datasets. GPM aims to enhance our understanding of global precipitation by utilizing a constellation of satellites equipped with advanced instruments to measure precipitation rates and patterns. It combines data from various satellite instruments and ground-based measurements to create a comprehensive picture of precipitation around the world. This combination of data sources helps to improve the accuracy and coverage of rainfall estimates. Global Precipitation Measurement product provides real-time rainfall data using dual-frequency precipitation radar (Ku/Ka-band) and a multi-channel GPM Microwave Imager (GMI). Dual-frequency precipitation radar provides three dimensional measurements of precipitation structure and characteristics. IMERG is a product of the GPM mission that provides high-resolution global precipitation estimates. IMERG uses a multi-satellite approach to capture precipitation data from various sources, including passive microwave sensors and infrared data. The IMERG algorithm incorporates a variety of data sources, including those from the GPM Core Observatory satellite and its partner satellites, as well as data from other geostationary and polar-orbiting satellites. The algorithm integrates these diverse data streams to generate precipitation estimates on a global grid with a temporal resolution of 30 min. This study used daily accumulated rainfall derived from the originally available rainfall product at the interval of 3 h.

Sentinel-2 data has been utilized for land use and land cover (LULC) classification. This dataset provides a global map depicting various categories of land use and land cover. The map is derived from ESA Sentinel-2 imagery with a 10-m resolution. It represents a composite of Land Use/Land Cover (LULC) predictions for ten classes throughout the year, generating a representative snapshot of 2018. This map was produced using a deep learning model trained on over 5 billion hand-labelled Sentinel-2 pixels, sampled from over 20,000 sites distributed across all major biomes. The underlying deep learning model utilizes six bands of Sentinel-2 surface reflectance data: visible blue, green, red, near infrared, and two shortwave infrared bands. To create the final map, the model was applied to multiple dates of imagery throughout the year, and the outputs were composited into a final representative map of 2018.

10.4 Results & Discussion

The prodigious situation of flood inundated area has been observed over Kerala state during period from 11th to 21st August 2018. This study used satellite-based SAR imagery in conjunction with the meteorological observations. Two sets of Sentinel-1 SAR data have been downloaded for mapping the spatial extent of the flooding in the Kerala region. Multi-temporal Sentinel-1 SAR images acquired between 15 to 30 July 2018 are considered as pre-flood image and those acquired between 11 to 21 August 2018 are considered as during flood image, a significant flooding event was observed in the Kerala state. Additionally, the study incorporated the GPM daily rainfall product to see the spatio-temporal variability of rainfall and its relationship with the occurrence of flood.

10.4.1 Spatio-Temporal Variability of Precipitation

The GPM daily precipitation data has been utilized to examine how rainfall patterns change over time and across different locations in the Kerala region, specifically during the monsoon season from June to September in year 2018. To understand the location, strength, intensity and frequency of rainfall during the flooding period, rainfall maps have been prepared for the study area. Data collected from GPM on a daily data basis reveals how rainfall is distributed across various districts within Kerala state (Fig. 10.4). The findings highlight that, between the 1st and 31st of August month in year 2018, Kerala state has received daily rainfall range between 0.13 mm to 29.87 mm with an average daily rainfall of 4.73 mm and the average cumulative rainfall recorded for august month is 146.71 mm (Fig. 10.4). The statistics of the diurnal rainfall during flood period from August 01 to 31, 2018 over different districts in Kerala State is shown in (Fig. 10.4) where the daily rainfall has been analysed for each district. The results indicated that Kerala state has received the maximum cumulative rainfall up to 557.04 mm and the minimum cumulative rainfall up to 238.73 mm with a mean cumulative rainfall of 427.07 mm during the period of August 2018 (Fig. 10.5). The maximum rainfall has been received by Wayanad district (557 $\pm$ 11) and minimum rainfall has been received by Thiruvananthapuram district (239 $\pm$ 4). According to GPM datasets, the Kerala region has experience monsoonal cumulative rainfall of 1221 ($\pm$106) mm out of which the maximum cumulative rainfall has been recorded in Kasaragod district 1420 ($\pm$145) mm and minimum has been recorded in Thiruvananthapuram district 758 ($\pm$ 50) mm during June to September (Fig. 10.6). The high frequency and intensity of rainfall, and heavy discharge from rivers during these periods causes widespread floods. Additionally, it was observed that during flood period, Kerala state has received a heavy rainfall as well as runoff from foothills as well as from adjoining states and affected Alappuzha, Ernakulam, Kottayam and Trissur district with unprecedented floods. The cumulative rainfall has been recorded at all four districts are 391 mm, 460 mm, 426 mm and 463 mm during august 2018.

10.4.2 Flood Inundation Mapping Using Sentinel-1 SAR Images

Sentinel-1 SAR images with dual polarizations prove invaluable for delineating flood extents in inundated regions, offering the advantage of all-weather image acquisition. In this study, two sets of SAR data were utilized: one captured before the flood and the other during or after the flood, covering the affected area for flood mapping and monitoring purposes (Fig. 10.7). Figure 10.7 displays images of the VH and VV polarizations before the flood in the top left and top right, respectively, and after the flood in the bottom left and bottom right. A widely employed technique like thresholding technique was employed to separate flood inundated areas in SAR

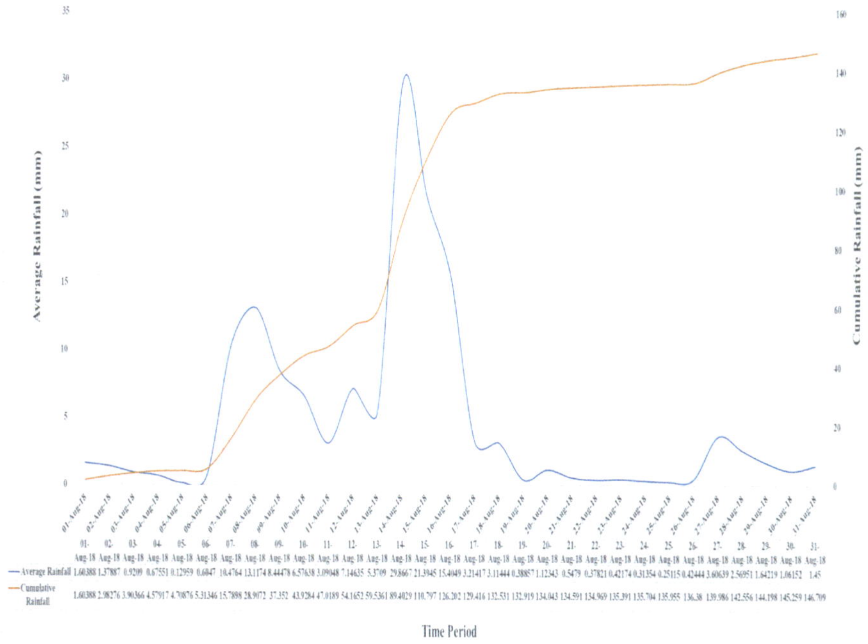

Fig. 10.4 Diurnal and cumulative rainfall during flood period from 01 to 31 August 2018

images. This study employed threshold values ranging from -2 to -24 dB for VH and VV polarizations to identify the extensive floodwater pixels in the study area. Both VH and VV polarizations were processed to assess flood extents, enabling a comparison of their accuracy in delineating floodwater pixels (Fig. 10.8). Pixels with intensities below the threshold value were classified as flood or open water, while those above the threshold value were deemed dry land.

Notably, VV polarization outperformed VH polarization in classifying flood pixels, as it exhibited higher backscatter values. The mapping of floodwaters in each district involved overlaying the extracted water pixels, with inundated areas during the flood period highlighted in red (Fig. 10.9). It's important to note that the choice of threshold values typically depends on the contrast between flooded and non-flooded classes and may need to be adjusted for each SAR image. It was observed that VV polarisation is most useful compared to VH polarisation for classify the flood pixels. VV polarizations showed high backscatter value as compared with VH polarizations. Flood water mapping for each district is done by superimposing the extracted water pixels. The area highlighted in red colour indicates the inundated areas during flooded period are shown in (Fig. 10.10).

The threshold value generally depends on the contrast between the flood and non-flooded class and generally needs to be set for each SAR images. It was also observed that flood extent was in its peak during August 15 to 18, 2018 because of the high

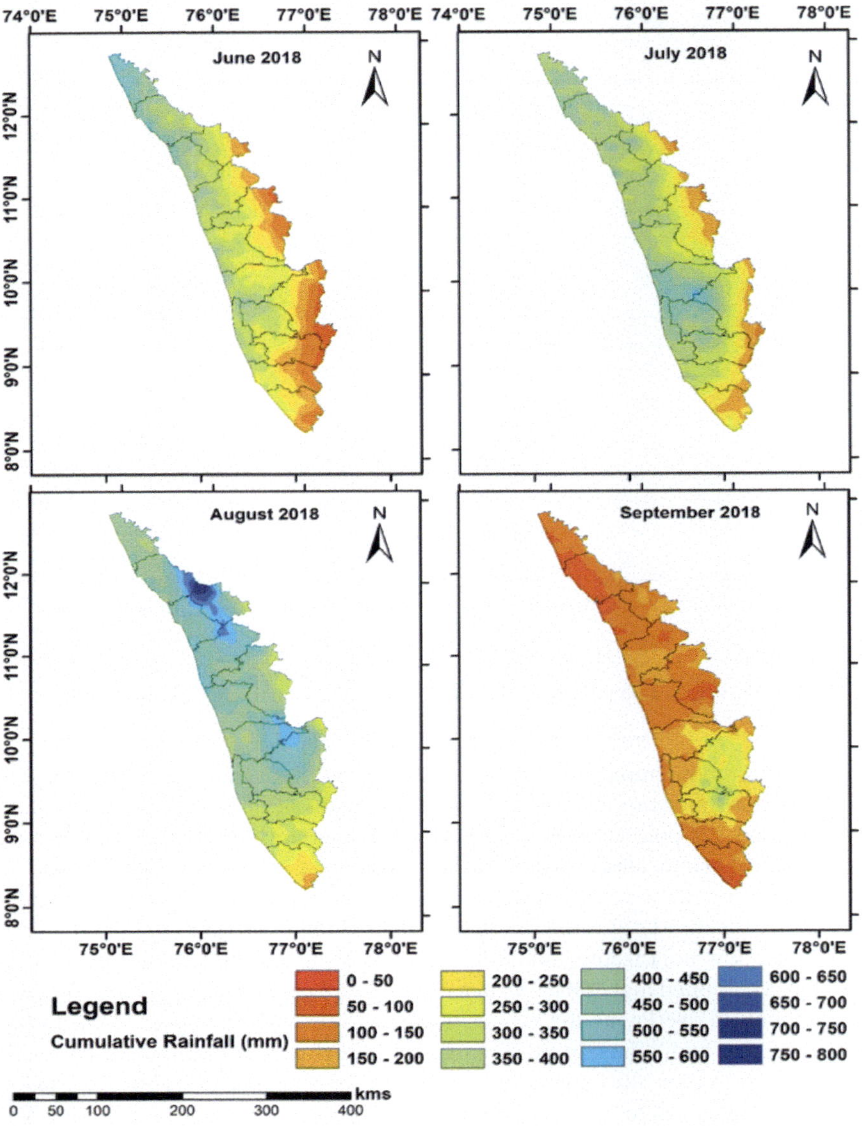

Fig. 10.5 Monthly cumulative rainfall during monsoon season (June to September) in year 2018

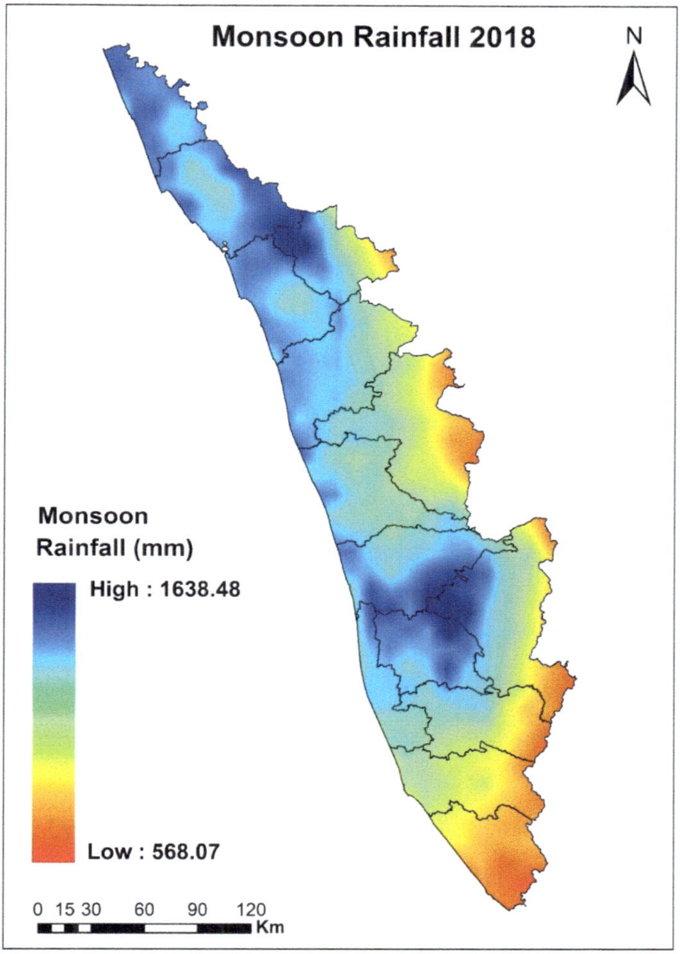

Fig. 10.6 Cumulative rainfall during monsoon season in year 2018

intensity rainfall, enormous amount of discharged water. The discharge amount of water crossed the danger level and impacted a large part of agricultural and urban patches during this time. The findings indicated that approximately 995.49 Square Kilometres (sq. km) of land in Kerala were inundated during the peak flooding stage.

District-wise zonal statistics, calculated using ArcGIS, provided insights into the total affected area. The results highlighted severe impacts on Alappuzha and Trissur districts, with approximately 171.48 sq. km (of the total geographic area 1453.7 sq. km) and 164.52 sq. km (of the total geographic area 3075.7 sq. km) under flood inundation. Most of the flooded areas were concentrated along the river basins. The area under flood inundation in each district over study area is shown in (Fig. 10.10). The

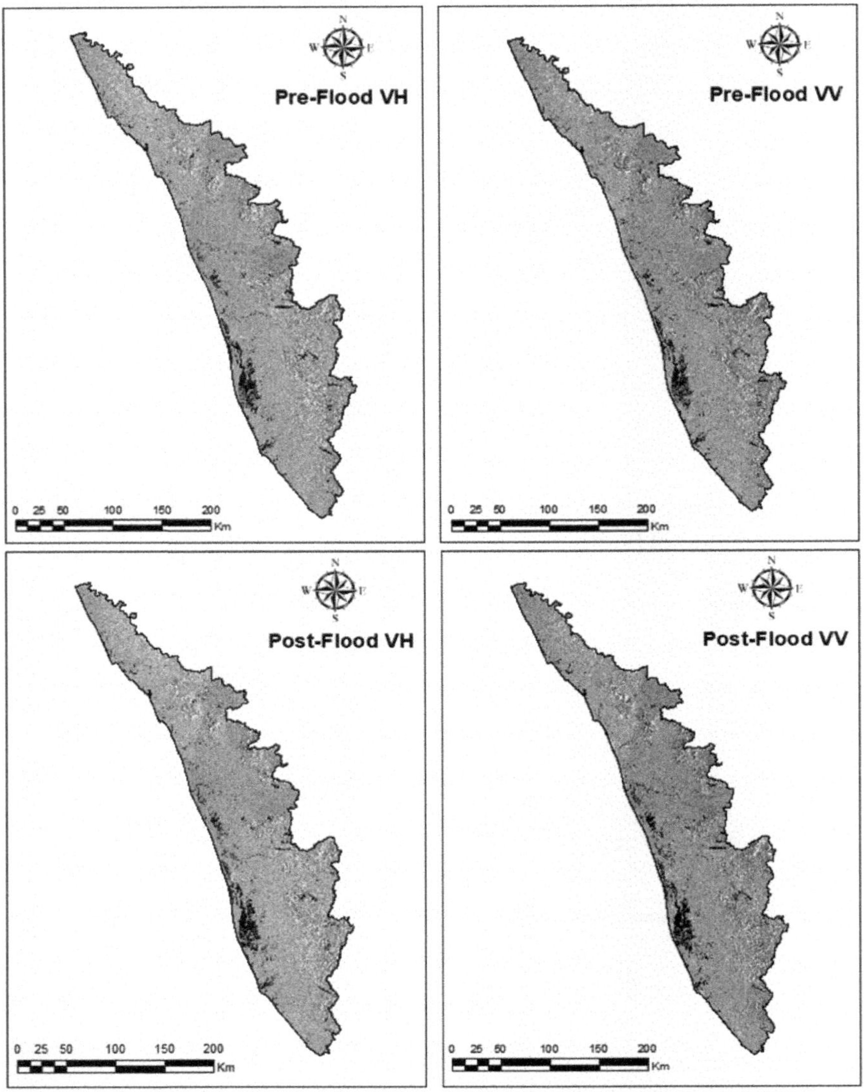

Fig. 10.7 Two sets of dual polarised SAR data: pre-flood and post-flood over study area

devastating floods damaged the state's agriculture production mainly the plantation and spice crops which are the backbone of the state's agriculture. The net area under cultivation during the year 2017–18 was 2462.52 sq km, which occupies 52.7% of the total area in the State and about 503.49 sq. km crop area have been affected by flood. It also affected the buildings and infrastructures. The total area coverage of built-up is 9648.44 sq. km out of which about 88.42 sq. km area has been affected

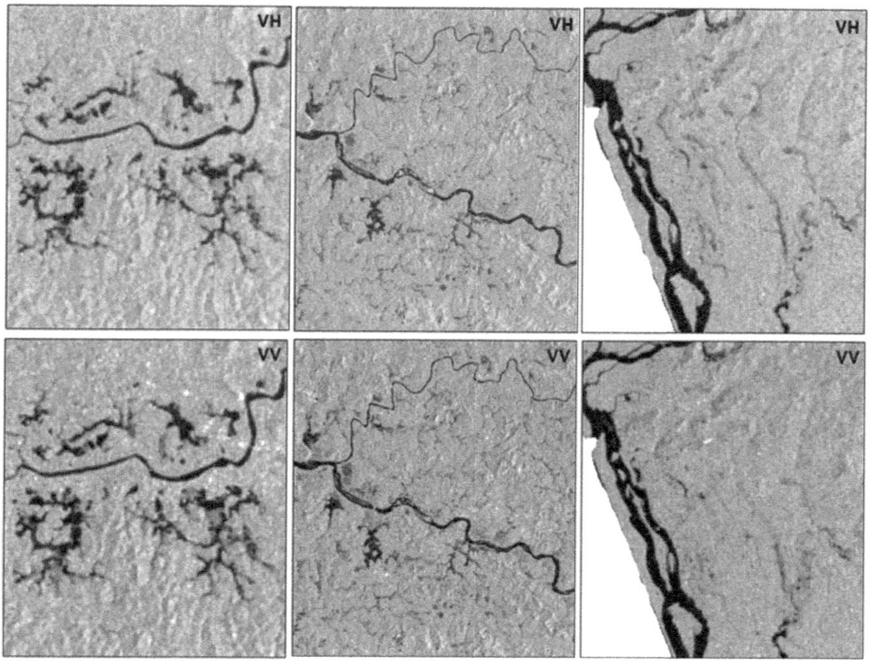

Fig. 10.8 Classified flood water in both polarizations (VH and VV) over study area

by flood (Fig. 10.11–10.12). Most flooded areas were found adjacent to the river and other water bodies because of its narrowness at several places passing from different districts. A decrease in flood extent was observed on August 25, 2018. This analysis underscores the importance of flood impact assessments, which can serve as valuable tools in supporting mitigation strategies. Numerous studies have indicated that extreme flooding significantly alters channel width and causes substantial sediment mobilization in river channels. The massive indiscriminate deforestation is also one of the major reasons for the floods which lead to large amounts of topsoil coming loose in the rains. Thus, the soil, instead of soaking up the rainfall, flows down into the river and in turn causes the riverbeds and its tributaries to rise.

The flood of 1924 unfolded against the backdrop of Kerala's unique geography, characterized by its intricate network of rivers, backwaters, and a coastal stretch that exposes the region to the full force of monsoon rains. During the monsoon season, these natural features can transform into agents of destruction when the rains exceed their regular patterns. Such was the case in 1924 when unprecedented levels of rainfall battered the region, triggering one of the worst flooding incidents in its history. Even as the years have passed, the memory of the 1924 Kerala flood continues to serve as a poignant reminder of the delicate balance between human settlements and the environment's formidable forces. From the proposed approach, it is found that SAR data along with GIS can be used efficiently for mapping and monitoring of

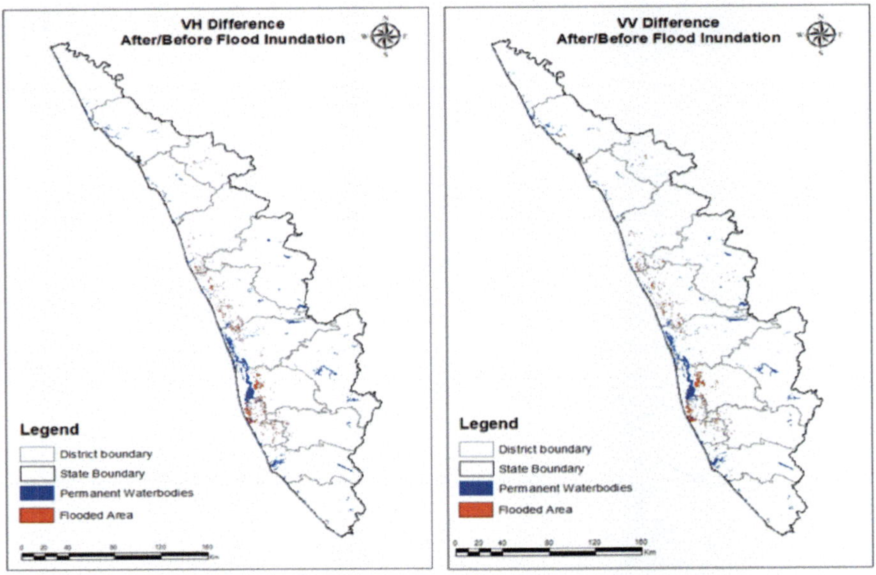

Fig. 10.9 Flood Inundated map over Kerala state and the area highlighted in red colour indicates the flooded area

flood inundated area and to be capable of overcoming some of the critical limitations (Mudi & Das, 2022; Yun & Hanssen, 2012). The real-time remote sensing data collection and efficient monitoring of flood events is fundamental to minimising the impact on life and property. The resultant maps could be utilized by the government officials, resource managers effectively for damage assessment and relief measures. The inundated area is integrated with district boundary layer for extracting the district-wise inundation statistics. The ability to create flood inundation maps in near real-time and monitor floods using remote sensing data is crucial in supporting disaster management authorities. This technology aids in devising rescue plans, quickly assessing property damage and loss of life, and ensuring food safety measures are implemented during flood emergencies. Additionally, integrating different hydrological and climatological data into various machine learning models has the potential to enhance forecasting accuracy and represents a promising direction for future research in this field.

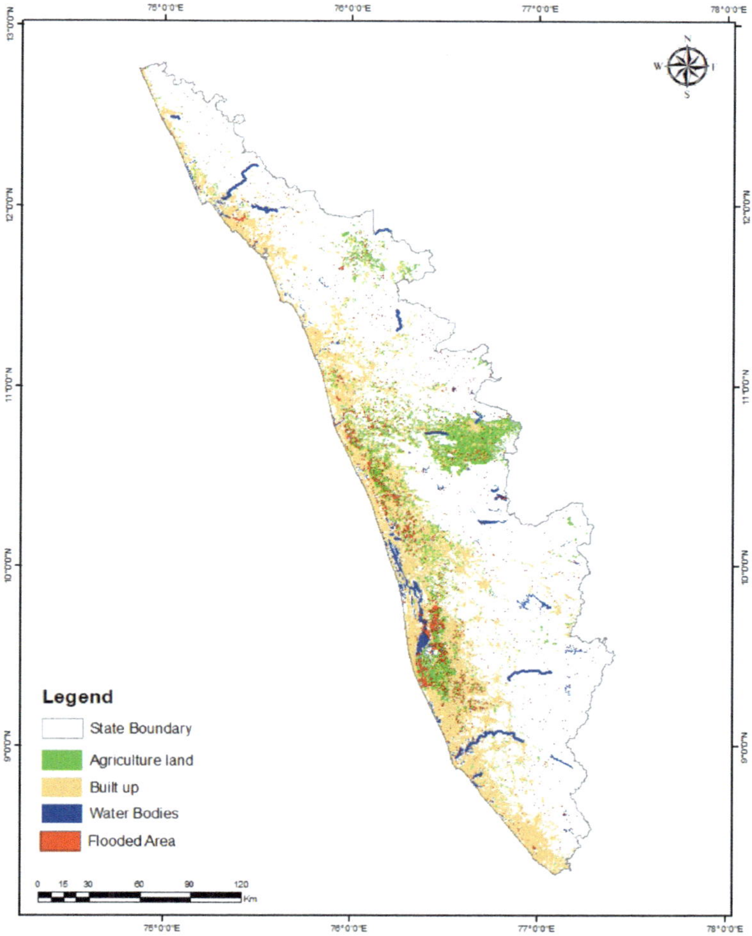

Fig. 10.10 Land use/land cover (LULC) of Kerala state and the area highlighted in red colour indicates the flooded area

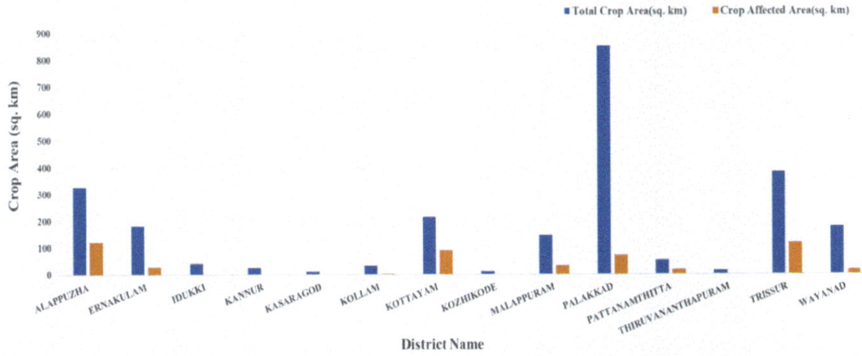

Fig. 10.11 Flood inundated agriculture area (District wise statistics)

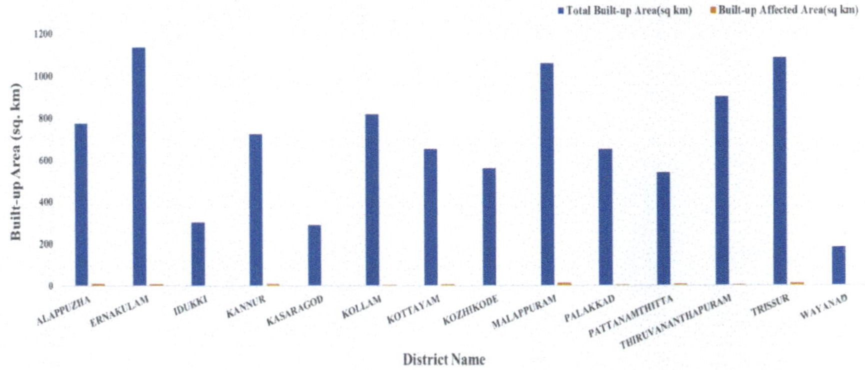

Fig. 10.12 Flood inundated built-up area (District wise statistics)

10.5 Conclusion

Rapid and precise identification of water-affected regions can offer crucial and timely insights during flood occurrences. This study conclude that remote sensing technology provides information on spatial flood extent and flood damage statistics like district-wise flood affected area, submerged crops, etc. A cloud computing approach Google Earth Engine (GEE) platform is utilized to derive the flood inundated areas. The GEE platform proves to be a valuable tool for processing and analysing inundation maps without the need for data downloads or high-end computing hardware. This study evaluated the effectiveness of the Otsu thresholding technique in Sentinel-1 SAR data for delineating flood pixels. Otsu's method exhibits the ability to automatically identify thresholds for distinguishing between water and non-water pixels without necessitating user-defined parameters or ground-based reference points, unlike supervised machine learning methods. Its automated and real-time functionality, coupled with an acceptable level of accuracy, underscores its potential for swift flood inundation mapping. The flood inundated area is computed by two different set of images. These results suggest that VV polarization is more useful than VH polarization to distinguish flooded areas. Based on the results of the present study, it can be concluded that satellite data with geospatial technology can provide effective solutions for better management of disasters like floods. Near real-time flood inundation mapping and monitoring during flood disaster using remote sensing data is a vital support to disaster management authority to plan for rescue operations and rapid assessment of property damage, life loses and to ensure food safety.

10.6 Data Availability Statement

Sentinel-1, Sentinel-2 and GPM data used for the study is publically available.

Acknowledgements The author would like to thanks the National Aeronautics and Space Administration (NASA) and European Space Agency (ESA) for providing high resolution satellite data in public domain to make this research possible. The anonymous reviewers and the editor are thanked for their insightful comments which helped to improve this paper.

Conflict of Interest The authors declare no potential conflicts of interest.

References

Abrams, M., Crippen, R., & Fujisada, H. (2020). ASTER global digital elevation model (GDEM) and ASTER global water body dataset (ASTWBD). *Remote Sensing, 12*(7), 1156.

Bernhofen, M. V., Whyman, C., Trigg, M. A., Sleigh, P. A., Smith, A. M., Sampson, C. C., ... & Winsemius, H. C. (2018). A first collective validation of global fluvial flood models for major floods in Nigeria and Mozambique. *Environmental Research Letters, 13*(10), 104007.

Bhaduri, S., & Ramachandran, R. (2013). Flood extent mapping using TerraSAR-X data and its comparison with MODIS in the context of rapid disaster response: A case study of 2010 Pakistan flood. *Remote Sensing, 5*(1), 184–206.

Brivio, P. A., Colombo, R., Maggi, M., & Tomasoni, R. (2002). Integration of remote sensing data and GIS for accurate mapping of flooded areas. *International Journal of Remote Sensing, 23*(3), 429–41.

Chen, L. C., & Lin, W. T. (2015). Flood mapping and change analysis using PALSAR data: A case study in northern Taiwan. *Sensors, 15*(6), 13911–13929.

Clement, M. A., Kilsby, C. G., & Moore, P. (2018). Multi-temporal synthetic aperture radar flood mapping using change detection. *Journal of Flood Risk Management, 11*(2), 152–168.

Comer, J., Olbert, A. I., Nash, S., & Hartnett, M. (2017). Development of high-resolution multi-scale modelling system for simulation of coastal-fluvial urban flooding. *Natural Hazards and Earth System Sciences, 17*(2), 205–224.

Cook, K. L., Andermann, C., Gimbert, F., Adhikari, B. R., & Hovius, N. (2018). Glacial lake outburst floods as drivers of fluvial erosion in the Himalaya. *Science, 362*(6410), 53–57.

Cozzolino, D., Parrilli, S., Scarpa, G., Poggi, G., & Verdoliva, L. (2013). Fast adaptive nonlocal SAR despeckling. *IEEE Geoscience and Remote Sensing Letters, 11*(2), 524–528.

Dube, A., Ashrit, R., Ashish, A., Sharma, K., Iyengar, G. R., Rajagopal, E. N., & Basu, S. (2014). Forecasting the heavy rainfall during Himalayan flooding—June 2013. *Weather and Climate Extremes, 4*, 22–34.

Gaume, E., Borga, M., Llasat, M. C., Maouche, S., Lang, M., & Diakakis, M. (2016). Sub-chapter 1.3.4. Mediterranean extreme floods and flash floods. In J. Moatti & S. Thiébault (Eds.), *The Mediterranean region under climate change: A scientific update*. IRD Éditions. https://doi.org/10.4000/books.irdeditions.23181

George, C., Kanakamma, L. G., John, J., Sunny, G., Cohen, A., & De Silva, M. J. (2012). Post-tsunami mental health: A cross-sectional survey of the predictors of common mental disorders in South India 9–11 months after the 2004 T sunami. *Asia-Pacific Psychiatry, 4*(2), 104–112.

Gorelick, N., Hancher, M., Dixon, M., Ilyushchenko, S., Thau, D., & Moore, R. (2017). Google earth engine: Planetary-scale geospatial analysis for everyone. *Remote Sensing Environment., 202*, 18–27.

Hinkel, J., Lincke, D., Vafeidis, A. T., Perrette, M., Nicholls, R. J., Tol, R. S., ... & Levermann, A. (2014). Coastal flood damage and adaptation costs under 21st century sea-level rise. *Proceedings of the National Academy of Sciences, 111*(9), 3292–3297.

Hofste, J. G., van der Velde, R., Wen, J., Wang, X., Wang, Z., Zheng, D., ... & Su, Z. (2021). Year-long, broad-band, microwave backscatter observations of an alpine meadow over the Tibetan Plateau with a ground-based scatterometer. *Earth System Science Data, 13*(6), 2819–2856. https://doi.org/10.5194/essd-13-2819-2021.

Khan, S. I., Hong, Y., & Choi, M. (2018). Comparative analysis of ALOS PALSAR-2 and Sentinel-1A data for flood mapping and damage assessment: A case study of Houston, USA. *Remote Sensing, 10*(8), 1292.

Kirkpatrick, J. I. M., & Olbert, A. I. (2020). Modelling the effects of climate change on urban coastal-fluvial flooding. *Journal of Water and Climate Change, 11*(S1), 270–288.

Lal, P., Prakash, A., Kumar, A., Srivastava, P. K., Saikia, P., & Pandey, A. C., et al. (2020). Evaluating the 2018 extreme flood hazard events in Kerala, India. *Remote Sensing Letters, 11*(5), 436–45.

Lee, J. S., Wen, J. H., Ainsworth, T. L., Chen, K. S., & Chen, A. J. (2008). Improved sigma filter for speckle filtering of SAR imagery. *IEEE Transactions on Geoscience and Remote Sensing, 47*(1), 202–213.

Martha, T. R., Roy, P., Govindharaj, K. B., Kumar, K. V., Diwakar, P. G., & Dadhwal, V. K. (2015). Landslides triggered by the june 2013 extreme rainfall event in parts of Uttarakhand State, India. *Landslides, 12*, 135–146.

Moskolaï, W., Abdou, W., Dipanda, A., & Kolyang, D. T. (2020). Application of lstm architectures for next frame forecasting in Sentinel-1 images time series. arXiv preprint arXiv:2009.00841.

Mudi, S., & Das, P. (2022). Flood hazard mapping in Assam using Sentinel-1 SAR data. *Geospatial Technology for Environmental Hazards: Modelling and Management in Asian Countries*, 459–473.

Mullissa, A. G., Marcos, D., Tuia, D., Herold, M., & Reiche, J. (2020). DeSpeckNet: Generalizing deep learning-based SAR image despeckling. *IEEE Transactions on Geoscience and Remote Sensing, 60*, 1–15.

Nabangchang, O., Allaire, M., Leangcharoen, P., Jarungrattanapong, R., & Whittington, D. (2015). Economic costs incurred by households in the 2011 greater Bangkok flood. *Water Resources Research, 51*(1), 58–77.

Nace, T. (2017). Fluvial and flooding processes: Geomorphology in Boreal regions. *Science Trends*.

Parthasarathy, B., Munot, A. A., & Kothawale, D. R. (1994). All-India monthly and seasonal rainfall series: 1871–1993. *Theoretical and Applied Climatology, 49*, 217–224.

Prasad, A. K., & Singh, R. P. (2005). Extreme rainfall event of July 25–27, 2005 over Mumbai, west coast, India. *Journal of the Indian Society of Remote Sensing, 33*, 365–370.

Ray, K., Pandey, P., Pandey, C., Dimri, A. P., & Kishore, K. (2019). On the recent floods in India. *Current Science, 117*(2), 204–218.

Roxy, M. K., Ghosh, S., Pathak, A., Athulya, R., Mujumdar, M., Murtugudde, R., ... & Rajeevan, M. (2017). A threefold rise in widespread extreme rain events over central India. *Nature communications, 8*(1), 708.

Rözer, V., & Surminski, S. (2020). New build homes, flood resilience and environmental justice-current and future trends under climate change across england and wales. *Grantham Research Institute on Climate Change and the Environment.*

Sengupta, N., & Kakran, S. P. (2011). *Report of working group on flood management and region-specific issues for xii plan, government of India planning commission.* New Delhi.

SK M. (2020). Flood inundation mapping using Synthetic Aperture Radar (SAR) data and its impact on Land Use /Land Cover (LULC): A case study of Kerala flood 2018, India. *Disaster Adv, 13*, 8.

Singh, O., & Kumar, M. (2017). Flood occurrences, damages, and management challenges in India: A geographical perspective. *Arabian Journal of Geosciences, 10*, 1–19.

Torres, R., Snoeij, P., Geudtner, D., Bibby, D., Davidson, M., Attema, E., ... & Rostan, F. (2012). GMES Sentinel-1 mission. *Remote Sensing of Environment, 120*, 9–24.

Tramblay, Y., Bouaicha, R., Brocca, L., Dorigo, W., Bouvier, C., Camici, S., & Servat, E. (2012). Estimation of antecedent wetness conditions for flood modelling in northern Morocco. *Hydrology and Earth SysTem Sciences, 16*(11), 4375–4386.

Vishnu, C. L., Sajinkumar, K. S., Oommen, T., Coffman, R. A., Thrivik-ramji, K. P., Rani, V. R., & Keerthy, S. (2019). Satellite-based assessment of the august 2018 flood in parts of Kerala, India. *Geomatics, Natural Hazards and Risk, 10*(1), 758–767.

Wahlstrom, M., & Guha-Sapir, D. (2015). The human costs of weather-related disasters, The United Nations office for diaster risk reduction.

Yun, S. H., & Hanssen, R. F. (2012). Quantitative analysis of the potential of multi-frequency multi-polarization SAR for monitoring flooded vegetation. *Remote Sensing of Environment, 117*, 149–161.

Chapter 11
A Study of Impact of Rainfall and Groundwater on Agricultural Crops and Drinking Water in the Bhandara District, India

Nanabhau Kudnar, Nand Lal Kushwaha, Madiga Rajashekhar, Varun Narayan Mishra, Rongali Mahesh, and Malkhan Singh Jatav

11.1 Introduction

Every culture on the planet has its roots in the presence of water. Water was the source of stability for the nomadic human population. He arrived near the river's banks because those were the places where the majority of his wants could be satisfied. It was there that he began to truly mature. He and his animals had uninterrupted access to water and his desire to settle by the river grew stronger (Anand et al., 2021; Meshram et al., 2018; Nair & Mirajkar, 2021; Rajasekhar et al., 2021; Subash & Sikka, 2014). After starting farming, he realized the true importance of water. As the use of water for personal hygiene, for animals, for agriculture, for communication began to increase, the water culture came into being, developed and became stable. He bestowed divinity on those things which gave him stability (Aher et al., 2019; Bessah et al., 2021; Taneja et al., 2019). Before India or the rest of the world, another

N. Kudnar (✉)
C. J. Patel College Tirora, Gondia, Maharashtra Pin-441911, India
e-mail: nanabhaukudnar@gmail.com

N. L. Kushwaha · R. Mahesh · M. S. Jatav
Division of Agricultural Engineering, ICAR-Indian Agricultural Research Institute, New Delhi 11001, India
e-mail: nand.kushwaha@icar.gov.in

M. Rajashekhar
UGC SRF- Research Scholar, Dept. of Geology, Yogi Vemana University, Kadapa, Andhra Pradesh, India

V. N. Mishra
Amity Institute of Geoinformatics & Remote Sensing, Noida, India

P. K. Rai and S. Rai (eds.), *Climate, Environment and Agricultural Development*, Advances in Geographical and Environmental Sciences,
https://doi.org/10.1007/978-981-97-8363-2_11

question has gained importance. Water availability per capita is declining as a result of population growth, and even the water that is available is becoming unfit for human use due to rising water pollution. Lakes and surface rivers have turned into sewage islands. Since there isn't much fresh water left on Earth, it begs the question of how humans can exist if it is available in such vast quantities in contaminated form. There is yet more to the question. Surface water pollution has caused contaminated water to begin seeping into the earth, which has led to the contamination of ground water as well (Hussein et al., 2017; Srivastava et al., 2011; Zolekar et al., 2021). Visible effects of this pollution are beginning to be seen. We are seeing various types of epidemics increasing day by day (Addisu et al., 2015; Jain & Kumar, 2012). Epidemics like cholera, jaundice, malaria, typhoid, dengue, various types of influenza, gastro, diarrhea are spreading on a large scale (Chandniha et al., 2016; Kharake et al., 2021; Kushwaha et al., 2022). Domestic waste and effluents from industrial units, which contain phosphorus and nitrogenous fertilizers, are discharged into the rivers. Due to the presence of elements like phosphorus and nitrogen in water, its fertility increases, but such water is not fit for drinking (Deshmukh & Aher, 2016; Gajbhiye et al., 2015).

Since humans eat fish and are impacted by eating fish, industrial pollutants and pesticides that are absorbed in fish have a very negative impact on human health. In areas with significant pesticide use, underground water becomes contaminated, which makes it easy for contaminated water to find its way into drinking supplies. The fact that the majority of historical civilizations developed along riverbanks suggests that water has not only been necessary for existence, but also a valuable resource. The demand for drinking water and irrigation has expanded over the past few decades due to fast urbanization, population growth that hasn't stopped, expansion of industrial activity, and other factors. On the one hand, due to the uncontrolled exploitation of surface and underground water to supply the increasing demand of water, the ground water level is declining, and on the other hand, the quality and usefulness of water is decreasing due to the increasing amount of pollutants. Disasters like erratic rainfall, drought and flood have greatly affected the underground water recharge. Today, in the blind race for development, expansion of industrial activities and rapid urbanization have polluted the country's major rivers, especially Ganga, Yamuna, Godavari, Kaveri, Narmada and Krishna. One of the main reasons for the deterioration of water quality is increasing water pollution (Jayadas & Ambujam, 2019; Kudnar, 2022; Rajasekhar et al., 2020).

11.2 Study Area

The Bhandara district is located between 20°38' and 21°36' north latitude and between 79°27' and 80°06' east longitude (Fig. 11.1). India's state of Maharashtra has the district of Bhandara (Bisen et al., 2022). Bhandaria has 11,98,810 residents. The district has a total area of 3716 square kilometers. The aromatic tandulachya caste and tala are the district's two main draws. The Talavacha district is Yala. In Bhandariat, there are 3648 Lahan Lahan Talis. This district has abundant forests and mineral resources. The district or city headquarters is called Bhandara (Kashiwar et al., 2021; Meshram & Ingle, 2012). The district is rich in iron ore, and the principal crops grown there are rice, wheat, and millet (a cereal grain). Notable castes include Tandalachya Chinnor, Dubhraj, Kali Kamod, and Suvasik. Known for being the largest rice-growing district in Maharashtra is Bhandara. The district's principal river is Wainganga (Kudnar, 2020a, 2020b; Kudnar & Rajshekhar, 2020; Kudnar et al., 2022). Alternatively, even in the summer, the district's river flow remains uncorroded. Alternatively, there are just Dharnas in the districts of Bawanthadi, Chulband, Kanhan, Bagh, Gose Dharan, and Kavalevda Dharan.

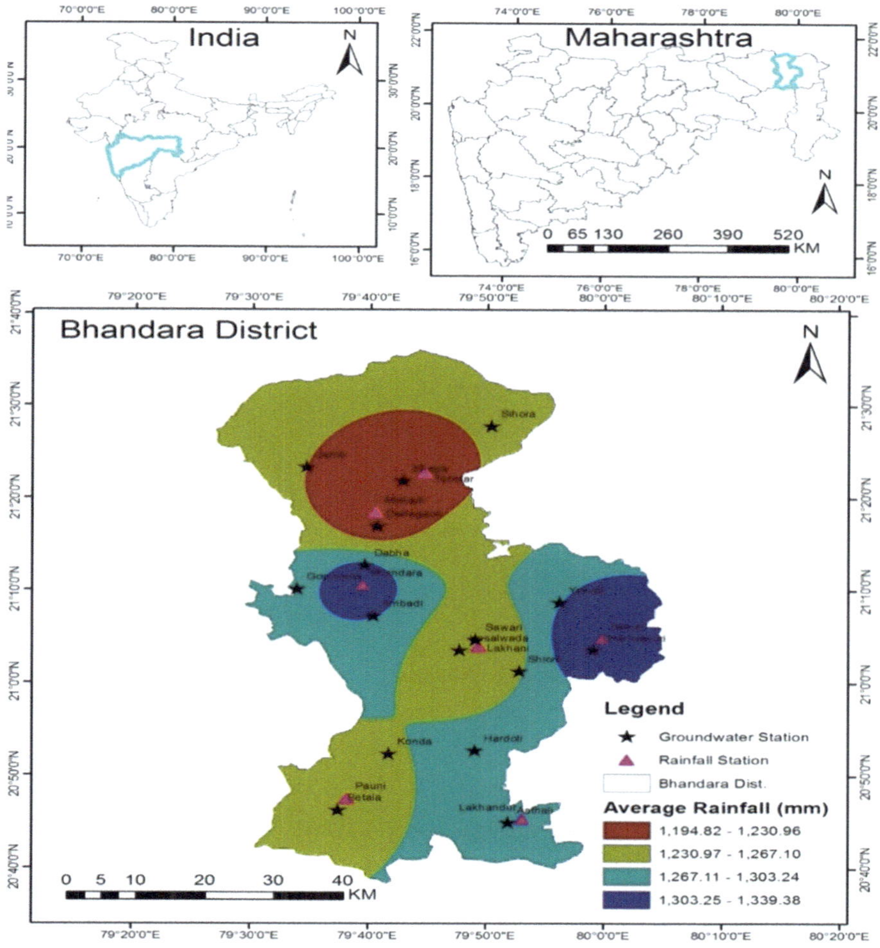

Fig. 11.1 Study area

11.3 Methodology

At first, rainwater is pure, but as land moves, both organic and inorganic minerals dissolve into it. In this sense, the minerals and other nutrients found in natural water are significant for the health of all living things, including people, in addition to being necessary. The following qualities should be present in drinking water from a quality perspective:

When such dangerous compounds are added, water's physical, chemical, and biological qualities change. Polluted water is unsuitable for use in residential, commercial, industrial, or agricultural settings, as well as for any other general purpose, due to the detrimental side effects of pollutants on water quality.

Apart from drinking, a lot of water is consumed for home tasks, gardening, farming, raising livestock, industrial and commercial operations, and so on. The water utilized for these purposes eventually becomes contaminated. In this contaminated water, traces of substances or chemicals that came in contact with water during commercial activities remain in the form of residues. Their presence makes water unfit for use. When this contaminated water is found in a clean water source, it also contaminates it. Contaminated water contains viruses, bacteria and other harmful micro-organisms along with organic and inorganic compounds and chemicals which pollute the water sources according to their nature.

11.3.1 Data Used

Data from seven rainfall stations in seven tehsils from 1971 to 2013 were obtained from IMD, Pune, and used in the analysis of the Bhandara district. In the process, it is separated into four periods: pre-monsoon, monsoon, post-monsoon, and winter monsoon (Fig. 11.1).

11.3.2 Descriptive Statistics Analysis

$$\mathrm{Sd} = \sqrt{\frac{\sum_{i=1}^{n}(xi - x)2}{n - 1}} \tag{11.1}$$

$$\mathrm{Sk} = \frac{n}{(n-1)(n-2)} \sum \frac{(xj - x)3}{s} \tag{11.2}$$

Kurtosis value is calculated by the following formula (Figs. 11.2 and 11.3)

$$\mathrm{Kr} = \{\frac{n(n+1)}{(n-1)(n-2)(n-3)} \sum \frac{(xj - x)4}{s}\} - \frac{3(n-1)2}{(n-2)(n-3)} \tag{11.3}$$

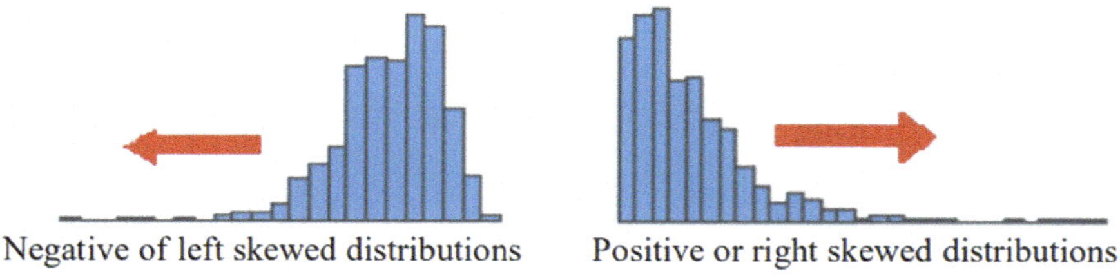

Fig. 11.2. Two type of distribution pattern with different skewnwess

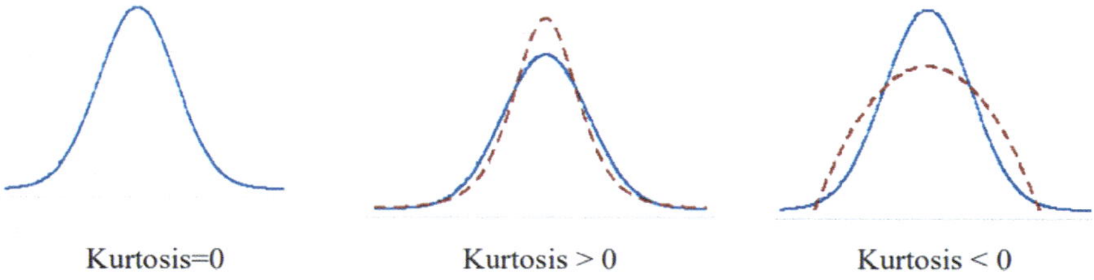

Fig. 11.3. Three type of distribution pattern with different Kurtosis

11.3.3 Mann–Kendall Test (Non-Parametric Test)

The non-parametric test is (Fathian et al., 2014; Hamed & Rao, 1998; Kendall, 1975; Mann, 1945) usind in study area. The following equation:

$$S = \sum_{i=1}^{n-1} \sum_{j=i+1}^{n} sgn\left(X_j - X_i\right) \tag{11.4}$$

$$Sgn(Y) = \left\{ \begin{Bmatrix} +1 \; if \;\; \mathrm{Y} > 0 \\ 0 \;\; if \;\; \mathrm{Y} = 0 \\ -1 \; if \;\; \mathrm{Y} < 0 \end{Bmatrix} \right. \tag{11.5}$$

$$Var(S) = \frac{\left[n(n-1)(2n+1) - \sum_t t(t-1)(2t+5)\right]}{18} \tag{11.6}$$

$$Z_{mk} = \begin{cases} \frac{S-1}{\sqrt{V(S)}} \; if \;\; S > 0 \\ 0 \, if \;\; S = 0 \\ \frac{S+1}{\sqrt{V(S)}} \; if \;\; S < 0 \end{cases} \tag{11.7}$$

11.3.4 Regression Model (Parametric Test)

The regression model apply a geographically rainfall pattern (Ahmad et al., 2020; Rupa et al., 1992).

$$y = mx + c \tag{11.8}$$

11.3.5 Spearman's Rank Correlation Coefficient

The Spearman's rank correlation coefficient appliy a two station (Misra & Tripathi, 2021).

$$\rho = 1 - 6\sum_{n-1}^{n} d^2 / n\left(n^2 - 1\right) \tag{11.9}$$

11.3.6 Inverse Distance Weighing

The researcher are Shepard, Mueller et al., Gong et al. defines IDW as,

$$z = \frac{\sum_{i=1}^{n} w_i z_i}{\sum_{i=1}^{n} w_i} \tag{11.10}$$

$$w_i = \frac{1}{d_i^p}$$

11.3.7 Ground Water Quality

A total of 16 ground water wells have been studied in this study area and this data is collected in pre-monsoon period and it is collected in 2011 (Fig. 11.4).

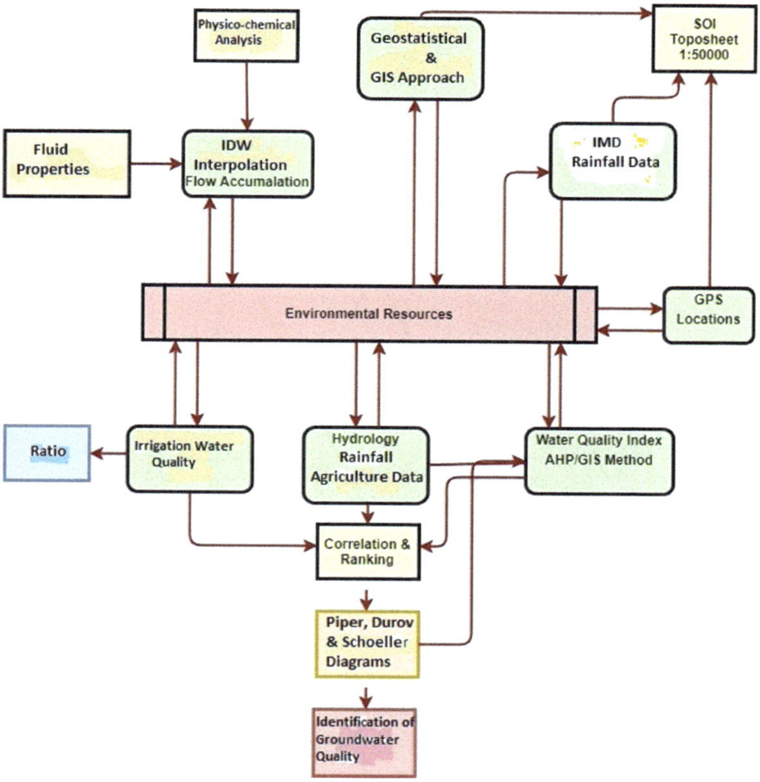

Fig. 11.4 Flow chart of methodology

11.3.8 Piper Diagram

This diagram uses graphics to explain the water sample's chemical characteristics. It displays data on sulfate, chloride, carbonate, and hydrogen carbonate by plotting calcium, magnesium, sodium, and potassium using a matrix (Fig. 11.5), which was created by Hilly and changed by Piper.

11.3.9 Groundwater for Drinking and Irrigation Purposes

Numerous scientists have examined the characteristics and water quality of groundwater to determine whether or not it is drinkable (Singh et al., 2011). When irrigating a farm, the electrical conductivity of the water provides information about whether or not it is acceptable for agriculture. This information is then used to determine the water's quality.

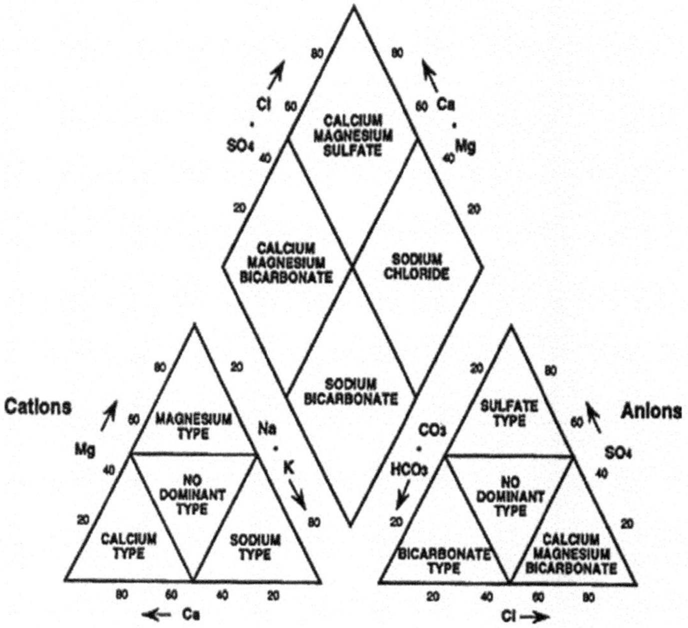

Fig. 11.5 Piper diagram

11.4 Results and Discussion

The eastern region of the state is home to the Bhandara district, which is also divided into the eastern districts of Gondia, Chandrapur, and Nagpur in the south, and the northern districts of Madhya Pradesh in the west. The largest river in the Bhandara district is the Wainganga; other rivers that pass through the area include the Wagh, Chulband, Bawanthadi, Ambagad, Bodalkasa, Gadhvi, Sur, and Maru. Talavancha district is another name for Bhandara district. This region is called the district of ponds because the residents of Kohli Jamati have been gathering water, draining, and creating ponds using a traditional way since 400 years ago. The primary crop grown in the Bhandara district is rice, while other crops grown there include oats, tur, soy, moong, wheat, harbhara, and barley. Some places only accept flying drinks. In terms of mineral riches, the state's Bhandara district is wealthy. Manganese is a plentiful mineral in the district's northern region. Mubalak, or pandhari soil, is located in Tumsar and Kavadsi. Other than this, the area only has limited access to other minerals. Manganese, chromite, quartzite, sillimanite, and kyanite are all mined.

Table 11.1 Rainfall statistics (1971–2013)

Tehsil	Maximum	Minimum	Mean/Average	Std. deviation	Kurtosis	Skewness
Mohadi	1845.1	636.5	1194.82	247.89	0.20	0.37
Bhandara	2057.5	787.7	1327.5	306.35	0.21	0.7
Pauni	2117.3	748.9	1257.54	334.72	0.07	0.86
Lakhandur	2174.9	659	1247.63	339.68	0.16	0.68
Tumsar	2208.7	433.1	1223.5	323.66	1.12	0.51
Lakhani	1994.4	630	1239.68	291.14	0.58	0.42
Sakoli	2041.3	836	1339.39	287.03	0.10	0.52

11.4.1 Statistical Characteristics

Because the Sun is close to the Tropic of Cancer, the belt of high pressure that forms in North-West India in the wintertime slowly advances southward. There is increased pressure in the southern region of India in April. Sagaravarhi makes up the majority of its territory. There is pressure in northwest India. I am still feeble, though. more pressure originates from the pressure area. They come from the Maharashtra region and rain in the months of April and May. For this reason, or monsoon, Embassy, This is how it is stated. The onset of the monsoon is typically a very difficult procedure. The monsoon mechanism is a body of water and the terrain in the summer. The features of the Indian subdivision are as follows. In Maharashtra, it typically rains. The southwest monsoon winds are responsible for about 85% of these precipitation. There is rising signs of increased rain in Maharashtra's eastern region. This is because the Bhandara region, which receives rain between August and September, is a branch of the monsoon winds over the Bay of Bengal. There is 110–160 m of rain in the study region. The mountains of Gaimukha continued to show signs of rain. 160 cm of rain fall in the southern region, or cold haven. This month has a lot of sunshine. Additionally, there is a feeling of nausea due to the abundance of evidence of fumes in the air (Table 11.1).

11.4.2 Trend Analysis

There is rainfall data available from 1971 to 2013. The study area's north, east, and northeast have had less rainfall, with the exception of Mohadi and other areas that receive little to no precipitation. Nevertheless, there isn't a noticeable monsoon trend when looking at the whole region. He also mentioned how the monsoon varies greatly and is not constant. There has been a rise in weather events including lightning strikes and deluges of rain as a result of climate change-related atmospheric instability. The Arabian Sea cyclone is becoming more intense, and forecasting the weather becomes more difficult as a result of the growing adversity. Meteorological agencies' capacity to forecast intense rainfall is impacted by climate change. The monitoring of weather reports has increased with the aid of radars, satellites, automated weather stations, and rain gauges. Over the past five years, there has been a 30–40% rise in the accuracy of forecasts for extreme rainfall, heat waves, cold waves, and smoke due to advancements in computing systems, monitoring, and modeling.

11.4.3 Linear Regression Analysis

The linear regression equations R^2 of each station were predicted. Bhandara ($R^2 = 0.009$) in Fig. 11.6a, Lakhandur ($R^2 = 1.914E-4$) in Fig. 11.6b, Mohadi ($R^2 = 0.007$) in Fig. 11.6d, Pauni ($R^2 = 0.001$) in Fig. 11.6e, and Tumsar ($R^2 = 0.001$)in Fig. 11.6g, increasing trend in Lakhani ($R^2 = 7.225E-4$) in Fig. 11.6band Sakoli ($R^2 = 0.001$) in Fig. 11.6f along with their R^2 values presented in above.

11.4.4 Spatial–temporal Rainfall Distribution

It abruptly turns into a frigid area when the water vapor builds and reaches higher levels. Hygroscopic condensation nuclei cause a decrease in evaporation and the tendency for water droplets to form. These could be dust or suspended particles that are present in the condensation center's surroundings. They cease to be affected by gravity when they harden because they become lighter. Shower by convection: These are the kinds of rain that occur when warm air that has been heated by the sun comes into touch with the earth's surface. Once upon a time, water drops fell, and as a result of the drips' absence, rain fell. You know you're ready when a mountain is lying in the dust. The opposite side of the mountain used to be absolutely dry due to the wind's ups and downs, which submerged all the moisture there. They are created when two dust particles of different temperatures collide. Usually, one trait predominates and the other is chilly. Usually, these rains are cyclonic or gusty. They are produced by variations in temperature over the earth's surface. These typically happen towards the end of the summer and are accompanied by strong storms that typically inflict significant damage, particularly to crops (Fig. 11.7).

The defining characteristics include a shortage of drinking water, inadequate animal shelter and food, a lack of jobs in rural regions, which leads to the migration of villagers to adjacent towns, and an annual drought. There would have been a significant drought even though the years 1972, 1974, 1979, 1984, 1996, 2002, and 2009 were included in the year of drought. Comparing the two negative aspects, it would have been acceptable to classify 1972 as either a drought or a drought. As a result, during the drought, the quality of the drinking water would have been adequate today (Fig. 11.8; the rainfall was less than 50% of the norm). Nonetheless, there were variations in the distribution of rainfall across the study area. Similarly, a large-scale coverage of Maharashtra would have been largely dependent on climate

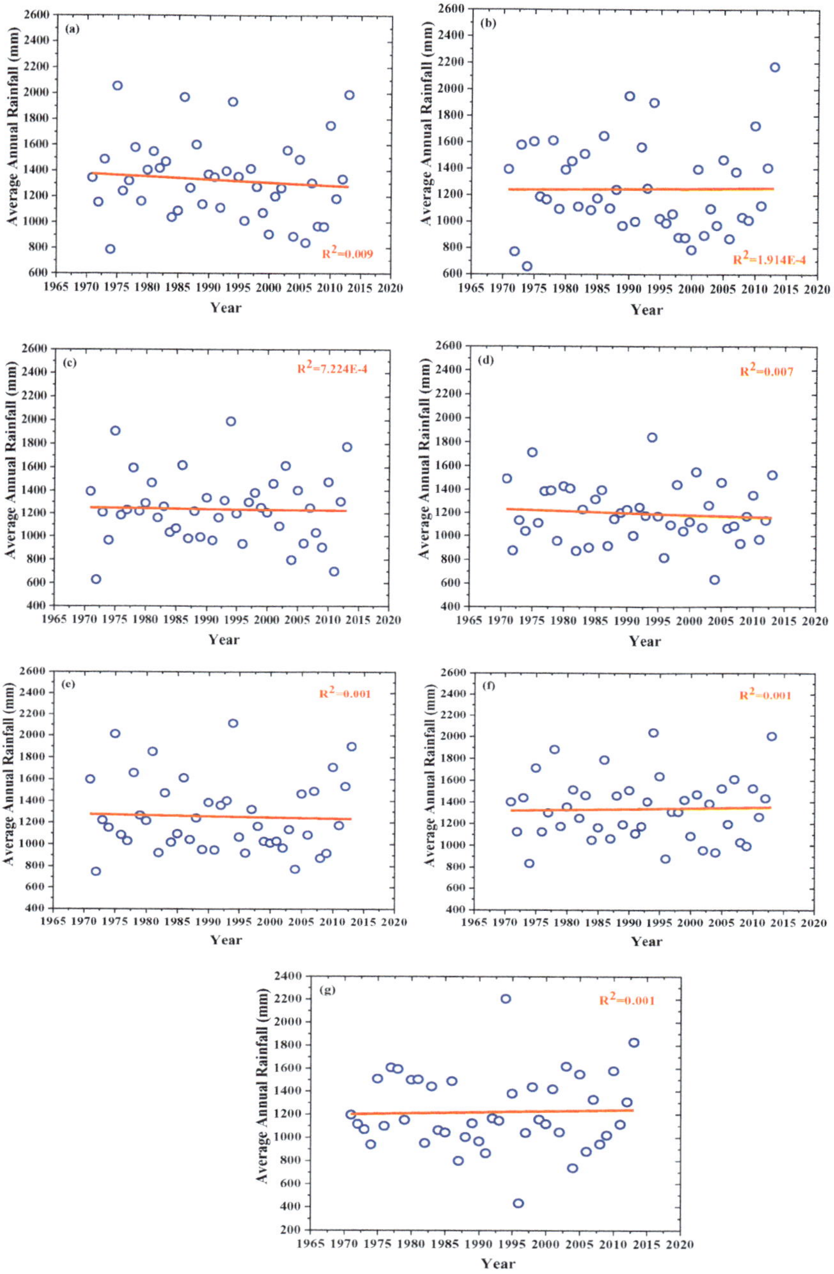

Fig. 11.6 Linear regression analysis (a) Bhandara, (b) Lakhandur, (c) Lakhani, (d) Mohadi, (e) Pauni, (f) Sakoli and (g) Tumsar stations over the period of 2071–2013

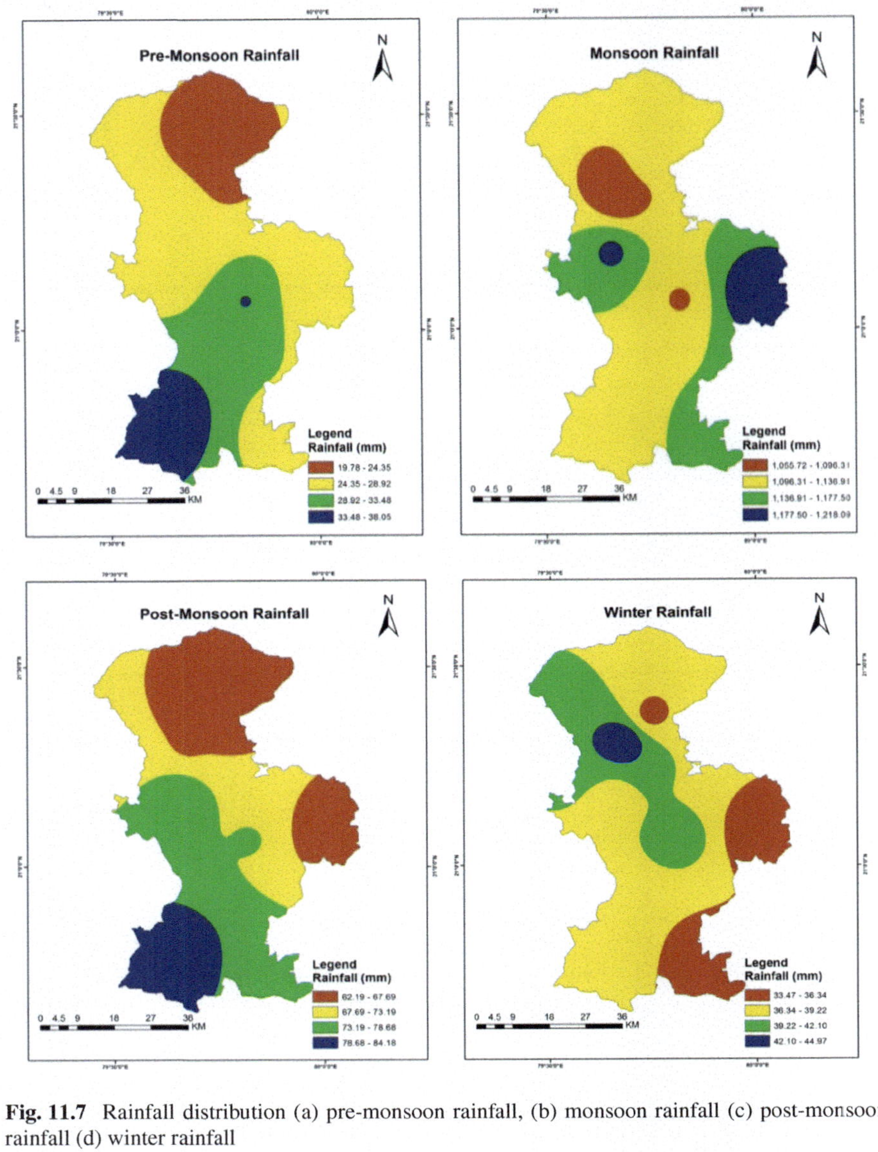

Fig. 11.7 Rainfall distribution (a) pre-monsoon rainfall, (b) monsoon rainfall (c) post-monsoon rainfall (d) winter rainfall

change which comprises the finished components? The local climatic circumstances maintained the monsoon's westerly winds flowing from the Arabian Sea at Nangarun and suggested the existence of a low pressure system in the Bay of Bengal. Or the excessive amount of moisture brought in from the western warm Arabian Sea caused heavy rains 1975, 1990, 1994, 2003, and 2013 in Fig. 11.9.

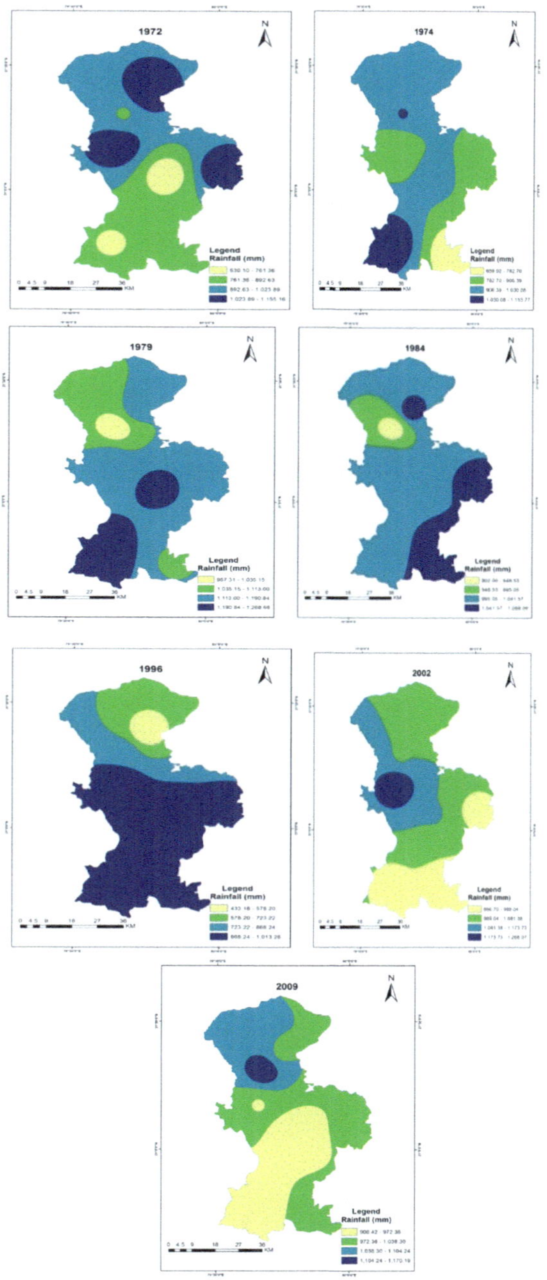

Fig. 11.8 Drought rainfall (a) 1972 (b) 1974 (c) 1979 (d) 1984 (e) 1996 (f) 2002 (g) 2009

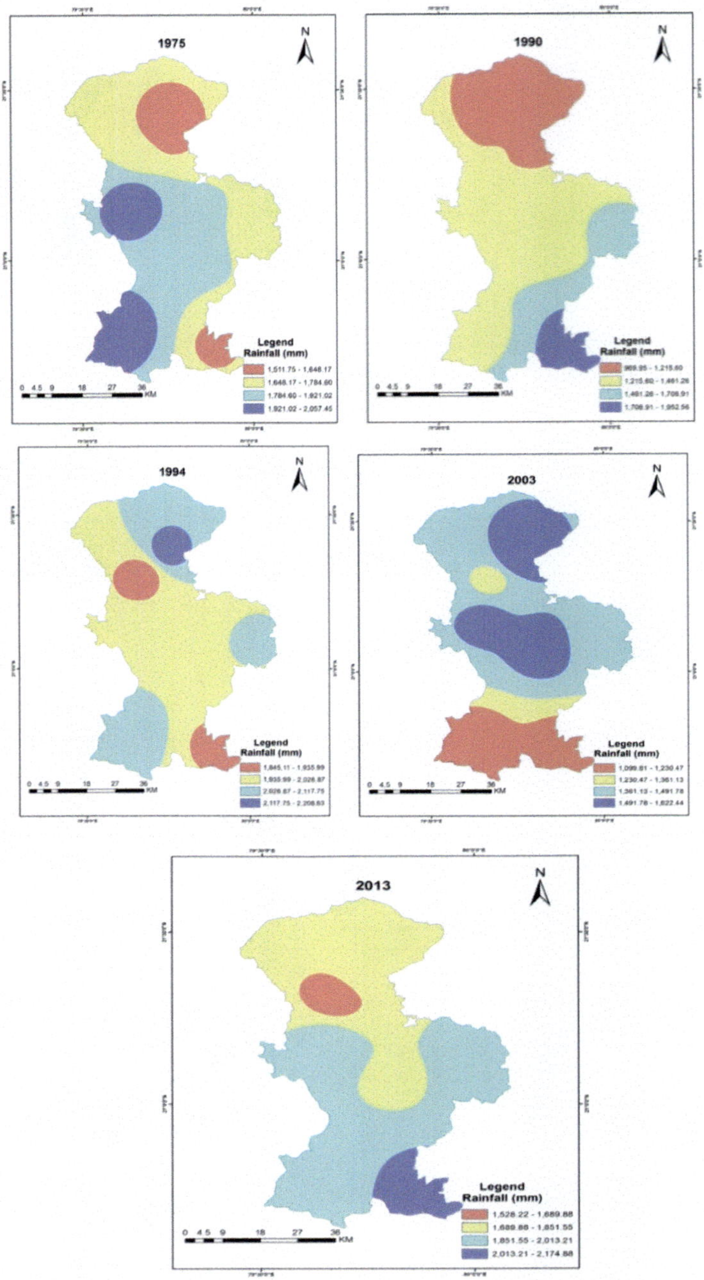

Fig. 11.9 Heavy rainfall (a) 1975, (b) 1990, (c) 1994, (d) 2003 and (e) 2013

11.4.5 Effects of Water Pollution

Humans can contract cholera, dysentery, rot, gastrointestinal ailments, and other illnesses by drinking contaminated water. Pregnant women who drink dirty water experience hormonal changes in their bodies, which may result in health issues or even death for the fetus. According to several studies, drinking contaminated water is one of the main causes of skin cancer. Drinking contaminated water is another significant factor contributing to polio-related disabilities. In order to prevent pollution of the nation's water supplies and maintain their quality, the Water Pollution Prevention and Control Act was passed in 1974. Water pollution can come from a variety of sources, thus in order to effectively regulate it, the sources must be stopped and the water must be purified through appropriate management and treatment. The following actions can be taken in order to resolve this issue. Given the value of water as a resource, all citizens of the nation should voluntarily contribute to maintaining it clean and uncontaminated. One of the best methods to reduce pollution is to recycle and reuse water, so try to include these practices into your daily routine. To ensure that the water quality is not impacted, industrial groupings should work together to establish shared effluent treatment facilities in locations where there are clusters of industrial units. In order to limit the number of pollutants at the source site, new solutions for the manufacture of other usable materials should be explored through recycling wastes such as ferrous and non-ferrous materials, paper, and plastic waste from treatment plants. By employing fertilizers in the right amounts based on the needs of the crops, pollution can be decreased through soil testing by minimizing their seepage in the reservoir. Commercial fertilizers and nutrient management should be applied by farmers in the right and proper amounts. For the purpose of controlling pollution brought on by industrial effluents, a thorough strategy should be created and put into action in a timely way. Licenses for polluting industrial facilities that release untreated industrial effluents ought to be revoked by limiting the availability of water. Using environmentally friendly toiletries and household goods can actively and positively contribute to the reduction of pollution, with minimal negative effects on the environment.

Sodium (Na):- Soil, plants, water, and food items are abundant sources of sodium. The majority of minerals that contain sodium contain sodium chloride. Ground water typically contains naturally occurring sodium. Although sodium does not smell when it is in ground water, its taste is detectable at concentrations of 200 mg or more per liter. Most rocks and soils that readily dissolve sodium include its constituents. The following are a few typical sources of too much sodium in groundwater: by natural erosive processes of minerals and rocks containing sodium, as well as brackish water present in some aquifers. See page of salinized surface water and sewage effluent poisoning of groundwater. See page of leachate from industrial or landfill sites.

Potassium (K)—It frequently occurs in rocks and soil. In water, potassium has no color or smell. However, as a result, the water could taste salty. The following are some of the main sources of potassium: fertilizer spills; sea water in locations susceptible to salt water invasion; and weathering and erosion of minerals based on potassium (such as feldspar, etc.).

Calcium (Ca):- It gets into fresh water by leaching, seepage of soil, and weathering of rocks, particularly limestone (marble, calcite, dolomite, gypsum, fluoride, and apatite, etc.). Surface water has less calcium than ground water does. Because calcium is present in water as the (Ca+ +) ion, it affects how hard the water is.

Magnesium (Mg): Due to this, the taste of drinking water can be undesirable. If its concentration is 100 mg per liter (Mg/liter) or more, the drinking water may taste unpleasant to sensitive people. The taste is unpleasant to the average person at concentrations above 500 mg per liter (Mg/L). Excess magnesium in drinking water can have a laxative effect, especially when the concentration of magnesium sulfate is greater than 700 Mg/L.

Carbonate and Bi-carbonate (Co3 & HCo3):- Carbon dioxide dissolved in rain water and snow is the major source of carbonate and bicarbonate ions found in ground water. Carbon dioxide (Co2) is released by the decay of organic matter. The pH (Table 11.2) of water itself gives an indication of the form of carbon dioxide present in it. Generally, the concentration of bicarbonate in groundwater ranges from 700 to 800 ppm. The parameters analyzed, include pH, Electrical Conductivity (EC), TDS, Total Alkalinity (TA), Total Hardness (TH), Na^+, DO, Nitrate (NO3) and Fluoride (F) etc.

The type of ground water presented in the study area is also analyzed using Pier chart. It is observed from the Piper plot (Fig. 11.10). This diagram explains the chemical properties of the water sample through a graphical method. It shows the information of sulphate, chloride, carbonate, hydrogen carbonate by creating different plots of calcium, magnesium, sodium, potassium, and all this information through a matrix.

11.4.6 Drinking and Irrigation Purposes

While studying whether groundwater is potable or not, a total of sixteen parameters were considered, out of which two samples were found to have excessive hardness and ten samples were found to have low nitrate content below in Table 11.3.

While determining whether the water is suitable for irrigation of the agricultural crop or not based on the parameter of electrical conductivity in the water, it was noticed that a total of 16 samples were taken and one place with the highest salinity was found, followed by nine sample with high salinity concentrations, six with moderate salinity and none with the lowest salinity Table 11.4. From this it can be seen that the water here is suitable for agriculture and 6.25% of the total parameter was found as very high salinity water followed by high salinity water as 56.25% followed by medium salinity water as 37.5%.

Table 11.2 Decision matrix of the selected water quality parameter

Parameter	pH	EC	TDS	TA	TH	Na^+	Do	K^+	COD	Ca^{2+}	Mg^{+2}	HCO_3	Cl^-	SO_4^{2-}	NO_3^-	F-
pH	1															
EC	0.5	1														
TDS	0.33	0.12	1													
TA	0.25	0.11	0.14	1												
TH	0.2	1	0.12	0.5	1											
Na^+	0.17	0.5	0.11	0.33	0.2	1										
Do	0.14	0.25	1	0.25	0.17	0.14	1									
K^+	0.12	0.2	0.5	0.2	0.14	0.12	0.12	1								
COD	0.11	0.14	0.33	0.17	0.12	0.11	0.11	0.12	1							
Ca^{2+}	1	0.12	0.25	0.14	0.11	1	1	0.11	0.14	1						
Mg^{+2}	0.5	1	0.2	0.12	1	0.5	0.5	1	0.12	0.2	1					
HCO_3	0.33	0.5	0.17	0.11	0.5	0.33	0.33	0.5	0.11	0.17	0.5	1				
Cl^-	0.25	0.33	0.14	1	0.33	0.25	0.25	0.33	1	0.14	0.33	0.14	1			
SO_4^{2-}	0.2	0.25	0.12	0.5	0.25	0.2	0.2	0.25	0.5	0.12	0.25	0.12	0.5	1		
NO_3^-	0.17	0.2	0.11	0.33	0.2	0.17	0.17	0.2	0.33	0.11	0.2	0.11	0.33	0.2	1	
F-	0.14	0.17	1	0.25	0.17	0.14	0.14	0.17	0.25	1	0.17	1	0.25	0.17	0.14	1

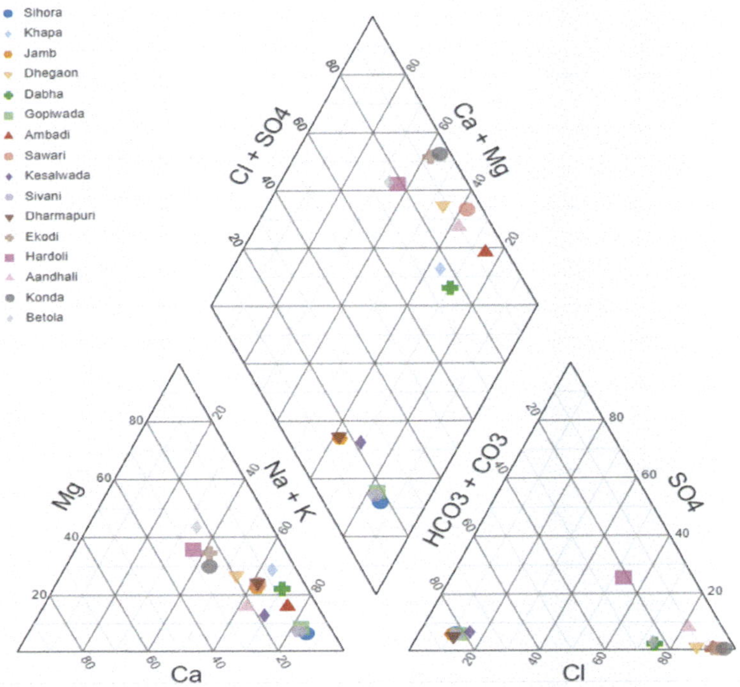

Fig. 11.10 Piper graph

Table 11.3 Ground Water Samples for Drinking based

Parameters	DL	MPL	Samples with conc. < DL	Samples with conc. inDL-MPL	Sampleswithconc. > MPL
TA(mg/L)	200	600	15	4	-
TH(mg/L)	300	600	8	9	2
NO_3(mg/L)	45	Norelaxation	8	-	10
F(mg/L)	1.0	1.5	18	-	-

Table 11.4 Ground water for irrigation based on EC

Type	EC (μS/cm)	No. of samples	% of samples
LowSalinityWater	<250	Nil	Nil
Medium SalinityWater	250–750	6	37.5
High SalinityWater	750–2250	9	56.25
VeryHighSalinityWater	>2250	1	6.25
Total		16	100.0

11.5 Conclusion

The rainfall trend and pattern from 1971 to 2013 are explained by the study conducted in the Bhandara district. Seven rainfall stations were examined for this study, and it was discovered that the distribution of rainfall in these stations was negative; the only site where a positive rainfall trend was observed was Sakoli. 16 ground water samples were examined during this ground water investigation in order to determine whether or not the ground water is drinkable and suitable for agriculture. Of the overall sample, 37.5% of the water is contaminated while the remaining 62.5% is potable. The primary cause of this is because the water's quality has been impacted by the chemical pesticides and fertilizers used in the fields. Through the use of graphics, the Piper diagram provides an explanation of the chemical properties of the water sample. It displays data on sulfate, chloride, carbonate, and hydrogen carbonate by generating various calcium, magnesium, sodium, and potassium plots and utilizing a matrix to display all of this data. Many researchers can utilize this data to inform their regional planning efforts.

References

Addisu, S., Selassie, Y. G., Fissha, G., & Gedif, B. (2015). Time series trend analysis of temperature and rainfall in lake Tana Sub-basin Ethiopia. *Environmental Systems Research, 4–25*(1), 1–12. https://doi.org/10.1186/s40068-015-0051-0

Aher, S., Shinde, S., Gawali, P., et al. (2019). Spatio-temporal analysis and estimation of rainfall variability in and around upper Godavari River basin India. *Arabian Journal Geosciences, 12*, 682. https://doi.org/10.1007/s12517-019-4869-z

Ahmad, S., Israr, M., et al. (2020). Spatio-temporal trends in snow extent and their linkage to hydro-climatological and topographical factors in the Chitral River basin (Hindukush, Pakistan). *Geocarto International, 35*(7), 711–734. https://doi.org/10.1080/10106049.2018.1524517

Anand, A., Singh, A. et al. (2021). Rainfall rate estimation over India using global precipitation measurement's microwave imager datasets and different variants of fuzzy information system. *Geocarto International, 37*, 6213. https://doi.org/10.1080/10106049.2021.1936208

Bessah, E., Boakye, E. A., Agodzo, S. K., et al. (2021). Increased seasonal rainfall in the twenty-first century over Ghana and its potential implications for agriculture productivity. *Environment, Development and Sustainability, 23*, 12342–12365. https://doi.org/10.1007/s10668-020-01171-5

Bisen, D. K., Kudnar, N. S. et al., (2022). Geo-Spatial modeling in the assessment of environmental resources for sustainable water resource management in a semi-arid region: A case study of Bhandara District, India. *International Journal of Scientific Research in Science, Engineering and Technology, 9*(4), 286–299, https://doi.org/10.32628/IJSRSET229445

Chandniha, S. K., Meshram, S. G., Adamowski, J. F., & Meshram, C. (2016). Trend analysis of precipitation in Jharkhand State India. *Theoretical and Applied Climatology, 130*(1–2), 261–274. https://doi.org/10.1007/s00704-016-1875-x

Deshmukh, K. K., & Aher, S. P. (2016). Assessment of the impact of municipal solid waste on groundwater quality near the Sangamner city using GIS approach. *Water Resources Management, 30*, 2425–2443. https://doi.org/10.1007/s11269-016-1299-5

Fathian, F., Dehghan, Z., Bazrkar, M. H., & Eslamian, S. (2014). Trends in hydrologic and climatic variables affected by four variations of mann-kendall approach in Urmia Lake basin

Iran. *Hydrological Sciences Journal, 61*(5), 892–904. https://doi.org/10.1080/02626667.2014.932911
Gajbhiye, S., Meshram, C., Mirabbasi, R., & Sharma, S. K. (2015). Trend analysis of rainfall time series for Sindh River basin in India. *Theoretical and Applied Climatology, 125*(3), 593–608.
Hamed, K. H., & Rao, A. R. (1998). A modified mann-kendall trend test for auto correlated data. *Journal of Hydrology, 204*, 182–196.
Hussein, S. J., Shattri, M., et al. (2017). Rainfall–runoff modelling and water balance analysis for Al-Hindiyah barrage Iraq using remote sensing and GIS. *Geocarto International, 32*(12), 1407–1420. https://doi.org/10.1080/10106049.2016.1213889
Jain, S. K., & Kumar, V. (2012). Trend analysis of rainfall and temperature data for India. *Current Science, 102*(1), 37–49.
Jayadas, A., & Ambujam, N. K. (2019). Observed trends in indices for daily rainfall extremes specific to the agriculture sector in lower Vellar River sub-basin India. *Journal of Earth System Science, 128*, 61. https://doi.org/10.1007/s12040-019-1074-0
Kashiwar, S. R., Kundu, M. C., & Dongarwar, U. R. (2021). Soil erosion estimation of Bhandara region of Maharashtra, India, by integrated use of RUSLE, remote sensing, and GIS. *Natural Hazards*. https://doi.org/10.1007/s11069-021-04974-5
Kendall, M. G. (1975). Rank correlation methods. Charles Griffin, London Khaliq MN, Ouarda TBMJ (2007) short communication on the critical values of the standard normal homogeneity test (SNHT). Int J Climatol 27(5):681–687. https://doi.org/10.1002/joc.1438
Kharake, A., Pathare, J., Deshmukh, P. (2021). Spatio-temporal variability of intra-monsoonal rainfall in Pravara-Mula River basin, India. *Arabian Journal of Geosciences, 14*, 890. https://doi.org/10.1007/s12517-021-07268-z
Kudnar, N. S. (2020a) GIS-based assessment of morphological and Hydrological parameters of Wainganga River basin, central India. *Modeling Earth Systems and Environment, 6*, 1933–1950. https://doi.org/10.1007/s40808-020-00804-y
Kudnar, N. S. (2020b). GIS-based investigation of topography, watershed, and Hydrological parameters of Wainganga River basin, central India. *Sustainable development practices using geoinformatics* (pp. 301–318). Scrivener Publishing LLC. https://doi.org/10.1002/9781119687160.ch19
Kudnar, N. S., & Rajasekhar, M. (2020). A study of the morphometric analysis and cycle of erosion in Waingangā basin India. *Modeling Earth Systems and Environment, 6*, 311–327. https://doi.org/10.1007/s40808-019-00680-1
Kudnar, N. S., Diwate, P., Mishra, V. N., et al. (2022). Spatio-temporal variability and trend analysis of rainfall in Wainganga River basin, central India, and forecasting using state-space models. *Theoretical and Applied Climatology, 150*, 469–488. https://doi.org/10.1007/s00704-022-04168-4
Kudnar, N. S. (2022). Geospatial modeling in the assessment of environmental resources for sustainable water resource management in a semiarid region: A GIS approach. *Current Directions in Water Scarcity Research, 7*, 133–151. https://doi.org/10.1016/B978-0-323-91910-4.00009-1
Kushwaha, N., Elbeltagi, A., Mehan, S., et al. (2022). Comparative study on morphometric analysis and RUSLE-based approaches for micro-watershed prioritization using remote sensing and GIS. *Arabian Journal of Geosciences, 15*, 564. https://doi.org/10.1007/s12517-022-09837-2
Mann, H. B. (1945). Nonparametric tests against trend. *Econometrica, 13*(3), 245–259. https://doi.org/10.2307/1907187
Meshram, S. G., Singh, S. K., Meshram, C., et al. (2018). Statistical evaluation of rainfall time series in concurrence with agriculture and water resources of Ken River basin, central India (1901–2010). *Theoretical and Applied Climatology, 134*, 1231–1243. https://doi.org/10.1007/s00704-017-2335-y
Meshram, R. R., & Ingle, K. B. (2012). Mineralogy and origin of dumortierite from Girola area, Bhandara district, Eastern Maharashtra. *Journal of the Geological Society of India, 79*, 181–188. https://doi.org/10.1007/s12594-012-0020-4

Misra, A. K., & Tripathi, A. (2021). Mathematical models of how the transpiration affects rainfall through agriculture crops. *Environmental Modeling and Assessment, 26*, 837–848. https://doi.org/10.1007/s10666-021-09769-3

Nair, S. C., & Mirajkar, A. B. (2021). Spatio–temporal rainfall trend anomalies in Vidarbha region using historic and predicted data: A case study. *Modeling Earth Systems and Environment, 7*, 503–510. https://doi.org/10.1007/s40808-020-00928-1

Rajasekhar, M., Gadhiraju, S. R., Kadam, A. et al., (2020) Identification of groundwater recharge-based potential rainwater harvesting sites for sustainable development of a semiarid region of southern India using geospatial, AHP, and SCS-CN approach. *Arabian Journal of Geosciences*, 13–24. https://doi.org/10.1007/s12517-019-4996-6

Rajasekhar, M., SudarsanaRaju, G. et al., (2021). Multi-criteria land suitability analysis for agriculture in semi-arid region of Kadapa district, Southern India: Geospatial approaches. *Remote Sensing of Land, 5*(2), 59–72. https://doi.org/10.21523/gcj1.2021050201

Rupa Kumar, K., Pant, G. B., Parthasarathy, B., Sontakke, N. A. (1992). Spatial and sub-seasonal patterns of the long term trends of Indian summer monsoon rainfall. Int J Climatol, 12(3), 257–268. https://doi.org/10.1002/joc.3370120303

Singh, P. K., Kumar, S., & Singh, U. C. (2011). Groundwater resource evaluation in the Gwalior area, India, using satellite data: An integrated geomorphological and geophysical approach. *Hydrogeology Journal, 19*, 1421–1429. https://doi.org/10.1007/s10040-011-0758-6

Srivastava, P. K., Mukherjee, S., Gupta, M., et al. (2011). Characterizing monsoonal variation on water quality index of River Mahi in India using geographical information system. *Water Quality Exposure and Health, 2*, 193–203. https://doi.org/10.1007/s12403-011-0038-7

Subash, N., & Sikka, A. K. (2014). Trend analysis of rainfall and temperature and its relationship over India. *Theoretical and Applied Climatology, 117*(3–4), 449–462. https://doi.org/10.1007/s00704-013-1015-9

Taneja, P., Labhasetwar, P., & Nagarnaik, P. (2019). Nitrate in drinking water and vegetables: Intake and risk assessment in rural and urban areas of Nagpur and Bhandara districts of India. *Environmental Science and Pollution Research, 26*, 2026–2037. https://doi.org/10.1007/s11356-017-9195-y

Zolekar, R. B., Todmal, R. S., Bhagat, V. S., et al. (2021). Hydro-chemical characterization and geospatial analysis of groundwater for drinking and agricultural usage in Nashik district in Maharashtra, India. *Environment, Development and Sustainability, 23*, 4433–4452. https://doi.org/10.1007/s10668-020-00782-2

Chapter 12
The Impact of Climate Change on the Environment, Water Resources, and Agriculture: A Comprehensive Review

Prerna Mehta

12.1 Introduction

Climate change, a defining challenge of our era, exerts profound impacts on ecosystems, water resources, and agricultural systems globally. Its intricate interplay with natural processes and human activities reshapes the very fabric of our planet's ecological balance. This exploration delves into the intricate web of consequences stemming from climate change, examining its multifaceted effects on the environment, water resources, and agriculture (Abbass et al., 2022; Anik et al., 2023). Through meticulous scientific inquiry and interdisciplinary analysis, we strive to unravel the underlying mechanisms driving environmental transformations and elucidate the pathways towards resilience and adaptation (Fig. 12.1).

At the heart of climate change lies a nexus of interconnected systems, each bearing the imprint of rising global temperatures and shifting climatic patterns (Bibi & Rahman, 2023). Ecosystems, ranging from the Arctic tundra with iconic species like the polar bear (*Ursus maritimus*) to the Amazon rainforest teeming with biodiversity, witness unprecedented disruptions in species distributions, phenological shifts, and habitat degradation. Emblematic species such as the jaguar (Panthera onca) face existential threats as their habitats erode under the weight of environmental stressors (Jagadish, 2020). In tandem, vegetation communities undergo profound transformations, with changes in temperature and precipitation altering species composition. From the boreal forests of Canada with species like spruce (*Picea* spp.) to the savannas of Africa, the specter of climate-induced shifts looms large, reshaping landscapes and

P. Mehta (✉)
Department of Biotechnology, GD Rungta College of Science & Technology Bhilai-490024, Bhilai, Chhattisgarh, India
e-mail: prernamehta326@gmail.com

P. K. Rai and S. Rai (eds.), *Climate, Environment and Agricultural Development*, Advances in Geographical and Environmental Sciences,
https://doi.org/10.1007/978-981-97-8363-2_12

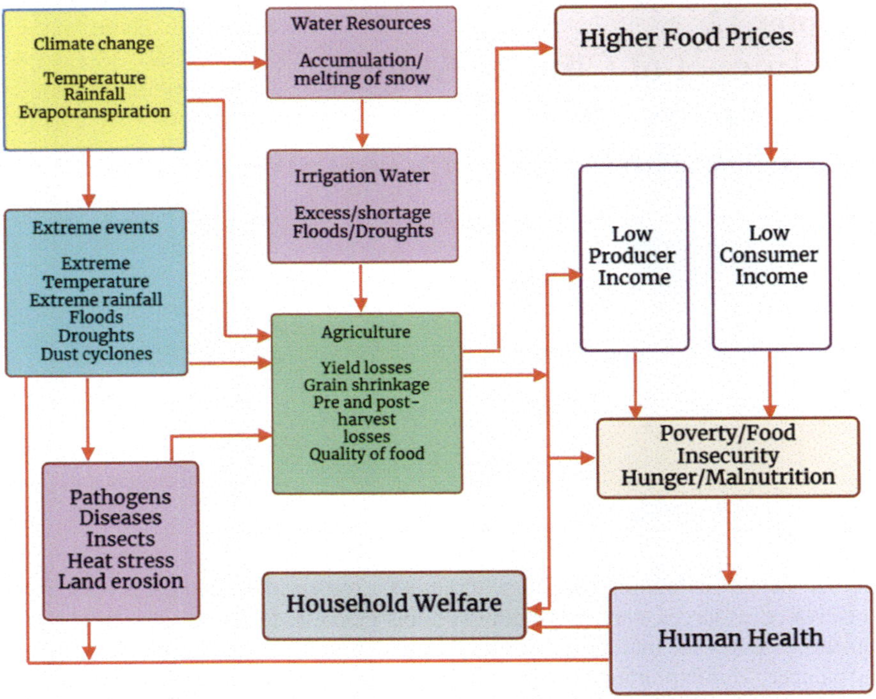

Fig. 12.1 Provides a concise depiction of the multifaceted impacts of climate change on water resources, agriculture, food security, and household welfare

challenging the resilience of ecosystems worldwide. Water resources, the lifeblood of ecosystems and human societies alike, bear the brunt of climate change's impact. Alterations in precipitation patterns, melting glaciers, and changing runoff regimes usher in an era of uncertainty, where water scarcity and quality concerns escalate across regions. From the Himalayan rivers to the deltas of Bangladesh, communities grapple with the specter of sea-level rise, salinity intrusion, and dwindling freshwater reserves, posing significant challenges to agricultural productivity, public health, and economic stability. In agricultural landscapes, where reliance on predictable weather patterns and stable water supplies is paramount, the effects of climate change reverberate through crop yields, pest dynamics, and food security (Oyewo et al., 2020). From smallholder farms in sub-Saharan Africa to industrialized agricultural systems in North America, adaptation strategies must be forged to withstand the onslaught of changing climatic norms and ensure sustainable livelihoods for future generations.

Through synthesis of scientific research and empirical evidence, this exploration seeks to shed light on the intricate interplay between climate change and its far-reaching consequences. By unraveling the complex dynamics at play and identifying areas of vulnerability and resilience, we endeavor to pave the way for evidence-based solutions and adaptive strategies. From enhancing ecosystem resilience to improving water resource management and promoting sustainable agricultural practices, the pathway to a resilient future hinges on collective action, informed decision-making, and unwavering commitment to safeguarding the health and integrity of our planet amidst the challenges of climate change.

12.2 Importance of Understanding Climate Change Impacts

Understanding the impacts of climate change is paramount for informed decision-making and effective adaptation strategies. It enables societies to anticipate and mitigate the adverse effects on ecosystems, water resources, agriculture, and human communities (Corami, 2023). Through interdisciplinary research and empirical evidence, comprehending these impacts fosters resilience and facilitates the transition to sustainable practices in a changing climate (Table 12.1).

12.3 Environmental Impacts of Climate Change

12.3.1 Effects on Ecosystems and Biodiversity

Climate change stands as a formidable force reshaping ecosystems and biodiversity worldwide. The intricate balance of life is under siege as rising temperatures, altered precipitation patterns, and extreme weather events disrupt habitats and species distributions (Butkevičienė et al., 2023). In the Arctic, the iconic polar bear (*Ursus maritimus*) faces unprecedented challenges as sea ice retreats, diminishing its hunting grounds and threatening its survival. Similarly, in tropical rainforests like the Amazon basin, the vibrant biodiversity, including species like the jaguar (*Panthera onca*) and poison dart frogs (*Dendrobatidae*), confronts escalating pressures from deforestation and shifting climatic conditions. These changes reverberate through intricate ecological networks, impacting species interactions, population dynamics, and the provision of critical ecosystem services. The ramifications of climate change extend beneath the ocean's surface, where coral reefs, among the most diverse ecosystems on Earth, confront existential threats. Species such as Acropora corals, integral to reef structures, are vulnerable to bleaching events triggered by rising sea temperatures. As corals decline, entire ecosystems, including fish species like the clownfish (*Amphiprioninae*) and parrotfish (*Scaridae*), face habitat loss and diminished resilience to environmental stressors. Furthermore, in temperate forests, iconic species such as the sugar maple (*Acer saccharum*) confront challenges as warming temperatures disrupt traditional growing seasons and alter species compositions. Shifts in phenology and distribution patterns are evident as species like the European robin (*Erithacus rubecula*) adjust their behaviors to adapt to changing climates.

Table 12.1 Different climatic changes and their impacts

Importance	Description	Reference
Ecological Dynamics	- Predicting how climate change influences species distributions, population dynamics, and ecosystem functioning informs conservation strategies - Studying species' responses to changing environmental conditions helps in identifying key indicators of ecosystem health and resilience	Åkesson et al. (2021)
Hydrological Processes	- Analyzing the impacts of climate change on precipitation patterns, snowmelt, and runoff is essential for water resource management and flood control - Understanding how changes in temperature and precipitation affect groundwater recharge rates and aquifer dynamics informs sustainable water use practices	Tilahun et al. (2023)
Agricultural Adaptation	- Assessing the effects of climate change on crop growth, phenology, and pest and disease outbreaks enables the development of resilient agricultural practices - Researching climate-resilient crop varieties and agronomic techniques helps farmers mitigate production risks and maintain food security in changing climates	Mall et al. (2017)
Socioeconomic Resilience	- Evaluating the socioeconomic implications of climate change, including impacts on livelihoods, food systems, and vulnerable communities, guides adaptation and mitigation strategies - Understanding the intersection of climate change with socio-economic factors, such as poverty, inequality, and access to resources, aids in promoting equitable and sustainable development	Carmen et al. (2022)
Policy and Decision-Making	- Integrating scientific knowledge of climate change impacts into policy decisions helps governments and organizations prioritize actions to mitigate greenhouse gas emissions and adapt to changing conditions - Incorporating climate risk assessments and projections into infrastructure planning, land-use management, and disaster preparedness enhances resilience and reduces vulnerability to climate-related hazards	Orlove et al. (2020)

The urgency of addressing climate change's impacts on ecosystems and biodiversity cannot be overstated. From the Arctic to the tropics, from forests to oceans, species are struggling to cope with the pace and scale of environmental change. Conservation efforts must be informed by scientific research and interdisciplinary collaboration to mitigate the loss of biodiversity and safeguard the resilience of ecosystems. This necessitates innovative approaches, including habitat restoration, protected area management, and sustainable land-use practices, aimed at preserving the rich tapestry of life on our planet for future generations. Only through concerted action and collective commitment can we hope to mitigate the far-reaching consequences of climate change and secure a sustainable future for all species and ecosystems.

12.4 Vegetation Dynamics and Species Vulnerability

The dynamics of vegetation and the vulnerability of species represent critical facets of ecosystem responses to climate change. Across diverse biomes, from forests to grasslands, alterations in temperature, precipitation patterns, and atmospheric carbon dioxide levels influence vegetation composition and distribution (Chaudhry & Sidhu, 2022). In boreal forests, species such as the spruce (*Picea* spp.) and fir (*Abies* spp.) exhibit shifts in distribution as warming temperatures extend their growing seasons and enable colonization of previously inhospitable habitats (Zhu et al., 2021). However, these changes also bring about increased susceptibility to pests and pathogens, posing threats to forest health and resilience (Au et al., 2021) (Table 12.2).

12.5 Climate Change and Water Resources

12.5.1 Altered Hydrological Cycles and Water Availability

Climate change triggers substantial shifts in hydrological cycles, fundamentally altering the availability and distribution of water resources. In the Arctic, the accelerated melting of ice caps and glaciers, including the Greenland Ice Sheet, intensifies freshwater discharge into the North Atlantic Ocean. This phenomenon, known as the Greenland Ice Sheet's meltwater pulse, contributes to rising sea levels and disrupts oceanic circulation patterns, impacting weather systems globally. Moreover, in mountainous regions such as the Himalayas, the retreat of glaciers like the Gangotri Glacier jeopardizes the freshwater supply for millions of people in downstream river basins, including the Ganges and Brahmaputra. The decline in glacial mass alters river flow dynamics, posing challenges for irrigation, hydropower generation, and aquatic ecosystems dependent on seasonal meltwater inputs (Fig. 12.2).

Concurrently, altered precipitation patterns exacerbate water availability challenges in arid and semi-arid regions, such as the southwestern United States and the Mediterranean Basin. Species like the saguaro cactus (*Carnegiea gigantea*) and the Joshua tree (*Yucca brevifolia*) in North American deserts face heightened stress as prolonged droughts and erratic rainfall patterns disrupt their water supply. Similarly, in the Mediterranean region, where water scarcity is already a pressing issue, climate change exacerbates the strain on freshwater resources, threatening agriculture, urban water supplies, and biodiversity. To mitigate the impacts of altered hydrological cycles, interdisciplinary approaches integrating hydrological modeling, ecosystem

Table 12.2 Provides comprehensive details about vegetation and species

Region/Forest	Vegetation/Species	Description	Specific species	Reference
Boreal Forests	Spruce (*Picea* spp.), Fir (*Abies* spp.)	These forests, found in northern regions, are experiencing shifts in distribution patterns due to warming temperatures. Increased temperatures extend the growing season, altering the ecology of these forests. However, these changes also render the forests more susceptible to pests and pathogens, leading to concerns about long-term forest health and resilience	*Picea glauca (white spruce), Picea mariana (black spruce), Abies balsamea (balsam fir)*	Pureswaran et al. (2015)
Arid and Semi-Arid Regions	*Saguaro Cactus (Carnegiea gigantea), Acacia Trees (Acacia spp.)*	Vegetation in arid and semi-arid regions displays remarkable adaptations to conserve water and withstand harsh conditions. Saguaro cacti, for example, have specialized tissues for water storage, while Acacia trees have deep root systems. However, prolonged droughts and increasing temperatures associated with climate change pose significant challenges to these species, potentially altering the structure and composition of these ecosystems	*Carnegiea gigantea (saguaro), Acacia tortilis (umbrella thorn acacia), Acacia karroo (sweet thorn)*	Renzi et al. (2019)

(continued)

Table 12.2 (continued)

Region/ Forest	Vegetation/ Species	Description	Specific species	Reference
Alpine Ecosystems	Alpine Meadow Grasses, Cushion Plants	Alpine ecosystems are characterized by their extreme conditions and unique biodiversity. Alpine meadow grasses and cushion plants are specially adapted to survive in these harsh environments, but they are vulnerable to changes in temperature and precipitation patterns. As temperatures rise, vegetation zones in alpine regions may shift upward in elevation, affecting species distributions and community composition. Challenges such as limited dispersal abilities and competition from invasive species further exacerbate the vulnerability of alpine flora to climate change	Festuca spp. (fescue grasses), Saxifraga spp. (stonecrop), Androsace spp. (rock jasmine)	Wang et al. (2021)
Boreal Forests	Spruce (Picea spp.), Fir (Abies spp.)	These forests, found in northern regions, are experiencing shifts in distribution patterns due to warming temperatures. Increased temperatures extend the growing season, altering the ecology of these forests. However, these changes also render the forests more susceptible to pests and pathogens, leading to concerns about long-term forest health and resilience	*Picea glauca (white spruce), Picea mariana (black spruce), Abies balsamea (balsam fir)*	D'Orangeville et al. (2018)

monitoring, and stakeholder engagement are imperative. By understanding the intricacies of water availability dynamics and species vulnerabilities, policymakers can develop adaptive strategies that enhance water resilience and ensure sustainable water management practices in the face of a changing climate.

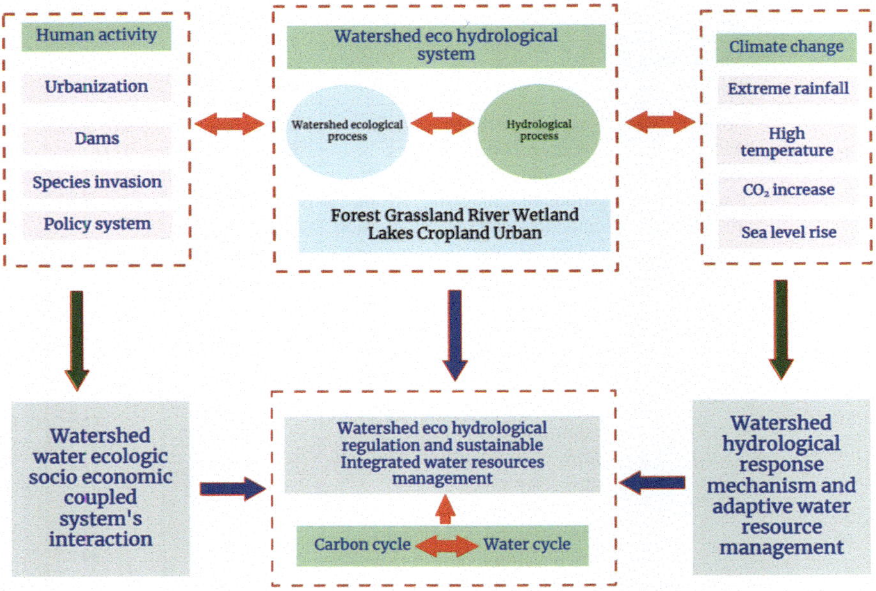

Fig. 12.2 Delineates the primary objectives of ecohydrological processes within watersheds

12.6 Implications for Freshwater Ecosystems and Human Communities

Freshwater ecosystems and human communities face intricate challenges due to altered hydrological cycles induced by climate change. Variations in precipitation patterns and river flow dynamics exacerbate water scarcity in semi-arid regions and areas dependent on glacial meltwater, threatening the survival of diverse species such as salmon (*Oncorhynchus* spp.) in Pacific Northwest rivers and freshwater fish communities in African lakes. Concurrently, coastal regions experience saltwater intrusion into freshwater sources, compromising water quality and jeopardizing the habitat of species like mangroves (*Rhizophora* spp.) and freshwater turtles (*Trachemys scripta*) in estuarine environments.

Addressing these implications demands integrated approaches that blend scientific research, community engagement, and policy interventions. Strategic investments in ecosystem restoration, sustainable water management practices, and climate-resilient infrastructure are imperative to safeguard freshwater resources and bolster the resilience of both ecological and human systems. Collaborative endeavors among governments, civil society organizations, and local communities are essential for devising and implementing adaptive strategies that foster water security, promote biodiversity conservation, and ensure equitable access to freshwater resources in the face of escalating climate change impacts.

12.7 Agricultural Challenges and Adaptations

12.7.1 Crop Yield Variability and Food Security

Agriculture faces mounting challenges due to climate change-induced variability in crop yields, posing significant threats to global food security. Shifts in temperature and precipitation patterns alter growing conditions, affecting crop growth, development, and productivity (Franzoni et al., 2022). Crops such as maize (*Zea mays*), wheat (*Triticum aestivum*), and rice (*Oryza sativa*) exhibit heightened sensitivity to changes in climate, with extreme weather events like droughts and heatwaves exacerbating yield losses and compromising harvests (Wessolek et al., 2023). Vulnerable regions, including sub-Saharan Africa and parts of Asia, experience heightened risks of food insecurity as smallholder farmers grapple with diminished crop yields and reduced income opportunities. Moreover, disruptions in global food supply chains further exacerbate the challenges, exacerbating hunger and malnutrition among vulnerable populations (Rai, 2020).

12.8 Strategies for Climate-Resilient Agriculture

To mitigate the impacts of climate change on agriculture and enhance resilience, innovative strategies for climate-resilient agriculture are imperative. Adopting drought-resistant crop varieties, such as drought-tolerant maize hybrids and heat-tolerant rice varieties (Sadok & Jagadish, 2020), helps buffer against climate-induced stresses and stabilize yields in water-limited environments (Sachdev et al., 2021). Furthermore, implementing agroforestry practices, cover cropping, and conservation agriculture techniques enhances soil health, water retention, and biodiversity, fostering resilient agricultural ecosystems (Li et al., 2023). Precision agriculture technologies, including remote sensing, precision irrigation, and crop modeling, enable farmers to optimize resource use, minimize input costs, and mitigate climate risks. Additionally, promoting diversified cropping systems and crop rotation schemes reduces vulnerability to pests, diseases, and adverse weather events, enhancing farm resilience and sustainability over the long term.

12.9 Interdisciplinary Perspectives and Solutions

12.9.1 Synthesis of Scientific Research and Empirical Evidence

Interdisciplinary collaboration is vital for synthesizing scientific research and empirical evidence to understand the complexities of climate change impacts and inform effective adaptation strategies. Integrating findings from climate science, ecology, hydrology, and social sciences provides comprehensive insights into the multifaceted nature of climate change challenges. Through systematic reviews, meta-analyses, and modeling approaches, researchers elucidate the underlying mechanisms driving environmental transformations, water resource dynamics, and agricultural vulnerabilities (Papageorgiou et al., 2021). By analyzing historical data, satellite imagery, and field observations, scientists discern patterns, trends, and emerging risks associated with climate change, enabling evidence-based decision-making and informed policy interventions.

12.10 Proposing Adaptive Strategies and Policy Recommendations

Proposing adaptive strategies and policy recommendations is paramount for building resilience and addressing the impacts of climate change across diverse sectors. Drawing upon interdisciplinary research and stakeholder engagement, policymakers develop adaptive management frameworks that prioritize risk reduction, resource conservation, and community resilience. Implementing nature-based solutions, such as ecosystem restoration, green infrastructure, and sustainable land management practices, enhances ecosystem resilience and buffers against climate-induced stresses (Cao et al., 2022). Moreover, integrating climate considerations into land-use planning, water resource management, and disaster risk reduction policies fosters adaptive capacity and reduces vulnerability to climate extremes.

Furthermore, fostering international cooperation and knowledge exchange is crucial for scaling up adaptation efforts and fostering collective action on a global scale. By promoting cross-sectoral collaboration, capacity-building initiatives, and technology transfer, countries can enhance adaptive governance structures and facilitate the implementation of climate-resilient policies and practices. Additionally, leveraging financial mechanisms, such as climate finance and green investment funds, mobilizes resources for climate adaptation projects and supports the transition towards sustainable and resilient development pathways.

12.11 Conclusion and Future Directions

12.11.1 Summary of Key Findings and Insights

The comprehensive analysis underscores the profound impact of climate change on ecosystems, water resources, and agriculture. Recent studies reveal alarming trends, including accelerated glacier melting and increased frequency of extreme weather events. Scientific data highlight the vulnerability of freshwater ecosystems, such as coral reefs and wetlands, to temperature rise and ocean acidification. Agricultural systems face escalating challenges, with shifting growing seasons and the spread of pests and diseases threatening global food security.

12.12 Call to Action for Sustainable Climate Solutions

Moving forward, governments must enact robust policies and allocate resources to address climate change effectively. Initiatives such as the Paris Agreement and the Green New Deal represent critical steps toward emissions reduction and renewable energy adoption. Moreover, new acts like the Climate Action Now Act in the United States emphasize the urgency of climate action and prioritize environmental sustainability in policymaking. It is imperative to integrate cutting-edge technologies, including artificial intelligence and satellite monitoring, to enhance climate resilience and adaptation strategies. Additionally, fostering international cooperation and knowledge-sharing initiatives will facilitate the implementation of sustainable climate solutions on a global scale. By embracing innovation and collective action, we can forge a path toward a more sustainable and resilient future for generations to come.

References

Abbass, K., Qasim, M. Z., Song, H., Murshed, M., Mahmood, H., & Younis, I. (2022). A review of the global climate change impacts, adaptation, and sustainable mitigation measures. *Environmental Science and Pollution Research, 29*, 42539–42559.

Anik, A. H., Sultan, M. B., Alam, M., Parvin, F., Ali, M. M., & Tareq, S. M. (2023). The impact of climate change on water resources and associated health risks in Bangladesh: A review. *Water Security, 18*, 100133.

Au, A., Lee, H., Ye, T., Dave, U., & Rahman, A. (2021). Bacteriophages: Combating antimicrobial resistance in food-borne bacteria prevalent in agriculture. *Microorganisms, 10*, 46.

Åkesson, A., Curtsdotter, A., Eklöf, A., Ebenman, B., Norberg, J., & Barabás, G. (2021). The importance of species interactions in eco-evolutionary community dynamics under climate change. *Nature Communications, 12*(1), 4759.

Butkevičienė, L. M., Steponavičienė, V., Pupalienė, R., Skinulienė, L., & Bogužas, V. (2023). Effect of different tillage systems and soil Biostimulants on agrochemical properties and intensity of soil CO_2 emission in wheat crop. *Agronomy, 13*, 338.

Bibi, F., & Rahman, A. (2023). An overview of climate change impacts on agriculture and their mitigation strategies. *Agriculture, 13*(8), 1508.

Cao, Z., Wang, S., Luo, P., Xie, D., & Zhu, W. (2022). Watershed ecohydrological processes in a changing environment: Opportunities and challenges. *Water, 14*(9), 1502.

Carmen, E., Fazey, I., Ross, H., Bedinger, M., Smith, F. M., Prager, K., ... & Morrison, D. (2022). Building community resilience in a context of climate change: The role of social capital. *Ambio, 51*(6), 1371–1387.

Corami, A. (2023). Nanotechnologies and phytoremediation: Pros and cons. In L. Newman, A. A. Ansari, S. S. Gill, M. Naeem & R. Gill (Eds.), *Phytoremediation: Management of environmental contaminants* (Vol 7, pp. 403–426) Springer International Publishing. ISBN 978-3-031-17988-4.

Chaudhry, S., & Sidhu, G. P. S. (2022). Climate change regulated abiotic stress mechanisms in plants: A comprehensive review. *Plant Cell Reports, 41*(1), 1–31.

Franzoni, G., Cocetta, G., Prinsi, B., Ferrante, A., & Espen, L. (2022). Biostimulants on crops: Their impact under abiotic stress conditions. *Horticulturae, 8*, 189.

Jagadish, S. V. K. (2020). Heat stress during flowering in cereals—Effects and adaptation strategies. *New Phytologist, 226*, 1567–1572.

Li, Q., Wang, L., Fu, Y., Lin, D., Hou, M., Li, X., Hu, D., & Wang, Z. (2023). Transformation of soil organic matter subjected to environmental disturbance and preservation of organic matter bound to soil minerals: A review. *Journal of Soils and Sediments, 23*, 1485–1500.

Mall, R. K., Gupta, A., & Sonkar, G. (2017). Effect of climate change on agricultural crops. In *Current developments in biotechnology and bioengineering* (pp. 23–46). Elsevier.

D'Orangeville, L., Houle, D., Duchesne, L., et al. (2018). Beneficial effects of climate warming on boreal tree growth may be transitory. *Nature CommunicAtions, 9*, 3213. https://doi.org/10.1038/s41467-018-05705-4

Orlove, B., Shwom, R., Markowitz, E., & Cheong, S. M. (2020). Climate decision-making. *Annual Review of Environment and Resources, 45*, 271–303.

Oyewo, O. A., Adeniyi, A., Bopape, M. F., Onyango, M. S. (2020). Chapter 4—Heavy metal mobility in surface water and soil, climate change, and soil interactions. In M. N. V. Prasad & M. Pietrzykowski (Eds.), *Climate change and soil interactions* (pp. 51–88). Elsevier. ISBN 978-0-12-818032-7.

Papageorgiou, A., Azzi, E. S., Enell, A., & Sundberg, C. (2021). Biochar produced from wood waste for soil remediation in Sweden: Carbon sequestration and other environmental impacts. *Science of the Total Environment, 776*, 145953.

Pureswaran, D. S., De Grandpré, L., Paré, D., Taylor, A., Barrette, M., Morin, H., ... & Kneeshaw, D. D. (2015). Climate-induced changes in host tree–insect phenology may drive ecological state-shift in boreal forests. *Ecology, 96*(6), 1480–1491.

Rai, R. (2020). Heat stress in crops: Driver of climate change impacting global food supply. In P. Singh, R. P. Singh & V. Srivastava, (Eds.), *Contemporary environmental issues and challenges in era of climate change* (pp. 99–117). Springer. ISBN 978-981-32-9595-7.

Renzi, J. J., Peachey, W. D., & Gerst, K. L. (2019). A decade of flowering phenology of the keystone saguaro cactus (Carnegiea gigantea). *American Journal of Botany, 106*(2), 199–210.

Sadok, W., & Jagadish, S. V. K. (2020). The hidden costs of nighttime warming on yields. *Trends in Plant Science, 25*, 644–651.

Sachdev, S., Ansari, S. A., Ansari, M. I., Fujita, M., & Hasanuzzaman, M. (2021). Abiotic stress and reactive oxygen species: Generation, signaling, and defense mechanisms. *Antioxidants, 10*, 277.

Tilahun, Z. A., Bizuneh, Y. K., & Mekonnen, A. G. (2023). The impacts of climate change on hydrological processes of Gilgel Gibe catchment, Southwest Ethiopia. *PLoS ONE, 18*(6), e0287314.

Wang, Y., Sun, J., Liu, B., Wang, J., & Zeng, T. (2021). Cushion plants as critical pioneers and engineers in alpine ecosystems across the Tibetan Plateau. *Ecology and Evolution, 11*(17), 11554–11558.

Wessolek, G., Bohne, K., & Trinks, S. (2023). Validation of soil thermal conductivity models. *International Journal of Thermophysics, 44*, 20.

Zhu, T., Fonseca De Lima, C.F., De Smet, I. (2021). The heat is on: How crop growth, development, and yield respond to high temperature. *Journal of Experimental Botany, 72*, 7359–7373.

Chapter 13
Spatio-Temporal Variability of Spectral Indices and Land Surface Temperature for Ecological Change Detection in Faridabad District India Between 1991 and 2021

Sarah Rahman and Swagata Ghosh

13.1 Introduction

The advent of the late twentieth century marked the age of modernization, during which societies witnessed tremendous metamorphosis in various aspects. The rapid pace of industrialization and urbanization has been the primary driver behind the transformation of natural landscapes into modified forms which prevents the natural cooling effect of vegetation and trees. Instead of reflecting sunlight, concrete absorbs and preserves heat and thus affects the comfort of city dwellers, alter rainfall patterns, aggravate the intensity of storms, and contribute to the overall warming of the planet. The alteration of land use patterns has disrupted the delicate balance of ecosystems, leading to the loss of biodiversity and ecological services (Levin & Levin, 1992).

Amidst the worldwide regions that suffered from ecological decline, constituent areas of urban agglomeration located in the arid and semi-arid regions are in acute danger, especially in the developing world. In the arid and semi-arid climatic zone, natural factors combined with anthropogenic activities trigger environmental degradation both in urban and its adjacent peri-urban and rural areas (da Silva et al., 2018; S. Kumar, Midya et al., 2021; Yongge Li et al., 2022). Majority of the published research focuses on an individual aspect of the environment and reported that climatic factors,

S. Rahman
Amity Institute of Geoinformatics and Remote Sensing (AIGIRS), Amity University, Sector 125, Noida, Uttar Pradesh 201313, India

S. Ghosh (✉)
Department of Geography, School of Earth Sciences, Central University of Karnataka, Kadaganchi, Aland Road, Kalaburagi, Karnataka 585367, India
e-mail: swagata.gis@gmail.com; swagataghosh@cuk.ac.in

P. K. Rai and S. Rai (eds.), *Climate, Environment and Agricultural Development*, Advances in Geographical and Environmental Sciences,
https://doi.org/10.1007/978-981-97-8363-2_13

soil salinity, reduced soil wetness, fallow land are responsible factors aggravating the temperature in the rural area during the daytime of summer causing environmental heating in the rural villages adjacent to various cities located in the arid and semi-arid part of the world (Abutaleb et al., 2021; Ghosh et al., 2021; S. Kumar et al., 2019; S. Kumar, Midya, et al., 2021; Mohammad et al., 2019; Pramanik & Punia, 2020; Rasul et al., 2015, 2016). However, published works of literature on the overall ecological status of urban agglomeration and adjacent regions in arid and semi-arid zones are still scarce. Therefore, to achieve a position of overall sustainability, a holistic approach is required to quantify the ecological change in urban and rural areas in the arid and semi-arid zone.

For the quantitative evaluation of the ecological change, spectral indices are effective tools. Normalized difference vegetation index (NDVI), fractional vegetation (FV), Enhanced vegetation index (EVI), Soil adjusted vegetation index (SAVI), Normalized difference moisture index (NDMI), Normalized difference soil index (NDSI), Bare soil index (BSI), Normalized difference build-up index (NDBI) are the major spectral indices which indicate several biophysical variables of land surface viz. greenness, wetness, dryness etc. (Abutaleb et al., 2021; Nejad & Zoratipour, 2019; Zha, 2003; Zhu et al., 2020). Such indices represent individually single biophysical variables and majority of the research focus on single aspect of ecology with the use of them in different parts of India (Badlani et al., 2017; V. Kumar et al., 2015; Richa Sharma & Joshi, 2016). Such single-factor based ecological evaluation is not enough for a holistic purview of ecological scenario due to the dynamicity of the ecological environment (Xu et al., 2019). To address such limitation, composite indicator based ecological evaluation is the preferred method. Composite indicator combines the benefit of all indices avoiding the hassle of assigning weights to each of the variable depending upon their characteristics and can be exhibited at a range of comparable spatiotemporal scale (Kerr & Ostrovsky, 2003). Composite indicator based ecological change detection have been attempted in different parts of the world mostly on the macro-scale (Alejandra et al., 2018; Karimi et al., 2021; Liao & Jiang, 2020; Li et al., 2021; Shi et al., 2021; Xu et al., 2019; Zhu et al., 2020). Yet, composite indicator based ecological analyses is limited on a local scale, especially in the Asian cities which have undergone rapid urbanization during 1990–2010 and share similar urban development processes resembling the urban growth pattern (Arfanuzzaman & Dahiya, 2019; Chen et al., 2020).

Therefore, in the present research, Faridabad having more than 1.2 million urban population, located in the semi-arid climatic zone of India is considered as the pilot study area to monitor the changes in the ecological status from 1991 to 2021. Faridabad has emerged as a satellite city of Delhi urban agglomeration with transit-oriented polycentric development (Kumar, Ghosh et al., 2022). Along with Faridabad, Gurugram, Ghaziabad and Gautam Budh Nagar have also been developed as constituent areas of Delhi urban agglomeration and well connected through commuting infrastructure. Similar, transportation networks centric development of urban agglomeration has been found in the Mumbai Metropolitan region of India and other Asian cities viz. Tokyo, Seoul and Hong Kong (Shaw, 1995; Sit, 1996). During this rapid urbanisation, proper planning is somewhere neglected which resulted in significant degradation in ecostate of such urban agglomeration with substantial land use land cover change (LULCC), air pollution, environmental heating etc. (Anwar et al., 2020; Ghosh et al., 2021; Hashmi et al., 2021; Kumar, Ghosh, et al., 2022; S. Kumar, Midya et al., 2021; Ranjan et al., 2022; Sabir & Gorus, 2019; R. Sharma et al., 2020; V. Sharma et al., 2021; V. Sharma, Ghosh, Kumari, et al., 2022; V. Sharma, Ghosh, Singh, et al., 2022; V. Sharma et al., 2023). Even though to the best of our knowledge, only a handful studies have been carried out for ecological change detection using the remote sensing-based composite ecological indicator in the Asian urban agglomeration. Though the composite indicator-based ecological change detection is the popular and extensively used research method in the field of environmental degradation, its potential and accuracy vary from place to place (Marchand et al., 2014; Yigitcanlar & Dur, 2010). Consequently, in the present research, a composite indicator-based method has been used in Faridabad where it has never been tested before to quantify the ecological degradation.

Besides composite indicator, LST is also an extremely important indicator of urban ecological status. Multi-temporal in-situ measurement of LST over large area is costly, cumbersome. However, thermal bands of Landsat satellite images with good spatial resolution fill the gap and successfully used in earlier research worldwide. Such research highlight the impact of LULC type over LST and indicated interrelationship of NDVI, NDMI, NDSI, NDBI, FV, impervious surface area (ISA), with change in LST which can assist in the prediction of LST at different scale (Guha & Govil, 2021a, 2021b; S. Kumar, Midya et al., 2021; Pramanik et al., 2021; Pramanik & Punia, 2020; Xu et al., 2019; Zhu et al., 2020). Earlier research conducted in Faridabad stated that besides urban-industrial concrete in the city area, barren land, open area, brick-kiln in the rural area are also contributing to a higher LST in Faridabad (Kumar, Ghosh et al., 2022; S. Kumar, Midya et al., 2021) while higher LST has been reported in urban built-up only in different urbanized areas of India (2021b; Guha & Govil, 2021a; Pramanik & Punia, 2020; Pramanik et al., 2021). Moreover, S. Kumar, Midya et al. (2021) has highlighted the rural–urban thermal disparity with increasing urban density and the interrelationship between LST and soil moisture in Faridabad. In light of these results, the present research might add significantly to existing findings regarding the present study area by examining the correlation between spectral indices indicating ecological condition and LST.

Overall, authors identified a gap of adequate research on the comprehensive analysis of ecological change in response to rapid urbanization in Asian urban agglomerations located in semi-arid and arid areas, with a specific focus on Faridabad. To address this research gap, the authors undertook a comprehensive study using a composite ecological indicator that integrates four spectral indices: Soil Adjusted Vegetation Index (SAVI), Normalized Difference Moisture Index (NDMI), Normalized Difference Snow Index (NDSI), and Normalized Difference Built-Up Index (NDBI). By utilizing these indices, the authors were able to monitor and assess changes in the ecological status of Faridabad over a span of 30 years, from 1991 to 2021. Further, to achieve deeper insight into the impact of the changing urban landscape composition over thermal environment during the study period, zonal statistics have been calculated from spectral indices and LST map and regression analysis has been performed between SAVI, NDBI, NDMI, NDSI and LST for both the years. This research contributes to a better understanding of the ecological consequences of urbanization and provides valuable insights for sustainable urban planning and management in similar regions.

13.2 Methodology

For carrying out the present research, Faridabad district (28°10′50"N to 28°29′04"N latitude and 77°6′49"E to 77°33′23"E longitude) covering an area of 741 km^2 is located on the south-eastern part of the state of Haryana has been selected as the study area (Fig. 13.1). Total population as per 2011 census is 1,809,733 which includes 370,878 rural and 1,438,855 urban population. Faridabad is the most populated and industrialized city in Haryana as people are migrated from different parts of Northern India (Ghosh & Ghosh, 2017).

Physiographically, Faridabad can be divided into two zones- The undulating plain and the Yamuna Khadar region with the Yamuna River flowing for 45 kms on the eastern side of the district. It is a semi-arid region that doesn't receive much rainfall. Still, the agricultural practices are a success in this region due to precipitation and irrigation. However, being a part of Delhi NCR Faridabad has 80% of the population residing in the urban area and are shifting their occupation from primary to the secondary or tertiary sector. Further, Faridabad emerged as a top industrialized area. Its closeness to the national capital helped the industries to grow and attract many MNCs and foreign investors with local investors. Faridabad Small Industries Association has introduced the project for the industries of Faridabad (Agrawal, 2021; Rasul et al., 2015). Cheap labour is also available from the nearby areas which boost profits. Moreover, broad gauge railway, highways, metro facilities, cycle tracks, installation of intelligent transport systems have further facilitated the mobility of labours from adjacent regions. History of land use reveals markedly varied patterns, various anthropogenic activities in the form of rapid industrialization, excessive mining, development of brick kilns are associated with a series of environmental degradation viz; overgrazing, particle pollution, water pollution etc. (Kumar, Ghosh et al., 2022; S. Kumar, Midya et al., 2021; Rai & Saha, 2015; V. Sharma et al., 2021).

The present study will cover a period of three decades. Such temporal scope of the study spanning three decades provides valuable insights into the long-term trends and patterns of ecological change in Faridabad. Therefore, Landsat 5 Thematic Mapper (TM) image (30 m spatial resolution for optical bands; 120 m resampled into 30 m spatial resolution for the thermal band) acquired on 14 March 1991 (image ID: LT05_L1TP_146040_19910314_20170127_01_T1) and Landsat 8 Operational Land Imager (OLI) (30 m spatial resolution for multispectral bands; 100 m (resampled into 30 m) spatial resolution for the thermal band) image acquired on 16 March 2021 (image ID: LC08_L1TP_146040_20210316_20210328_01_T1) downloaded from the United States Geological Survey (USGS) server has been used in the present study. Both these images have been acquired in March and at a similar local time to

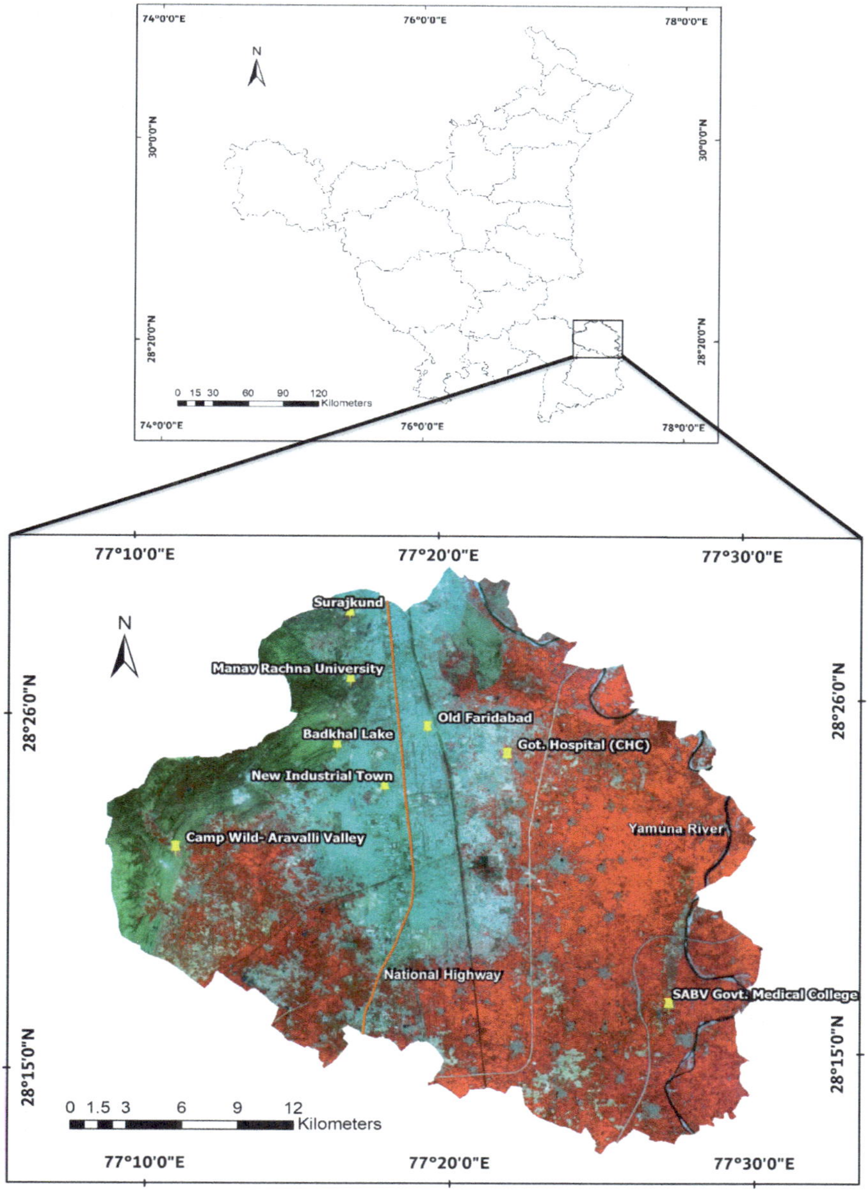

Fig. 13.1 Location map of the study area

nullify the effect of season on temperature. All the data sets are of Landsat Collection 1 Level-1 Tier 1 with standard terrain correction (L1TP) which includes radiometric correction and orthorectification with Universal Transverse Mercator (UTM) coordinate system, 43 North Zone and WGS 84 Datum useful for multitemporal analysis. By examining data from 1991 and 2021, we are able to analyze the cumulative effects of urbanization on the ecological balance. The chosen timeframe of 1991 and 2021 allows for a comprehensive analysis of the ecological changes and we can better understand the long-term impact of human activity on the natural ecology of the region.

For the calculation of temperature, thermal infrared (TIR) bands of images of 1991 and 2021 (band 6 of Landsat 5 TM; band 10 of Landsat 8 OLI) have been used. A detailed method of extracting LST from the thermal band has been explained in our previous publications (Ghosh et al., 2021; Kumar et al., 2019).

A composite ecological index has been calculated by combining SAVI (Huete, 1988), NDMI (Chandra et al., 2010), NDBI (Zha, 2003), NDSI [Calculated by combining bare soil index (BSI) and normalized difference impervious surface index (NDISI)] (Xu et al., 2012). SAVI, NDMI, NDBI and NDISI (representing greenness, wetness, dryness contributed by built-up and bare soil respectively) have been calculated using the following equations:

$$SAVI = \frac{1.5(NIR - RED)}{NIR + RED + 0.5} \tag{13.1}$$

$$NDMI = \frac{NIR - SWIR_2}{NIR + SWIR_2} \tag{13.2}$$

$$NDBI = \frac{SWIR_1 - NIR}{SWIR_1 + NIR} \tag{13.3}$$

$$BSI = \frac{(SWIR_1 + RED) - (BLUE + NIR)}{SWIR_1 + RED + BLUE + NIR} \tag{13.4}$$

$$NDISI = \frac{TIR - (MNDWI + NIR + SWIR_2)/3}{TIR + (MNDWI + NIR + SWIR_2)/3} \tag{13.5}$$

$$MNDWI = \frac{GREEN - SWIR_1}{GREEN + SWIR_2} \tag{13.6}$$

$$NDSI = \frac{BSI + NDISI}{2} \tag{13.7}$$

where,

$NIR \cong$ near-infrared band;
$RED \cong$ red band;
$SWIR_2 \cong$ middle-infrared band;
$SWIR_1 \cong$ short-wave near-infrared band;

$BLUE \cong$ blue band;
$GREEN \cong$ green band;
$TIR \cong$ thermal infrared band;

$$MNDWI \cong ModifiedNormalizedDifferenceWaterIndex$$

Most single indicator based ecological analysis use NDVI to quantify greenness, however, in the present study SAVI has been used as it is the favoured method to quantify the greenness of semiarid regions with thorny and discontinuous vegetation (Fatiha et al., 2013; Vani, 2017). NDMI is the best-suited spectral index for indirect determination of soil moisture in the drought-prone areas (Insan et al., 2021; Lin et al., 2010; Moazenzadeh et al., 2022). NDMI has been used in the present study as the soil water content is not adequate for growth of crops and plants in Faridabad. NDSI and NDBI are the standard indices to quantify the dryness caused by bare soil and built-up (Waqar & Hussain, 2012). For a holistic ecological change assessment all these spectral indices representing land surface biophysical variables (SAVI, NDMI, NDISI and NDBI) have been used to generate a composite ecological index.

Firstly, SAVI, NDMI, NDISI and NDBI have been stacked to make a four-band composite image. Further, taking the composite image as the input, principal component analysis (PCA) has been performed. PCA has been utilized as the factor weight allotment method. Unlike the other weighting method, individual understanding does not affect the weight assignment to the associated factors (in the present case SAVI, NDMI, NDISI and NDBI are considered as factors). Variance and factor loadings (the weight or importance of each spectral index in each principal component is factor loading) is calculated using the following equations (Jensen, 1986)

$$\%variance = \frac{eigenvalue\lambda_p \times 100}{\sum_{p=1}^{4} eigenvalue\lambda_p} \quad (13.8)$$

$$R_{kp} = \frac{a_{kp} \times \sqrt{\lambda_p}}{S_k} \quad (13.9)$$

where,

R_{kp} is factor loading;
a_{kp} is eigenvector for each factor and component p;
λ_p is the eigenvalue for the p'th component;
S_k is the standard deviation of each factor.

The factor loading of each spectral index to PC1 is considered as the importance of each ecological parameter to the composite index and thereby PC1 is considered as the composite ecological index

$$Compositeecologicalindex = f(SAVI, NDMI, NDSI, NDBI) \quad (13.10)$$

Such composite ecological index has been utilized to classify the study area based on the ecological condition. The high composite index indicated good ecological

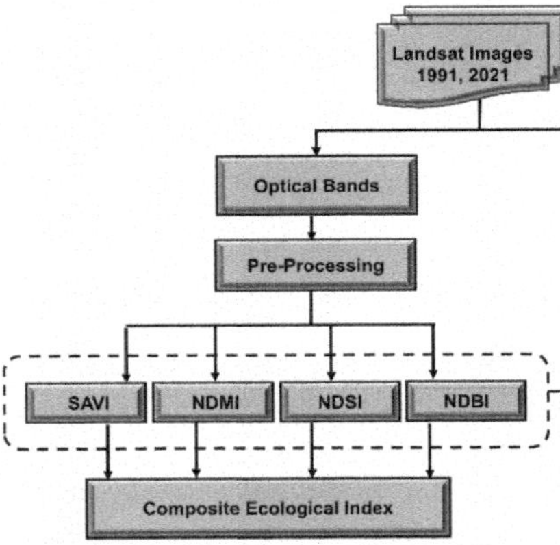

Fig. 13.2 Flowchart of the methodology

conditions and low values represent the poor ecological status. Further, the indices (SAVI, NDMI, NDISI and NDBI) which have been used to generate the composite ecological index has been correlated with LST. Multitemporal assessment of the indices represents change in several biophysical variables (greenness, wetness, and dryness) of the land surfaces which cause alteration in LST of the study area. Therefore, correlations analysis has been performed between SAVI, NDSI, NDMI and NDBI with LST for both years using linear regression by selecting the values of variables. The entire methodology has been described in Fig. 13.2.

13.3 Results

Spectral indices maps generated for both the years 1991 and 2021 have depicted the overall status of urban ecology in the study region (Fig. 13.3a, b, Fig. 13.4a, b, Fig. 13.5a, b, Fig. 13.6a, b). All indices values are limited from -0.74 to + 0.74. Lower SAVI and higher NDBI values indicate the expansion of built-up in the form of the residential, industrial and commercial areas especially in the northern, eastern, central, southern and south-western directions. In the western direction, the Aravalli region covered with degraded forest, open scrub and in the eastern corner of Faridabad riverine sand along Yamuna river, has SAVI value of 0.1 to 0.2. Besides the concrete built-up in the city area, barren land, open land covered with sandy soil has lesser moisture content and therefore lower NDMI. From 1991 to 2021, natural land covers and open spaces along the national highway, railway, the metro railway is transformed into buildings resulting in the expansion of the Faridabad city which cause the reduction in the NDMI on the eastern side of the canal. NDSI values are higher for the built-up, roadways highways in the city area, as well as it is higher in the open spaces in the rural–urban fringe, around the brick kiln in the rural zone, rocky Aravalli region.

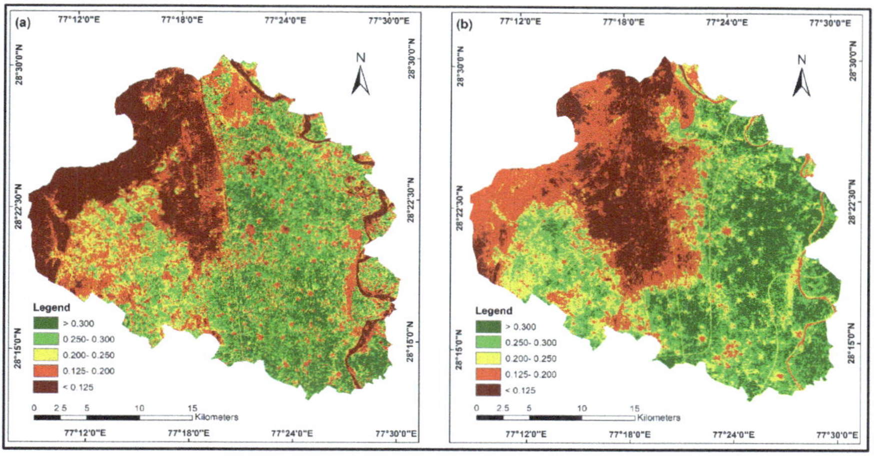

Fig. 13.3 SAVI map of (a) 1991 and (b) 2021

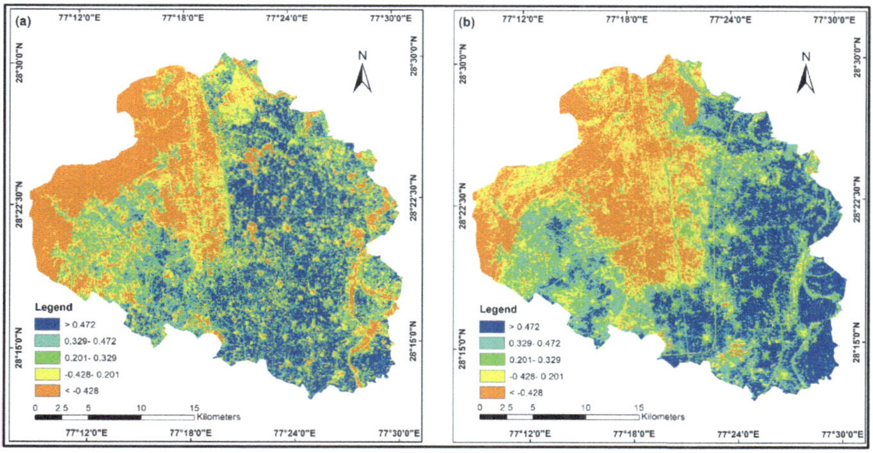

Fig. 13.4 NDMI map of (a) 1991 and (b) 2021

A composite ecological index has been formed by processing the statistics of PCs generated by the PC analysis of the integrated spectral indices image. Eigenvectors, eigenvalues, percentage of data covered in each of the four principal components, and factor loading of each index representing ecological indicators are displayed in Table 13.1–13.2. PC1 having the higher eigenvalues compared to PC2, PC3 and PC4 with a variance of above 95%, indicated that PC1 contains maximum information of the four indicators. Therefore, PC1 is represented as the composite ecological index. The contribution of each indicator to the composite ecological index is assessed by its loading to PC1 (Table 13.2). Based on the (+)-sign with SAVI and NDMI and (-)

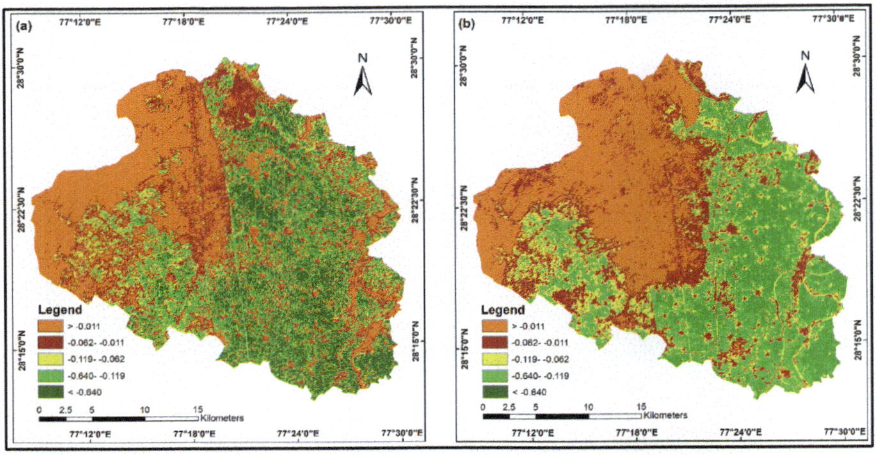

Fig. 13.5 NDSI map of (a) 1991 and (b) 2021

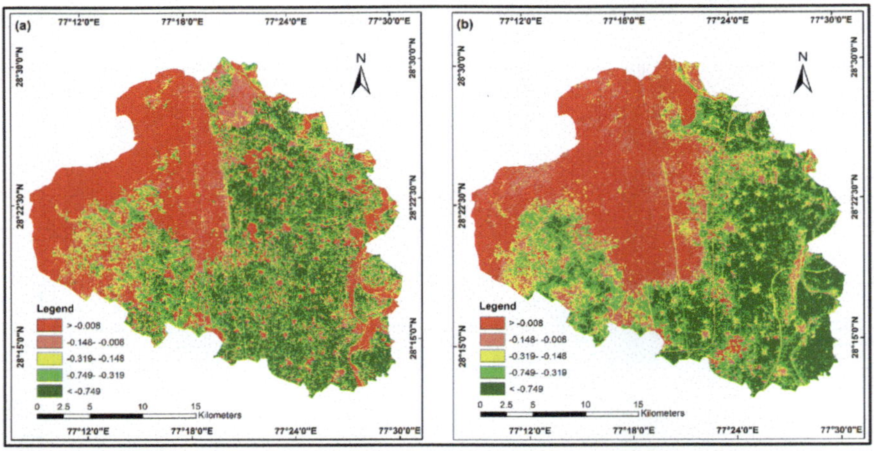

Fig. 13.6 NDBI map of (a) 1991 and (b) 2021

-sign with NDSI and NDBI, it can be remarked that these indicators contributed in opposite ways to the ecological status of the study region (Xu et al., 2019). Based on the composite ecological index, the ecological status of different parts of the study region has been categorized as very poor (< 0.2), poor (0.2–0.4), normal (0.4–0.6), good (0.6–0.8), and excellent (> 0.8) respectively (Fig. 13.7). A high composite ecological index indicates the most acceptable ecological environments having a higher degree of greenness, good soil moisture condition and tolerable dryness of the land.

Table 13.1 Eigenvector matrix, Eigenvalues and percentage of scene variance by each PC

	1991			
	PC1	PC2	PC3	PC4
SAVI	0.291	-0.883	-0.365	0.303
NDMI	0.586	0.031	0.393	0.609
NDSI	-0.475	-0.465	0.746	-0.493
NDBI	-0.586	-0.031	-0.393	-0.609
Eigen values	0.140	0.004	0.000	0.000
Sum of eigenvalues	0.144			
Percentage variance	97.087	2.773	0.138	0.000
	2021			
	PC1	PC2	PC3	PC4
SAVI	0.550	-0.549	-0.561	-0.284
NDMI	0.536	-0.343	0.704	0.312
NDSI	-0.310	-0.335	0.420	-0.783
NDBI	-0.559	-0.683	-0.110	0.455
Eigen values	0.049	0.002	0.000	0.000
Sum of eigenvalues	0.051			
Percentage variance	95.640	3.903	0.390	0.058

Table 13.2 Factor loadings of four ecological parameters to the four principal components

	1991			
	PC1	PC2	PC3	PC4
SAVI	0.871	-0.447	-0.041	0.000
NDMI	0.990	0.009	0.025	0.000
NDSI	-0.982	-0.162	0.058	0.000
NDBI	-0.990	-0.009	-0.025	0.000
	2021			
	PC1	PC2	PC3	PC4
SAVI	0.966	-0.195	-0.063	-0.012
NDMI	0.973	-0.126	0.082	0.014
NDSI	-0.953	-0.208	0.082	-0.060
NDBI	-0.952	-0.235	-0.012	0.019

Figure 13.8 (a-b) exhibits calculated LST map for the year 1991 (with LST ranging from 16.8 to 27.9 °C) and 2021 (with LST ranging from 20.4 °C to 31.2 °C) respectively. Comparative analysis shows a noteworthy change in the LST distribution pattern, in the study area, from 1991 to 2021. In 1991 eastern part has been identified as a cooler zone and the western zone.

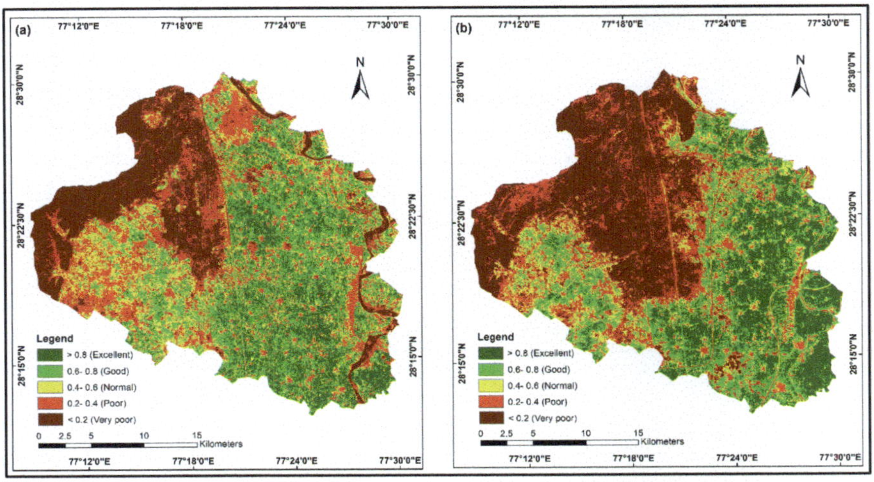

Fig. 13.7 Composite ecological index of (a) 1991 and (b) 2021

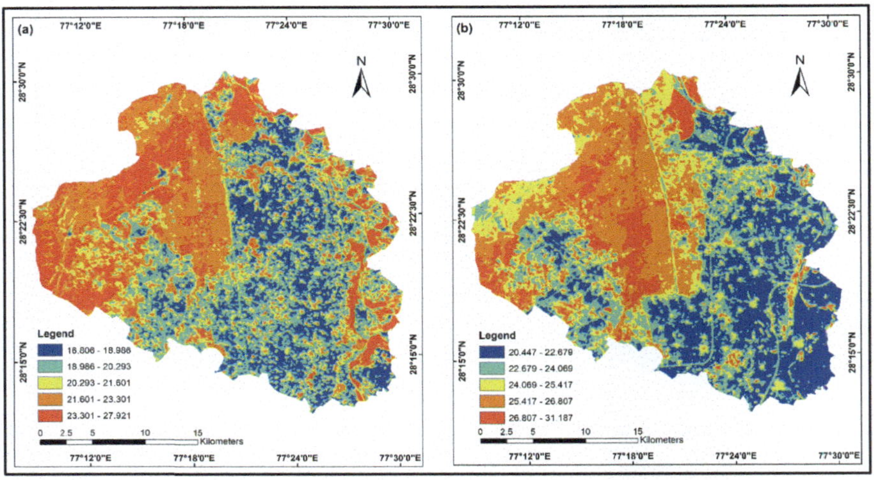

Fig. 13.8 LST map of (a) 1991 and (b) 2021

with higher average LST. On the contrary, in 2021 regions with lower LST are concentrated mostly in the eastern zone whereas the western, northern, and central regions exhibit higher temperatures ranging from 24 °C to 31 °C. Such modified LST distribution with higher LST (in 2021) at the central zone supports the evidence of ecological deterioration in that zone, as stated earlier.

In the regression analysis between mean SAVI, mean NDMI, mean NDSI, mean NDBI and mean LST, negative value of coefficients of SAVI and NDMI, has

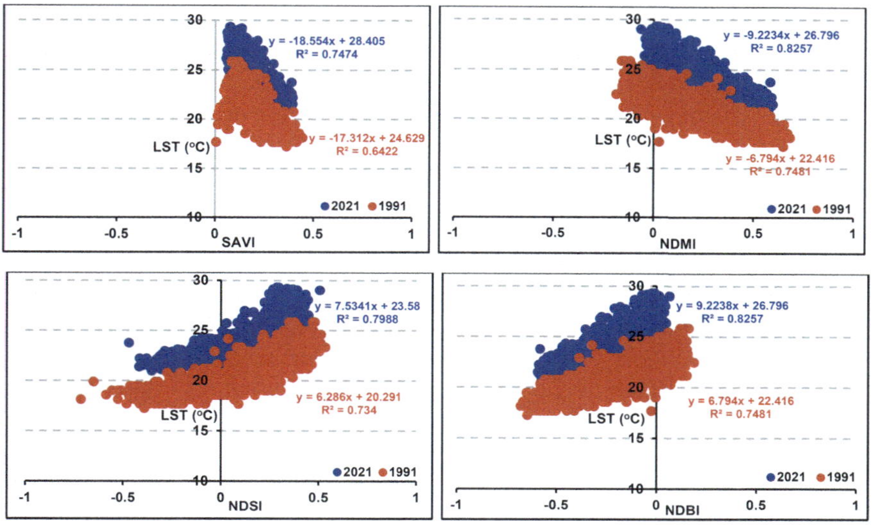

Fig. 13.9 Correlation between spectral indices and LST

explained the decrease in LST with increase in SAVI and NDMI, while positive value of coefficients of NDSI and NDBI depicted increase in LST with increase in NDSI and NDBI. For both 1991 and 2021, a strong positive correlation has been observed between LST with NDSI ($R^2 = 0.7$ in 1991 and $R^2 = 0.79$ in 2021) and NDBI ($R^2 = 0.74$ in 1991 and $R^2 = 0.82$ in 2021). While negative correlation has been identified between LST and SAVI ($R^2 = 0.64$ in 1991 and $R^2 = 0.74$ in 2021) and NDMI ($R^2 = 0.74$ in 1991 and $R^2 = 0.82$ in 2021) (Fig. 13.9).

13.4 Discussion

Identification of areas with poor ecological status is crucial for effective planning of sustainable developmental programs, particularly for rapidly growing regions like Faridabad. Present study identified 30% of the present study area with poor ecological status located in and around the northern and central parts, around Old Faridabad due to the high density built-up in the form of residential buildings, industrial buildings, power plants and increased anthropogenic activities. What is concerning is that this poor ecological status is not limited to Old Faridabad alone. It continues to spread towards the southern and southwestern directions, particularly towards Ballabhgarh and the New Industrial Town (NIT). These areas have emerged as newer urban centers due to the polycentric urban structure of Faridabad city, which revolves around Old Faridabad, Ballabhgarh, and NIT (Kumar, Ghosh, et al., 2022). However, the ecological condition is still better around the newer urban centres than the northern and central parts, as the built-up around Ballabhgarh and NIT are interspaced with green vegetation. In the eastern direction, wide distribution of agricultural land with scattered rural built-up maintain the ecological status from excellent to moderate category. Due to the emergence of Greater Faridabad as the expansion of the existing urban area of Faridabad towards the eastern side of national highway leads to the deterioration of the ecological status (Kumar, Ghosh et al., 2022; S. Kumar, Midya et al., 2021).

However, the western part of the district, despite being covered by the Aravalli range, has shown a minor improvement in 2021. Such improvement might be resulted from imposing the ban on mining in this range (Dhanwantri et al., 2021). The surrounding region of Yamuna River on the eastern boundary of the district is also showing an improved ecological status in 2021 in comparison to 1991. The probable reason might be the Yamuna River protection programmes which focus on treating the Yamuna and preserving the banks (Tribunal, 2013). Such river rejuvenating programmes can be initiated in other regions to improve the ecological status. Reduced pollution due to repeated lockdown in the covid scenario and consequent improvement of water quality could be another potential reason.

Due to the extension of Faridabad city boundary and consequent development of Greater Faridabad higher LST has been recorded. Improvement in ecological status around Yamuna River is associated with lowering in LST in the south-eastern zone. Degradation of the forest cover in the Aravalli region causes higher temperature despite minor ecological improvement due to the prohibition of mining in the western zone. Further, fallow land and open spaces occupied for future development in the rural area are responsible for higher LST. Similar observations have been reported in neighbouring areas of present study area (Chakraborti et al., 2019; Guha & Govil, 2021a; S. Kumar et al., 2019; Pramanik & Punia, 2019). Fallow and open space trigger the LST because of the low soil moisture content and reduced evapotranspiration in the semi-arid zone (S. Kumar, Midya et al., 2021; Son et al., 2012).

As average LST increases with the years, consequently a stronger relationship has been observed in 2021 compared to 2019. However, a different trend has been reported in other parts of India (Guha & Govil, 2021b). Further, in the linear regression equations (Fig. 13.9), coefficient values (-17 and -16 in 1991 and 2001 respectively) of SAVI is highest amongst all the indices. It indicates that vegetation is the factor which influence the LST most. With per unit decrease in vegetation cover, it would result in the most significant increase in LST. Higher SAVI is associated with a dense healthy green cover facilitates a high rate of evapotranspiration and lower LST (S. Kumar et al., 2019; Yue et al., 2007). It is also well reported that LST and green cover indices have a stronger relationship in the daytime compared to nighttime (Sun & Kafatos, 2007). However, it can also be remarked that due to long stretch of homogeneous agricultural region, SAVI and LST maintain strong correlation. While, a moderate correlation has been found between NDBI and LST, as built-up area is always heterogenous and interspaced with urban green cover in places (Guha & Govil, 2021b).

By considering the findings of present assessment, urban planners can make informed decisions and implement measures that promote sustainability and resilience in the face of climate change. Combination of high-rise buildings with appropriate vegetation, implementing sustainable infrastructure, and preserving open spaces ensure a sustainable and resilient future for urban areas.

13.5 Conclusion

Despite the limited data availability and the challenges posed by the dearth of a composite index, the present study successfully assessed the ecological status of Faridabad district over a period of thirty years using a composite ecological index combining several spectral indices serving as relevant ecological indicators. The temporal scope of the study, spanning three decades, provides valuable insights into the long-term trends and patterns of ecological change in Faridabad. Strong correlations between ecological change with the increasing concretization and change in LST over the time frame has indicated the need proper balance of built-up, vegetation and waterbodies in the study area as smart growth strategy. The poor ecological status in the northern and central part of the district can be attributed to the dense residential and industrial built-up. However, the rural area in the eastern part of the district offers a better ecological situation. To improve the ecological status in the region, it is essential to manage fallow land with plantations and consider green roofs for future industrial or residential buildings. The river rejuvenation initiatives undertaken in the Yamuna River protection programs and the ban on mining in the Aravalli region have brought about significant improvements, to the ecological status of these areas. The success of these projects reinforces the notion that targeted efforts and government-driven actions can yield tangible benefits in improving the overall environmental health and sustainability of a region.

Although the present research study focuses on a specific region, its findings carry global significance. The implications of these findings extend beyond the boundaries of the specific region and can contribute to the planning and development of satellite cities in the vicinity of Delhi. Gurugram, Ghaziabad, and Noida are among the satellite cities that have grown alongside Faridabad, sharing similar patterns of urban-industrial growth and associated ecological degradation. Therefore, the methodology employed in this research can be replicated in these cities as well as in other Indian and Asian cities that have undergone similar spatiotemporal urbanization as part of an urban agglomeration. Moreover, the findings of this research on the existing million-plus city can be valuable for anticipating and addressing future ecological scenarios in the incipient million-plus city. By understanding the ecological challenges faced by the current city, appropriate conservation measures can be planned and implemented in the future city to mitigate potential ecological degradation.

By implementing sustainable urban planning and management strategies, it is possible to strike a balance between urban development and ecological conservation. This would facilitate the restoration and preservation of the ecological health in the identified areas with poor ecological status. It is essential to prioritize environmental protection and incorporate ecological considerations into future urban development plans to ensure a sustainable and harmonious coexistence of human activities and the natural environment.

13.5.1 Disclosure Statement

No potential conflict of interest was reported by the authors.

References

Abutaleb, K., Freddy Mudede, M., Nkongolo, N., & Newete, S. W. (2021). Estimating urban greenness index using remote sensing data: A case study of an affluent vs poor suburbs in the city of Johannesburg. *Egyptian Journal of Remote Sensing and Space Science, 24*(3), 343–351. https://doi.org/10.1016/j.ejrs.2020.07.002

Agrawal, B. (2021). Case study : An effort of Faridabad entrepreneurs for energy case study—An effort of Faridabad entrepreneurs for energy efficiency. January 2016. https://doi.org/10.51976/gla.prastuti.v5i2.521607

Alejandra, M., Alberto, D., Marino, L., & Rodriguez, S. (2018). Urban environmental quality assessment using remote sensing and census data. *International Journal of Applied Earth Observation and Geoinformation, 71*(October 2017), 95–108. https://doi.org/10.1016/j.jag.2018.05.010

Anwar, A., Younis, M., & Ullah, I. (2020). Impact of urbanization and economic growth on CO_2 emission: A case of far east Asian countries. *International Journal of Environmental Research and Public Health, 17*(7). https://doi.org/10.3390/ijerph17072531

Arfanuzzaman, M., & Dahiya, B. (2019). Sustainable urbanization in Southeast Asia and beyond: Challenges of population growth, land use change, and environmental health. *Growth and Change, 50*(2), 725–744. https://doi.org/10.1111/grow.12297

Authors, F. O. R., Up, S., Alerts, F. O. R., & Organizers, F. O. R. (n.d.). Analysis of soil moisture as an indicator of land quality using vegetation index (SAVI and NDMI) retrieved from remote sensing data in Jember—Indonesia. March 2022.

Badlani, B., Patel, A. N., Patel, K., & Kalubarme, M. H. (2017). Urban growth monitoring using remote sensing and geo-informatics : Case study of Gandhinagar, Gujarat State (India). *International Journal of Geosciences, 08*, 563–576. https://doi.org/10.4236/ijg.2017.84030

Chakraborti, S., Banerjee, A., Sannigrahi, S., Maiti, A., & Jha, S. (2019). Assessing the dynamic relationship among land use pattern and land surface temperature : A spatial regression approach. Asian Geographer, 36(2), 5706. https://doi.org/10.1080/10225706.2019.1623054

Chandra, P. R., Nidhi, P., Stutee, C. S. R., Rajan, G. K. S., & Dutt, S. H. R. C. B. S. (2010). Assessing forest canopy closure in a geospatial medium to address management concerns for tropical Islands—Southeast Asia. *Environmental Monitoring and Assessment, 16*, 541–553. https://doi.org/10.1007/s10661-008-0717-4

Chen, T. L., Chiu, H. W., & Lin, Y. F. (2020). How do East and Southeast Asian cities differ from western cities? A systematic review of the urban form characteristics. *Sustainability (Switzerland), 12*(6). https://doi.org/10.3390/su12062423

Dhanwantri, K.,Dhote, M., Yadav, K. K., Rajendra, K. (2021). Chapter 30—Appraisal of ecosystem services and forest resource conservation in a tropical dry deciduous forest of Aravallis, India. *Modern Cartography Series, 10*, 659–678.

da Silva, R. M., Santos, C. A. G., Maranhão, K. U. de A., Silva, A. M., & de Lima, V. R. P. (2018). Geospatial assessment of eco-environmental changes in desertification area of the Brazilian semi-arid region. *Earth Sciences Research Journal,22*(3), 175–186. https://doi.org/10.15446/esrj.v22n3.69904

Fatiha, B., Abdelkader, A., & Latifa, H. (2013). Spatio temporal analysis of vegetation by vegetation indices from multi-dates satellite images: Application to a Semi-arid area in ALGERIA. *Energy Procedia, 36*, 667–675. https://doi.org/10.1016/j.egypro.2013.07.077

Ghosh, S., & Ghosh, S. (2017). Evaluating patterns of urban growth in Faridabad. sub-region of NCR, India using remote sensing, GIS and entropy approach. *In: Indian Cartographer, 37*, 258–261.

Ghosh, S., N., K. V., Kumar, S., & Midya, K. (2021). Seasonal contrast of land surface temperature in Faridabad: An urbanized district of Haryana, India. In *Methods and applications of geospatial technology in sustainable urbanism* (pp. 217–250). IGI Global. https://doi.org/10.4018/978-1-7998-2249-3.ch008

Guha, S., & Govil, H. (2021a). A long-term monthly analytical study on the relationship of LST with normalized difference spectral indices. *European Journal of Remote Sensing, 54*(1), 487–512. https://doi.org/10.1080/22797254.2021.1965496

Guha, S., & Govil, H. (2021b). Annual assessment on the relationship between land surface temperature and six remote sensing indices using landsat data from 1988 to 2019. *Geocarto International, 0*(0), 1–20. https://doi.org/10.1080/10106049.2021.1886339

Hashmi, S. H., Fan, H., Habib, Y., & Riaz, A. (2021). Non-linear relationship between urbanization paths and CO_2 emissions: A case of South, South-East and East Asian economies. *Urban Climate, 37*(February), 100814. https://doi.org/10.1016/j.uclim.2021.100814

Huete, A. R. (1988). A soil-adjusted vegetation index (SAVI). *Remote Sensing of Environment, 25*(3), 295–309.

Insan, K., Rahmi, N., & Dimyati, M. (2021). Remote sensing and GIS application for monitoring drought vulnerability in Indonesia : A review. *Bulletin of Electrical Engineering and Informatics, 10*(6), 3507–3518. https://doi.org/10.11591/eei.v10i6.3249

Jensen. (1986). *DIP*.

Karimi, M., Fathololoumi, S., Kiavarz, M., & Biswas, A. (2021). Land Surface Ecological Status Composition Index (LSESCI): A novel remote sensing-based technique for modeling land surface ecological status. *Ecological Indicators, 123*(August 2020), 107375. https://doi.org/10.1016/j.ecolind.2021.107375

Kerr, J. T., & Ostrovsky, M. (2003). From space to species : Ecological applications for remote sensing. *Trends in Ecology & Evolution, 18*(6), 299–305. https://doi.org/10.1016/S0169-5347(03)00071-5

Kumar, S., Ghosh, S., Hooda, R. S., & Singh, S. (2019). Monitoring and prediction of land use land cover changes and its impact on land surface temperature in the central part of hisar district, Haryana under semi-arid zone of India. *Journal of Landscape Ecology(Czech Republic), 12*(3), 117–140. https://doi.org/10.2478/jlecol-2019-0020

Kumar, S., Ghosh, S., & Singh, S. (2022). Polycentric urban growth and identification of urban hot spots in Faridabad, the million—plus metropolitan city of Haryana, India : A zonal assessment using spatial metrics. In *Environment, Development and Sustainability* (Issue 0123456789). Springer Netherlands. https://doi.org/10.1007/s10668-021-01782-6

Kumar, S., Midya, K., Ghosh, S., & Singh, S. (2021). Pixel-based vs. object-based anthropogenic impervious surface detection: Driver for urban-rural thermal disparity in Faridabad, Haryana, India. *Geocarto International, 0*(0), 1–23. https://doi.org/10.1080/10106049.2021.2002429

Kumar, V., Kumar, P., Murugan, P., Annadurai, M., & Red, G. (2015). Assessment of surface water dynamicsin Bangalore using WRI, NDWI, MNDWI, supervised classification and K-T. *Aquatic Procedia, 4*(Icwrcoe), 739–746. https://doi.org/10.1016/j.aqpro.2015.02.095

Levin, S. A., & Levin, S. A. (1992). The problem of pattern and scale in ecology. *Ecology, 73*(August 1989), 1943–1967.

Liao, W., & Jiang, W. (2020). Evaluation of the spatiotemporal variations in the eco-environmental quality in China based on the remote sensing ecological index. *Remote Sensing, 12*(15), 2462.

Li, Y., Wu, L., Han, Q., Wang, X., Zou, T., & Fan, C. (2021). Estimation of remote sensing based ecological index along the grand canal based on PCA-AHP-TOPSIS methodology. *Ecological Indicators, 122*, 107214. https://doi.org/10.1016/j.ecolind.2020.107214

Li, Y., Liu, W., Feng, Q., Zhu, M., Yang, L., & Zhang, J. (2022). Quantitative assessment for the spatiotemporal changes of ecosystem services, tradeoff-synergy relationships and drivers in the semi-arid regions of China. *Remote Sensing, 14*(1), 1–28. https://doi.org/10.3390/rs14010239

Lin, M., Wang, Q., Sun, F., Chu, T., & Shiu, Y. (2010). Quick Spatial Assessment of Drought Information Derived from MODIS Imagery Using Amplitude Analysis. *International Journal of Geological and Environmental Engineering, 4*(7), 271–275.

Marchand, F., Debruyne, L., Triste, L., Gerrard, C., Padel, S., & Lauwers, L. (2014). Key characteristics for tool choice in indicator-based sustainability assessment at farm level. *Ecology and Society, 19*(3). https://doi.org/10.5751/ES-06876-190346

Moazenzadeh, R., Mohammadi, B., Jafar, M., & Safari, S. (2022). Mechanics Soil moisture estimation using novel bio-inspired soft computing approaches. *Engineering Applications of Computational Fluid Mechanics, 16*(1), 826–840. https://doi.org/10.1080/19942060.2022.2037467

Mohammad, P., Goswami, A., & Bonafoni, S. (2019). The impact of the land cover dynamics on surface urban heat island variations in semi-arid cities: A case study in Ahmedabad City, India, using multi-sensor/source data. *Sensors (Switzerland), 19*(17). https://doi.org/10.3390/s19173701

Nejad, M. F., & Zoratipour, A. (2019). Assessment of LST and NDMI indices using MODIS and landsat images in Karun riparian forest. *Journal of Forest Science, 65*(1), 27–32. https://doi.org/10.17221/121/2018-JFS

Pramanik, S., Areendran, G., Punia, M., & Sahoo, S. (2021). Spatio-temporal pattern of urban eco-environmental quality of Indian megacities using geo-spatial techniques. *Geocarto International, 0*(0), 1–24. https://doi.org/10.1080/10106049.2021.1903578

Pramanik, S., & Punia, M. (2019). intensity : Source—Sink landscape—Based study in Delhi, India. *Environment, Development and Sustainability*, 0123456789. https://doi.org/10.1007/s10668-019-00515-0

Pramanik, S., & Punia, M. (2020). Land use/land cover change and surface urban heat island intensity: Source–sink landscape-based study in Delhi, India. *Environment, Development and Sustainability, 22*(8), 7331–7356. https://doi.org/10.1007/s10668-019-00515-0

Rai, S. C., & Saha, A. K. (2015). Impact of urban sprawl on groundwater quality: A case study of Faridabad city. *National Capital Region of Delhi.* https://doi.org/10.1007/s12517-015-1811-x

Ranjan, K., Sharma, V., & Ghosh, S. (2022). Assessment of urban growth and variation of aerosol optical depth in Faridabad district, Haryana, India. *Pollution, 8*(2), 447–461. https://doi.org/10.22059/POLL.2021.329185.1163

Rasul, A., Balzter, H., & Smith, C. (2015). Spatial variation of the daytime surface urban cool Island during the dry season in Erbil, Iraqi Kurdistan, from landsat 8. *Urban Climate, 14*(August 2016), 176–186. https://doi.org/10.1016/j.uclim.2015.09.001

Rasul, A., Balzter, H., & Smith, C. (2016). Diurnal and seasonal variation of surface urban cool and heat Islands in the semi-arid city of Erbil, Iraq. *Climate, 4*(3). https://doi.org/10.3390/cli4030042

Sabir, S., & Gorus, M. S. (2019). The impact of globalization on ecological footprint: Empirical evidence from the South Asian countries. *Environmental Science and Pollution Research, 26*(32), 33387–33398. https://doi.org/10.1007/s11356-019-06458-3

Sharma, R., Sinha, A., & Kautish, P. (2020). Examining the impacts of economic and demographic aspects on the ecological footprint in South and Southeast Asian countries. *Environmental Science and Pollution Research, 27*(29), 36970–36982. https://doi.org/10.1007/s11356-020-09659-3

Sharma, R., & Joshi, P. K. (2016). Mapping environmental impacts of rapid urbanization in the National Capital Region of India using remote sensing inputs. *Urban Climate, 15*, 70–82. https://doi.org/10.1016/j.uclim.2016.01.004

Sharma, V., Ghosh, S., Dey, S., & Singh, S. (2023). Modelling PM2.5 for data-scarce zone of Northwestern India using multi linear regression and random forest approaches. *Annals of GIS, 29*(3), 415–427. https://doi.org/10.1080/19475683.2023.2183523

Sharma, V., Ghosh, S., Kumari, M., Taloor, A. K., Singh, S., Arola, A., & Devara, P. C. S. (2022). Analysis and variation of the maiac aerosol optical depth in underexplored urbanized area of National Capital Region, India. *Journal of Landscape Ecology(Czech Republic), 15*(3), 82–101. https://doi.org/10.2478/jlecol-2022-0019

Sharma, V., Ghosh, S., Singh, S., Vishwakarma, D. K., Al-Ansari, N., Tiwari, R. K., & Kuriqi, A. (2022). Spatial variation and relation of aerosol optical depth with LULC and spectral indices. *Atmosphere, 13*(12), 1992. https://doi.org/10.3390/atmos13121992

Sharma, V., Ghosh, S., Bilal, M., Dey, S., & Singh, S. (2021). Performance of MODIS C6.1 dark target and deep blue aerosol products in Delhi National Capital Region, India: Application for aerosol studies. *Atmospheric Pollution Research, 12*(3), 65–74. https://doi.org/10.1016/j.apr.2021.01.023

Shaw, A. (1995). Satellite town development in Asia: The case of new Bombay, India. *Urban Geography, 16*(3), 254–271. https://doi.org/10.2747/0272-3638.16.3.254

Shi, H., Shi, T., Liu, Q., & Wang, Z. (2021). Ecological vulnerability of tourism scenic spots : Based on remote sensing ecological index. *Polish Journal of Environmental Studies, 30*(4), 3231–3248. https://doi.org/10.15244/pjoes/129916

Sit, V. F. S. (1996). Mega-city, extended metropolitan region, Desakota, and exo-urbanization: An introduction. *Asian Geographer, 15*(1–2), 1–14. https://doi.org/10.1080/10225706.1996.9684009

Son, N. T., Chen, C. F., Chen, C. R., Chang, L. Y., & Minh, V. Q. (2012). Monitoring agricultural drought in the lower Mekong basin using MODIS NDVI and land surface temperature data. *International Journal of Applied Earth Observations and Geoinformation, 18*(June 2006), 417–427. https://doi.org/10.1016/j.jag.2012.03.014

Sun, D., & Kafatos, M. (2007). Note on the NDVI-LST relationship and the use of temperature-related drought indices over North America. *Geophysical Research Letters, 34*(24), 1–4. https://doi.org/10.1029/2007GL031485

Tribunal, N. G. (2013). Restoration and conservation of river yamuna final report. 06.

Vani, V. (2017). Comparative study of NDVI and SAVI vegetation indices in anantapur district semi-arid areas. *International Journal of Civil Engineering and Technology, 8*(4), 559–566.

Waqar, M. M., & Hussain, E. (2012, January). Development of new indices for extraction of built-up area and bare soil from landsat. *Open Acess Scientific Reports, 1*(1).

Wei, J., Huang, W., Li, Z., Xue, W., Peng, Y., Sun, L., & Cribb, M. (2019). Estimating 1-km-resolution PM2.5 concentrations across China using the space-time random forest approach. *Remote Sensing of Environment, 231*(April), 111221. https://doi.org/10.1016/j.rse.2019.111221

Xu, H., Lin, D., & Tang, F. (2012). The impact of impervious surface development on land surface temperature in a subtropical city: Xiamen. *China*. https://doi.org/10.1002/joc.3554

Xu, H., Wang, Y., Guan, H., & Shi, T. (2019). Detecting ecological changes with a remote sensing based ecological index (RSEI) produced time series and change vector analysis. *Remote Sensing, 11*(20)1–24.

Yigitcanlar, T., & Dur, F. (2010). Developing a sustainability assessment model: The sustainable infrastructure, Land-use, environment and transport model. *Sustainability, 2*(1), 321–340. https://doi.org/10.3390/su2010321

Yue, W., Xu, J., Tan, W., & Xu, L. (2007). The relationship between land surface temperature and NDVI with remote sensing: Application to Shanghai landsat 7 ETM+ data. *International Journal of Remote Sensing, 28*(15), 3205–3226. https://doi.org/10.1080/01431160500306906

Zha, Y. (2003). Use of normalized di ff erence built-up index in automatically mapping urban areas from TM imagery. *International Journal of Remote Sensing, 24*(3), 583–594.

Zhu, X., Wang, X., Yan, D., Liu, Z., & Zhou, Y. (2020). Analysis of remotely-sensed ecological indexes' influence on urban thermal environment dynamic using an integrated ecological index: A case stud of Xi' an, China. *International Journal of Remote Sensing, 40*(9), 3421–3447. https://doi.org/10.1080/01431161.2018.1547448

Chapter 14
Bioenergy and Value Added Products from Biowaste and Its Role in Circular Economy

Swapnil Gaur, Nandini Dhyani, and Susheel Kumar Singh

14.1 Introduction

Reducing the amount of solid biowaste and limiting reliance on fossil fuel resources are two of the most important issues facing modern society. Lignocellulosic biomasses, which are biowaste and include mash and paper, horticulture, nourishment, forestry, and municipal solid waste, are thought to be the most capable renewable resources because they are widely available and help us avoid the conflicts that come with using consumable crops (Rai et al., 2021). A large amount of municipal waste is made up of biowaste, especially in rural and underdeveloped nations (Leow et al., 2018). According to Bhuvaneshwari et al. (2019), biowaste is often composed of decomposable plants, industrial biowaste, food-related biowaste from family cafeterias, and waste from botanical gardens. Furthermore, people are being encouraged to recycle waste and consume more efficiently by the era of growing rubbish, which is expected to double between now and 2030 (Ochieng et al., 2022). The global shift to post-carbon fossil social regimes may be facilitated by advancements and improvements in cleaner biomass production and processing (Mofijur et al., 2020a). Because of this, scientists have recently focused their attention on creating bioproducts and bioenergy from an affordable renewable resource obtained from biowaste from agriculture and industry (Muhammad et al., 2021; Sharma et al., 2023). Researchers are interested in using microalgae as a substitute feedstock for producing different types of bioenergy, including biodiesel, biohydrogen, and bioethanol (Lee et al.,

S. Gaur · N. Dhyani
School of Commerce, Graphic Era Hill University, Dehradun, UK, India

S. K. Singh (✉)
Department of Physics, DSMNRU, Lucknow, UP, India
e-mail: susheelsingh489@gmail.com

N. Dhyani
Graduate School of Business, Tula's Dehradun, Dehradun, India

P. K. Rai and S. Rai (eds.), *Climate, Environment and Agricultural Development*,
Advances in Geographical and Environmental Sciences,
https://doi.org/10.1007/978-981-97-8363-2_14

2019). Beyond that, developing a low-carbon, resource-efficient, and sustainable culture depends on the shift to a circular economy. For instance, a circular economy maximises asset effectiveness in terms of usage and output by reusing, reprocessing, repairing, and refurbishing assets (Alhawari et al., 2021; Stephenson & Damerell, 2022).

Biological products have emerged as a significant player in the development of a circular modern economy, being one of the biggest producers and consumers of environmental resources in both industrialised and developing nations (Belmonte-Urena et al., 2021). The development of biological products has positive effects on the environment and the economy. For example, it can replace biomaterials made from fossil fuels or chemicals, which is essential to the circular economy idea (Perea-Moreno et al., 2019). The development of the circular economy has put a priority on resource utilisation and the formation of biowaste during a bioproduct's whole life cycle, as well as on product life cycles, design, and generation strategies. The goals of resource efficiency and maintainability are complementary for the circular economy and the bioeconomy frameworks. In order to maintain asset viability and biomass sustainability, sustainable management of organic resources is a key component of global bioeconomy strategies. In addition to improved sustainability measures, the creation of biorefineries will be necessary for a circular bioeconomy (Barañano et al., 2021). A simple method for evaluating the effects of energy use, toxic chemical emissions, and natural resource use at every stage of a product or activity's life cycle is the life cycle assessment. Applying environmental sustainability to biowaste management agreements might help characterize the policies and predict future developments.

A sustainable method of managing degradable biowaste (such as that from forestry residues, cattle and agricultural waste, food waste, kitchen waste, green waste, seaweed, algae, sewage sludge, agro industries, and other organic industrial waste) is to convert biomass into biological products and biofuels (Uddin et al., 2021). Coordination of biorefineries is the backbone of the circular economy; these facilities use all of the waste components from biomass to produce energy, chemicals, and life. Conversely, processed solid and liquid waste streams, like pomace, peels, seeds, and byproduct streams, are rich in biomass and include a range of essential elements, including minerals, proteins, and carbohydrates (Zamri et al., 2021). According to Rai et al. (2020), these wastes have the potential to be an excellent and reasonably priced raw material for the production of value-added products with a wide range of industrial and medicinal applications, such as enzymes, pigments, bioactive compounds, and biofuels.

Since biobased treatment is seen as a feasible substitute for physico-chemical procedures for extending the recovery of organic products from wastes and producing noticeably better bioproduct yields, it is receiving a lot of attention for the expansion of the bioproducts refinery sector. Even while certain biowaste can be burned straight to create fuel, doing so pollutes the air (Mofijur et al., 2020b). Thus, it is crucial to valorize biowaste and turn it into other high-value goods like biofuel. Precisely, the circular economy progresses through green product planning, strategy adjustments, prolonging the life of consumer goods, waste control, and growing the market for auxiliary materials like food, pulverisation and development, organic products, basic crude materials, and biomass like biohydrogen (Ma & Liu, 2019). A deeper understanding of the modern circular economy has been provided by this chapter, which also summarises the final developments in biobased processes for the lignocellulosic waste utilisation and offers an outline of feasible biorefinery chemicals derived from biowaste. It also highlights the importance of the most recent limitations and projections to hinder the progress of the associated circular economy.

14.2 Global Scenario of Biowaste, Bioenergy and Bioproducts

Research and development are now being conducted in the field of producing bioenergy from biowaste. Because to changes in consumption patterns, urbanisation, industrialization, and population increase, a vast amount of municipal biomass waste is dispersed annually around the world (IEA, 2021). Based on data from the World Bank, the world's supply of municipal solid waste (MSW) was 2.01 billion tonnes in 2016. It is projected that annual MSW generation worldwide would reach 2.2 billion tonnes in 2025 and 3.40 billion tonnes in 2050 (IPCC, 2018). Biowaste, or organic waste from living things, can be processed to produce bioethanol, biodiesel, biohydrogen, and biogas, among other types of bioenergy. One long-term, promising solution to problems with traditional energy sources, climate change, and greenhouse gas emissions is the creation of bioenergy from biowaste. Anaerobic digestion, pyrolysis, thermal gasification, composting, transesterification, hydrothermal liquefaction, and fermentation are some of the processes utilised to produce bioenergy from biowaste (Khan & Kabir, 2020; Li et al., 2017; Yin et al., 2016). By using these techniques, organic matter in biowaste is broken down to create biogas, which can be utilised as an alternative to natural gas or as a fuel for power plants. The development of more sustainable and effective processes for producing bioenergy from biowaste is still being researched. An example of a recent study that uses biowaste to produce bioenergy is the holistic zero waste macroalgae-based third-generation bioethanol biorefinery system (Tiwari & Pal, 2021). Data-driven circular economy models for the conversion of biowaste to bioenergy are also being developed using machine learning and prediction modelling (Liao & Yao, 2021). In conclusion, the production of bioenergy from biowaste is a crucial field for study and development, and several techniques and tools are being employed to do so (D'Amato et al., 2019). The production of bioenergy from biowaste offers the opportunity to alleviate energy and environmental issues while advancing environmentally friendly waste management techniques.

14.3 Major Types of Bioenergy Generated from Biowastes

Agro-wastes produced by human activity are less harmful than chemical wastes, but they nevertheless pose a threat to the environment since they are improperly treated, which causes them to accumulate and eventually become a pollution hazard. The high level of organic matter in agro-wastes creates an ideal environment for microbial growth, which can have a number of negative effects on human health and the environment. Conversely, these same qualities indicate these wastes' potential for use in the manufacturing of biofuels. The different agro-food sectors' waste types and compositions are shown in (Fig. 14.1) (Table 14.1) and described in the sections that follow.

It is a well-known fact that agro-wastes are produced in agro-based companies' product processing units, and these wastes can be used to produce various kinds of biofuels (Nair et al., 2022). It's also noteworthy that, in recent years, there has been a lot of interest in using biocatalysts to produce biofuel (Atabani et al., 2021). In bioconversion procedures, a variety of microorganism-derived enzymes are also employed. Certain agro-industrial leftovers, like straw wastes, are challenging for biochemical breakdown, despite the fact that they may be sources of bioenergy. Under these circumstances, the material is pretreated chemically with strong acids or bases, or heated, which typically increases the substance's digestibility. Alkaline peroxide, NH_3 treatment, and bicarbonate are examples of chemical pretreatment. Being a N source for biodegradation and not producing distinct wastewater streams from the pretreatment process are two benefits of NH_3 treatment over other treatments (Khan & Kabir, 2020; Li et al., 2017). The following sections have covered the various wastes' roles in the creation of biofuels, which are alternatives to traditional energy.

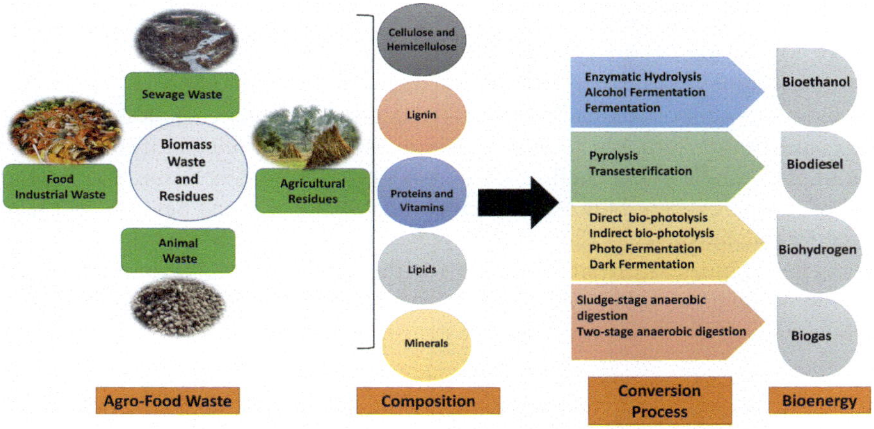

Fig. 14.1 Illustration of different type of bioenergy generated from agro-food waste

Table 14.1 List of major studies illustrating generation of bioenergy from agro-waste

S. No	Types of bioenergy	Biomass waste	Conversion process	References
1	Bioethanol	Food waste	Consolidated bioprocessing	Bibra et al. (2020)
2	Bioethanol	Bread waste	Separate Hydrolysis and Fermentation	Mihajlovski et al. (2020)
3	Bioethanol	Biowaste	Separate Hydrolysis and Fermentation	Tsafara et al. (2022)
4	Bioethanol and Biomethane	Bread waste	Separate Hydrolysis and Fermentation	Narisetty et al. (2022)
5	Bioethanol	Food waste	Consolidated bioprocessing	Bibra et al. (2023)
6	Biodiesel	Waste Cooking Oil and Grease	Transesterification	Park et al. (2019)
7	Biodiesel	Waste Cooking Oil	Transesterification	Velmurugan and Warrier (2022)
8	Biodiesel	Waste Cooking Oil	Transesterification	Cerón Ferrusca et al. (2023)
9	Biodiesel	Banana peel waste	Transesterification	Devasan et al. (2023)
10	Biodiesel	Vegetable waste	Transesterification	Sathish et al. (2023)
11	Biohydrogen	Organic wastewater	Dark Anaerobic Fermentation	Qu et al. (2022)
12	Biohydrogen and Methane	Biomass and Organic Waste	Dark Fermentation	Sittijunda et al. (2022)
13	Biohydrogen	Organic biowaste	Dark Fermentation	Kanwal and Torriero (2022)
14	Biohydrogen	Lignocellulosic biomass	Dark Fermentation	Jain et al. (2022)
15	Biohydrogen	Lignocellulosic biomass	Dark Fermentation	Talapko et al. (2023)
16	Biogas	Organic waste	Anaerobic digestion	Atelge et al. (2020)
17	Biogas	Organic waste and Cattle manure	Anaerobic digestion	Ellacuriaga et al. (2021)
18	Biogas	Lignocellulosic and organic waste	Anaerobic Digestion	Uddin et al. (2021)
19	Biogas	Organic waste and Cattle manure	Anaerobic digestion	Singh et al. (2023)
20	Biogas	Sugarcane Wastewater and Municipal Solid Waste	Anaerobic digestion	Tshemese et al. (2023)

Bioethanol: One kind of biofuel that can be made from biowaste, such as food scraps, energy crops, and agricultural wastes, is bioethanol. It can be combined with petrol and is frequently used as a fuel for transportation.

Biodiesel: Another type of biofuel that can be produced from biowaste is biodiesel, which is primarily made from plant, animal, and recycled cooking oil. It replaces traditional diesel fuel with a renewable fuel.

Biohydrogen: Dark fermentation and photo-fermentation are two methods that can be used to create biohydrogen, a clean, sustainable energy source from biowaste. It might be applied to fuel cells, which produce electricity.

Biogas: Anaerobic digestion of biowaste, which includes organic matter from industrial, agricultural, and municipal sources, produces biogas, a renewable energy source. It may be utilised to generate power and heat and is mostly composed of carbon dioxide and methane.

Through the use of organic waste for energy production, these forms of bioenergy support a circular economy, diversify energy sources, and cut greenhouse gas emissions.

14.4 Significance of Circular Economy of Bioenergy and Bioproducts

With varying degrees of emphasis on bioproducts and bioenergy, the circular economy and bioeconomy objectives are combined to form the term "circular economy.“ The circular bioeconomy's sustainability is in jeopardy due to its recent definition, which has led to the marginalisation of numerous important social, ethical, and ecological components in research papers, policy documents, and industrial practices (Gawel et al., 2019). The circular economy idea has gained significant traction both locally and globally. An increasing number of governments, businesses, and international organisations are embracing the bioeconomy and circular economy strategies in an effort to improve sustainability (Fig. 14.2). There have been concerns expressed on the consequences for biodiversity. In order to understand the two methods' relationships to other economic models, to each other, to sustainability, and to biodiversity, we have reviewed the most recent studies on them. Neither the bioeconomy nor the circular economy offer a clear road to sustainability and are both ill-defined, inconsistently applied, and inadequately quantified. A lot of actors prioritise economic expansion over environmental concerns. Inadequate or indirect approaches are frequently taken to biodiversity. Moreover, a number of historically marginalised groups—such as women and Native Americans—may be disregarded and get no assistance or engagement. These difficulties are made worse by capacity shortages as well as implementation-related legal and governance issues that are impacted by conventional wisdom. Governments and businesses must more clearly consider how their actions affect biodiversity when developing their sustainability plans. Opportunities include the work that already exists and may be expanded

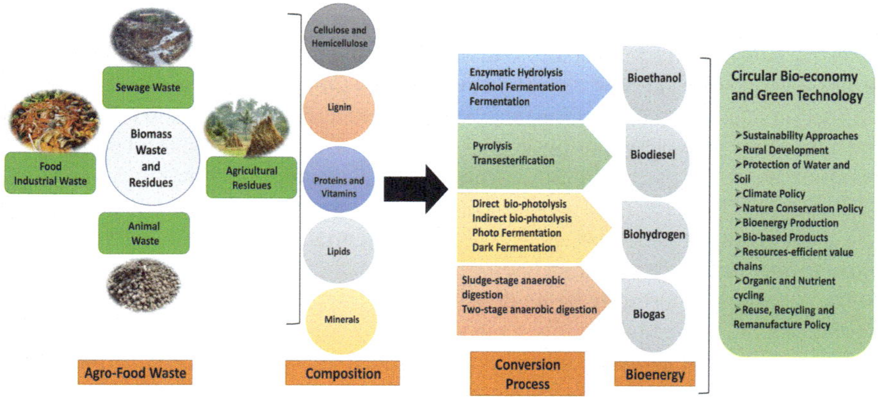

Fig. 14.2 Illustration of demonstration of circular economy

upon, the variety of partners that are already involved, and the relevance and timeliness of sustainable economics for achieving Sustainable Development Goals in a post-COVID future. To guarantee the sustainability of economic approaches, we suggest five key actions. In the end, the only way we can guarantee sustainability is if we begin to change people's perspectives and create a more concentrated agenda for the bioeconomy and circular economy that prioritises the health of local populations, ecosystems, and species.

14.5 Limitation and Future Aspects

The political objectives of the modern world are largely determined by fuel efficiency. Because these nations are supported by rare fuel resources, they are seen as being extremely powerful. Other nations that depend on these nations for power and energy would face severe consequences if their resources were to run out or if they stopped sharing them. In this case, having a fuel and electrical source that is sustainable is crucial. Utilising waste biomass for the manufacture of biofuels is a potential method that lowers environmental pollution while simultaneously increasing the pace of energy generation in the nation. One of the main benefits is the range of substrates that can be used to produce biofuel. Numerous nations have already set up facilities for producing biofuel, and there is a great deal of global research being done on the subject.

The qualities of biofuels such as enhanced performance and time improved emission characteristics favour them to replace conventional fuels on a quick scale (Pattanaik et al., 2019). At the same time, some constraints that impede large-scale applications make direct fossil fuel replacement impossible. Even after refinement, bioethanol production systems are inefficient because they may still contain contaminants that cause engine metal parts to dissolve, raising the engine's maintenance costs (Bibra et al., 2020). The high cost of manufacture is another significant obstacle to the production of ethanol. In addition, technology related to hydrolysis and pretreatment optimisation is crucial (Mihajlovski et al., 2020). Seasonal variations and geographic location can impact the availability of the selected feedstock, thus creating challenges for its supply. The cost of producing bioethanol can also be significantly impacted by variations in feedstock prices. It would be essential to raise the bioethanol yield because they make up more than one-third of the manufacturing costs (Tsafara et al., 2022). Research is currently being done to create an efficient bioethanol production system that has to be high yielding and reasonably priced.

While European nations concentrate on both dark and photo fermentation, Asian nations primarily focus on dark fermentation for the creation of biohydrogen. The current biohydrogen production system is deemed suitable for decentralised small-scale systems that are integrated with residues from waste processing facilities, industrial and agricultural wastes, or pure cultures from natural sources using reactors operating with mixed microflora, such as aerobic, anaerobic, thermophilic, photo non-sulfur producing bacteria, etc. (Qu et al., 2022). The poor yield and the fact that just one-third as much H_2 is produced from carbs as from lipids are some of the main drawbacks of producing biohydrogen via dark fermentation (Sittijunda et al., 2022). A lot of research is being done to improve the production of biohydrogen by merging different techniques. Economic viability, thermodynamic constraints, unintended fermentation, etc. are some of the main obstacles to the generation of biohydrogen (Ochieng et al., 2022). There are restrictions on how biohydrogen can be transported and stored. Because pressurised hydrogen is so explosive, it requires more space than other traditional petrol fuels and raises issues about transportation safety. More research is being done to enhance the production of biohydrogen, such as genetically engineering microorganisms to increase efficiency and designing intricate bioreactor designs. However, fuel economics and the social adaptation of H_2 energy systems also play a significant role in this process. The immobilisation of the microbial culture, bioreactor modifications, optimisation of necessary conditions (temperature, pH, etc.), selection and enrichment of the culture, substrate selection, and metabolic engineering by the biohydrogen specialists are among the strategies that are monitored to increase the productivity of H2 (Kanwal & Torriero, 2022; Talapko et al., 2023).

Waste availability, logistics, process efficiency, and end-product qualities are all factors in the biogas economy. Industry participants believe that the federal government's research funding programmes along with private industry investments could help expedite and bring these fuels to market at competitive prices. However, the modification of engines for the biogas combustion should be given significance, as the biogas-based engines are not evolved enough to deal with the technical challenges of biogas use (Tshemese et al., 2023). Potentially a green fuel, biodiesel has significant obstacles due to its high production costs, lower energy content than regular fossil diesel, and the emission of nitrogen oxide during combustion (Park et al., 2019). Because of the high cost, it is not profitable. Research is being done on a number of topics to increase the production of biodiesel, such as finding catalysts, alternative feedstocks, production processes, etc., to lower the cost of production (Velmurugan & Warrier, 2022). According to Rai et al. (2020), it was recommended that an integrated biorefinery method be used to maximise feedstock utilisation and improve production economics.

Some technological obstacles must be removed in order to realise better biofuel production. Given that feedstock accounts for more than 45% of the expenditures associated with the fermentation process, feedstock prices are crucial to the process's economics (Bhatia et al., 2015). Presently employed pretreatment procedures for biomass are more expensive and less effective. To boost sugar yield and lower inhibitory by-products, more research needs to be done (Nair et al., 2022). Additionally, new microorganisms must be discovered, and existing consortia of bacteria must be engineered to increase biofuel production, increase their tolerance to inhibitors derived from biomass, and make them efficient for the utilisation of a wide variety of carbon sources (Tshemese et al., 2023). The economics of producing biofuel could be enhanced by co-producing other valuable byproducts (Nair et al., 2022).

An essential stage in the bio-conversion process is optimising the process for producing biofuel from biowaste. It is important to ensure that the production process is economical in order to preserve the biofuels' economic sustainability. Given that the cost of fossil fuels is already rising globally, most consumers would not choose more expensive biofuels over conventional ones. The future promise of lower feedstock costs should also be pursued. Agro-industrial wastes are inexpensive to acquire since, in a technical sense, they are waste residues. To encourage the agro-industries to provide agro-waste, fuel price incentives may be provided. The bioconversion system will gain from the application of metabolic engineering techniques as well. The productivity and efficiency of the process would rise if effective microorganisms for enzymatic hydrolysis and saccharification during bioconversion were developed or found. All of these areas need more investigation in order to build effective systems for producing bioenergy from agro-industrial wastes.

The impact of issues such as energy security, climate change, and rising oil prices on the global population is significant. Furthermore, the needs of the world's growing population place a strain on energy supplies. It was evident that the value-adding of agricultural and associated wastes could result in the production of beneficial products like bioenergy. These techniques offer a sustainable energy source that is affordable for everyone in addition to controlling the amount of garbage that is deposited on the planet. Biotechnological advances are also aided by the circular economy approach. In the long run, biofuels made from agricultural wastes can be beneficial because they pollute less than traditional fuels. Environmental conservation has advanced thanks to the decrease in biowaste deposition and the substitution of biofuels for conventional gasoline. Through intensive study, the cost and efficiency of producing various biofuels, such as biogas, bioethanol, biodiesel, and biohydrogen, can be reduced and the environment can become more sustainable.

14.6 Conclusion

Energy is a fundamental component of socioeconomic growth, human welfare, and quality of life. The main causes of the rise in energy consumption over the past few decades have been the rapid urbanisation, industrialization, and modernization of society. A number of well-known problems, including global warming, climate change, rising greenhouse gas emissions, rising and falling fossil fuel prices, and rising energy consumption, compelled scientists to look into alternate energy sources. The need for alternative energy sources has grown in importance in order to keep up with the rapid rate of pollution, industrialization, and population development. Waste generation and the energy problem have made waste to energy for biofuels more feasible. In order to promote environmental and economic sustainability and lessen reliance on imported petroleum fuels, biofuels are made up of a variety of fuels. One important source of organic waste for biofuels is the waste produced by homes and businesses. Since they don't contain lignin and have a high productivity rate and oil content, algal and yeast biomasses have become increasingly important. Biorefineries can make advantage of the high-productivity lignocellulosic feedstock. Biofuels from a variety of substrates, including bioethanol, biodiesel, biogas, bioelectricity, and biohydrogen, can meet the world's energy needs. The use of bioenergy in international transportation has increased research activity even more. For most nations to fulfil their commitments to cut greenhouse gas emissions, researching the usage of these alternative energy sources is crucial. The various biomass transformation technologies for waste-to-bioenergy are the main topic of this chapter. There is discussion of the technology, opportunities, and application. The article describes the newly developed biochemical routes for biofuels, including microbial fuel cells, alcoholic fermentation, extraction-transesterification, and anaerobic digestion. We cover the process of transesterification to create biodiesel and waste to bioelectricity in microbial fuel cells.

References

Alhawari, O., Awan, U., Bhutta, M. K. S., & Ali Ülkü, M. (2021). Insights from circular economy literature: A review of extant definitions and unravelling paths to future research. *Sustainability, 13*, 859.

Atabani, A. E., Al-Muhtaseb, A. H., Kumar, G., Saratale, G. D., Aslam, M., Khan, H. A., Said, Z., & Mahmoud, E. (2019). Valorization of spent coffee grounds into biofuels and value-added products: Pathway towards integrated bio-refinery. *Fuel, 254*, 115640.

Atelge, M. R., Krisa, D., Kumar, G., et al. (2020). Biogas production from organic waste: Recent progress and perspectives. *Waste and Biomass Valorization, 11*, 1019–1040.

Barañano, L., Garbisu, N., Alkorta, I., Araujo, A., & Garbisu, C. (2021). Contextualization of the bioeconomy concept through Its links with related concepts and the challenges facing humanity. *Sustainability, 13*, 7746.

Belmonte-Urena, L. J., Plaza-Úbeda, J. A., Vazquez-Brust, D., & Yakovleva, N. (2021). Circular economy, degrowth and green growth as pathways for research on sustainable development goals: A global analysis and future agenda. *Ecological Economics, 185*, 107050.

Bhatia, S. K., Shim, Y. H., Jeon, J. M., Brigham, C. J., Kim, Y. H., Kim, H. J., Seo, H. M., Lee, J. H., Kim, J. H., Yi, D. H., Lee, Y. K., & Yang, Y. H. (2015). Starch based Polyhydroxybutyrate production in engineered *Escherichia coli*. *Bioprocess and Biosystems Engineering, 38*, 1479–1484.

Bibra, M., Samanta, D., Sharma, N. K., Singh, G., Johnson, G. R., & Sani, R. K. (2023). Food waste to bioethanol: Opportunities and challenges. *Fermentation, 9*(1), 8.

Bibra, M., Rathinam, N. K., Johnson, G. R., & Sani, R. K. (2020). Single pot biovalorization of food waste to ethanol by Geobacillus and Thermoanaerobacter spp. *Renewable Energy, 155*, 1032–1041.

Cerón Ferrusca, M., Romero, R., Martínez, S. L., Ramírez-Serrano, A., Natividad, R. (2023). Biodiesel production from waste cooking oil: A perspective on catalytic processes. *Processes, 11*(7), 1952.

D'Amato, D., Korhonen, J., & Toppinen, A. (2019). Circular, green, and bio economy: How do companies in land-use intensive sectors align with sustainability concepts? *Ecological Economics, 158*, 116–133.

Devasan, R., Ruatpuia, J. V. L., Gouda, S. P., et al. (2023). Microwave-assisted biodiesel production using bio-waste catalyst and process optimization using response surface methodology and kinetic study. *Science and Reports, 13*, 2570.

Ellacuriaga, M., García-Cascallana, J., & Gómez, X. (2021). Biogas production from organic wastes: Integrating concepts of circular economy. *Fuels, 2*, 144–167.

Gawel, E., Pannicke, N., & Hagemann, N. (2019). A path transition towards a bioeconomy: The crucial role of sustainability. *Sustainability, 11*, 3005.

IEA. (2021). *Net zero by 2050: A roadmap for the global energy sector* (pp. 18–19). International Energy Agency.

IPCC. (2018). *Global Warming of 1.5 ◦C. An IPCC special report on the impacts of global warming of 1.5 ◦C above pre-industrial levels and related global greenhouse gas emission pathways, in the context of strengthening the global response to the threat of climate change, sustainable development, and efforts to eradicate poverty*. V. Masson-Delmotte, P. Zhai, H.-O. Pörtner, D. Roberts, J. Skea, P. R. Shukla, A. Pirani, W. Moufouma-Okia, C. Péan, R. Pidcock et al. (Eds.), IPCC: Geneva, Switzerland, 2019.

Jain, R., Panwar, N. L., Jain, S. K., Gupta, T., Agarwal, C., & Meena, S. S. (2022). Bio-hydrogen production through dark fermentation: An overview. *Biomass Conversion and Biorefinery, 14*, 1–26.

Kanwal, F., & Torriero, A. A. J. (2022). Biohydrogen-A green fuel for sustainable energy solutions. *Energies, 15*, 7783.

Khan, I., & Kabir, Z. (2020). Waste-to-energy generation technologies and the developing economies: A multi-criteria analysis for sustainability assessment. *Renewable Energy, 150*, 320–333.

Lee, S. Y., Sankaran, R., Chew, K. W., Tan C. H., Krishnamoorthy, R., Chu, D. T., & Show, P. L. (2019). Waste to bioenergy: A review on the recent conversion technologies. *BMC Energy, 1*, 4.

Leow, C. W., Fan, Y. V., Chua, L. S., Muhamad, I. I., Klemeš, J. J., & Lee, C. T. (2018). A review on application of microorganisms for organic waste management. *Chemical Engineering Transactions, 63*, 85–90.

Li, Y., Liu, H., Yan, F., Su, D., Wang, Y., & Zhou, H. (2017). High-calorific biogas production from anaerobic digestion of food waste using a two-phase pressurized biofilm (TPPB) system. *Bioresource Technology, 224*, 56–62.

Liao, M., & Yao, Y. (2021). Applications of artificial intelligence-based modeling for bioenergy systems: A review. *GCB Bioenergy, 13*, 774–802.

Ma, Y., & Liu, Y. (2019). Turning food waste to energy and resources towards a great environmental and economic sustainability: An innovative integrated biological approach. *Biotechnology Advances, 37*(7), 107414.

Mihajlovski, K., Rajilić-Stojanović, M., & Dimitrijević-Branković, S. (2020). Enzymatic hydrolysis of waste bread by newly isolated Hymenobacter sp. CKS3, statistical optimization and bioethanol production. *Renewable Energy, 152*, 627–633.

Mofijur, M., Arafat Siddiki, S. Y., Ahmed, M. B., Djavanroodi, F., Fattah, I. M. R., Ong, H. C., et al. (2020a). Effect of Nanocatalysts on the transesterification reaction of first, second and third generation biodiesel sources—A mini-review. *Chemosphere, 4*, 128642.

Mofijur, M., Kusumo, F., Fattah, I. M. R., Mahmudul, H. M., Rasul, M. G., Shamsuddin, A. H., et al. (2020b). Resource recovery from waste coffee grounds using ultrasonic-assisted technology for bioenergy production. *Energies, 13*(7), 1770.

Muhammad, G., Alam, M. A., Mofijur, M., Jahirul, M. I., Lv, Y., Xiong, W., et al. (2021). Modern developmental aspects in the field of economical harvesting and biodiesel production from microalgae biomass. *Renewable and Sustainable Energy Reviews, 135*, 110209.

Nair, L. G., Agrawal, K., & Verma, P. (2022). An overview of sustainable approaches for bioenergy production from agro-industrial wastes. *Energy Nexus, 6*, 100086.

Narisetty, V., Nagarajan, S., Gadkari, S., Ranade, V. V., Zhang, J., Patchigolla, K., Bhatnagar, A., Awasthi, M. K., Pandey, A., & Kumar, V. (2022). Process optimization for recycling of bread waste into bioethanol and biomethane: A circular economy approach. *Energy Conversion and Management, 266*, 115784.

Ochieng, R., Gebremedhin, A., & Sarker, S. (2022). Integration of waste to bioenergy conversion systems: A critical review. *Energies, 15*, 2697.

Park, S. H., Khan, N., Lee, S., Zimmermann, K., DeRosa, M., Hamilton, L., Hudson, W., Hyder, S., Serratos, M., Sheffield, E., Veludhandi, A., & Pursell, D. P. (2019). Biodiesel production from locally sourced restaurant waste cooking oil and grease: Synthesis, characterization, and performance evaluation. *ACS Omega, 4*(4), 7775–7784.

Pattanaik, L., Pattnaik, F., Saxena, D. K., & Naik, S. N. (2019). Biofuels from agricultural wastes. In *Second and third generation of feedstocks* (pp. 103–142). Elsevier Inc. https://doi.org/10.1016/B978-0-12-815162-4.00005-7

Perea-Moreno, M. A., Samerón-Manzano, E., & Perea-Moreno, A. J. (2019). Biomass as renewable energy: Worldwide research trends. *Sustainability, 11*, 863.

Qu, X., Zeng, H., Gao, Y., Mo, T., & Li, Y. (2022). Bio-hydrogen production by dark anaerobic fermentation of organic wastewater. *Frontiers in Chemistry, 10*, 1095.

Rai, S., Solanki, M. K., Anal, A. K. D., Sagar, A., Solanki, A. C., Kashyap, B. K., Pandey, A. K. (2020). Emerging frontiers of microbes as agro-waste recycler. In B. K. Kashyap, M. K. Solanki, D. V. Kamboj & A. K. Pandey (Eds.), *Waste to energy: Prospects and applications* (pp. 3–27). Springer.

Rai, S., Solanki, M. K., Anal, A. K. D., Sagar, A., Solanki, A. C., Kashyap, B. K., Pandey, A. K., & A.K.,. (2021). Emerging frontiers of microbes as agro-waste recycler. In B. K. Kashyap, M. K. Solanki, D. V. Kamboj, & A. K. Pandey (Eds.), *Waste to energy: Prospects and applications* (pp. 3–27). Springer.

Sathish, T., Saravanan, R., Depoures, M. V., et al. (2023). Environmental remediation at vegetable marketplaces through production of biowaste catalysts for biofuel generation. *Science and Reports, 13*, 5067.

Sharma, S., Sagar, A., Yadav, S. S., & Rai, S. (2023). Production of value added products from fruit juice residues using enzyme technology. In M. Kuddus & P. W. Ramteke (Eds.), *Value-Addition in agri-food industry waste through Enzyme technology* (pp. 221–236). Academic Press.

Singh, A. K., Pal, P., Rathore, S. S., Sahoo, U. K., Sarangi, P. K., Prus, P., & Dziekański, P. (2023). Sustainable utilization of biowaste resources for biogas production to meet rural bioenergy requirements. *Energies, 16*(14), 5409.

Sittijunda, S., Baka, S., Jariyaboon, R., Reungsang, A., Imai, T., & Kongjan, P. (2022). Integration of dark fermentation with microbial electrolysis cells for biohydrogen and methane production from distillery wastewater and glycerol waste co-digestion. *Fermentation, 8*, 537.

Stephenson, P. J., & Damerell, A. (2022). Bioeconomy and circular economy approaches need to enhance the focus on biodiversity to achieve sustainability. *Sustainability, 14*(17), 10643.

Talapko, D., Talapko, J., Erić, I., & Škrlec, I. (2023). Biological hydrogen production from biowaste using dark fermentation Storage and Transportation. *Energies, 16*(8), 3321.

Tiwari, A. K., & Pal, D. B. (2021). Bio-processing: Biomass to commercial alcohol. In M. Srivastava, N. Srivastava & R. Singh (Eds.), *Bioenergy research biomass waste to energy* (pp. 149–168). Springer.

Tsafara, P., Passadis, K., Christianides, D., Chatziangelakis, E., Bousoulas, I., Malamis, D., Mai, S., Barampouti, E. M., & Moustakas, K. (2022). Advanced bioethanol production from source-separated bio-waste in pilot scale. *Sustainability., 14*(19), 12127.

Tshemese, Z., Deenadayalu, N., Linganiso, L. Z., & Chetty, M. (2023). An overview of biogas production from anaerobic digestion and the possibility of using sugarcane wastewater and municipal solid waste in a South African context. *Applied System Innovation., 6*(1), 13.

Uddin, M. N., Siddiki, S. Y. A., Mofijur, M., Djavanroodi, F., Hazrat, M. A., Show, P. L., Ahmed, S. F., & Chu, Y.-M. (2021). Prospects of bioenergy production from organic waste using anaerobic digestion technology: A mini review. *Frontiers in Energy Research, 9*, 627093.

Velmurugan, A., & Warrier, A. R. (2022). Production of biodiesel from waste cooking oil using mesoporous MgO-SnO_2 nanocomposite. *Journal of Engineering and Applied Science, 69*, 92.

Yin, Q., Zhu, X., Zhan, G., Bo, T., Yang, Y., & Tao, Y., et al., (2016). Enhanced methane production in an anaerobic digestion and microbial electrolysis cell coupled system with co-cultivation of Geobacter and Methanosarcina. *Journal of Environmental Sciences, 42*, 210–214.

Zamri, M. F. M. A., Hasmady, S., Akhiar, A., Ideris, F., Shamsuddin, A. H., Mofijur, M., et al. (2021). A comprehensive review on anaerobic digestion of organic fraction of municipal solid waste. *Renewable and Sustainable Energy Reviews, 137*, 110637.

Chapter 15
Gendering the Impact of Climate Change on Agriculture in Asia

Roli Misra and Nidhi Tewari

15.1 Introduction

In 2022, the worldwide repercussions of climate change intensified, leading to severe outcomes such as extreme weather occurrences, disruptions in ecosystems, and heightened socioeconomic difficulties on a global scale. Countries like India, China, Pakistan, and Bangladesh collectively experienced losses of around US$36 billion, with approximately 50 million individuals directly impacted by climate-related disasters. These events resulted in extensive damage amounting to more than US$36 billion in terms of infrastructure loss and livelihood disruptions (Economic and Social Commission for Asia and the Pacific, 2022). Amidst these alarming statistics, meeting the 2 °C target set forth in the 2015 Paris Agreement and attaining the Sustainable Development Goals by 2030 necessitates substantial upscaling of climate-resilient practices within agriculture and food systems. Achieving this transformational shift in agricultural systems requires coordinated efforts across various levels, ranging from global to local, and involving research, policy-making, and investment endeavours. Collaboration among stakeholders from both public and private sectors, as well as civil society, is crucial in this endeavour. Gender norms, socio-cultural contexts, religious beliefs, ethnicity, and other factors contribute to the distinct experiences of male and female farmers in confronting the impacts of climate change. These variations shape their roles within the agricultural sector, as well as their entitlements and opportunities. The interconnectedness of climate change impacts agriculture, communities, and the broader socio-economic fabric underscores the vulnerability of those dependent on agricultural livelihoods. Amidst these challenges, the intricate

R. Misra (✉)
Department of Economics, University of Lucknow, Lucknow, India
e-mail: misra_roli@lkouniv.ac.in

N. Tewari
Arthan Foundation, New Delhi, India

P. K. Rai and S. Rai (eds.), *Climate, Environment and Agricultural Development*,
Advances in Geographical and Environmental Sciences,
https://doi.org/10.1007/978-981-97-8363-2_15

social structure, particularly affecting women engaged in agriculture, accentuates their vulnerability. As climate change intensifies, it not only erodes the foundations of agriculture but also amplifies the existing socio-economic disparities, disproportionately affecting women who bear the brunt of increased care responsibilities and face the depletion of physical and mental well-being.

In this backdrop this paper aims to unravel these intricate dynamics, shedding light on the multifaceted impacts of climate change on agriculture and the vulnerable social strata, with a particular focus on the heightened vulnerabilities faced by women in agricultural communities. While economic impacts can look like loss of income opportunity and financial agency, the non-economic repercussions of climate change include altercation of social ties, inter and intra-household dynamics, social norms, and knowledge systems. Likewise, the impacts of increased care responsibilities and the subsequent depletion in carers' physical and mental health, are typically ignored. The non-economic impacts of climate change are frequently overlooked by conventional approaches to ascertaining the impact of climate change. In conclusion, this paper bridges the gap in studying the multifaceted impacts of climate change on women in agricultural communities. By unravelling these intricate dynamics, it aims to foster a deeper understanding of the vulnerabilities faced by women in the face of climate change, ultimately contributing to more gender-responsive approaches to climate adaptation and resilience-building efforts.

15.2 Theoretical Framework: Impact of Climate Change on Agriculture Through Feminist Perspectives and Intersectionality

Feminist perspectives on agriculture and climate change are integral for unravelling the intricate gendered dimensions within the realm of environmental challenges. In the agricultural sphere, traditional gender roles often assign women significant responsibilities in farming and food production, setting the stage for a gendered analysis of the impacts of climate change. These perspectives critically examine how climate change acts as an amplifier of existing gender disparities, disproportionately affecting women. From resource distribution to access to technological advancements and decision-making power, feminist analyses illuminate the multifaceted challenges women face in sustaining agricultural practices amid evolving climatic conditions. This scrutiny underscores the urgent need for gender-sensitive policies and interventions that acknowledge and address the unique vulnerabilities and contributions of women in agriculture.

Complementing the feminist lens, intersectionality emerges as a vital framework in understanding the layered impacts of climate change, particularly in the context of gender. According to Symington, intersectionality is a feminist theory that can be used as an analytical tool to study and comprehend the intersection of multiple identities with gender, as well as to appropriately respond to alleviate discrimination against oppressed classes. Rather than viewing gender in isolation, intersectional analyses consider how intersecting factors like race, class, and socioeconomic status intersect to shape distinct vulnerabilities within diverse social groups. Within the agricultural landscape, applying an intersectional approach allows for a nuanced exploration of the varied challenges faced by different groups of women, recognizing the non-uniformity of their experiences. This methodology unveils differential impacts and emphasizes the necessity of tailored, intersectionality-sensitive strategies for mitigating and adapting to the gendered consequences of climate change in agriculture. Gender equality plays a pivotal role in fostering fairness within the agricultural sector. Embracing a feminist perspective that acknowledges and values the equitable involvement of all stakeholders fosters a more innovative, adaptable, and ultimately effective transformative movement. As Lopes and Jomalinis (2011) highlight, the marginalization of women directly undermines agricultural progress. The dominance of men often acts as a barrier to transitioning to agroecology, stifling women's ability to freely express themselves, impeding their creative growth, and ultimately limiting their contributions to productive units. The integration of feminist perspectives and intersectionality not only enriches our understanding of the complex interplay between social dimensions and climate change but also lays the groundwork for inclusive policies and interventions that effectively address the specific needs of women across diverse agricultural settings grappling with the challenges of a changing climate.

15.3 Impact of Climate Change on Agriculture

Research indicates that the impact of temperature and precipitation changes on crop production and food security is expected to worsen in South Asia. These impacts have been studied by analyzing the relationship between crop yields and soil water availability during various stages of crop growth. Water and heat stresses during crucial periods such as crop establishment and key growth stages like flowering, pollination, and grain filling can severely affect many crops. If left unaddressed, the changes in temperature and precipitation patterns in South Asia will significantly affect agriculture in the long term. However, the actual effects of climate change on agriculture vary depending on the type of crops, specific locations, and the ability of communities to adapt to climatic risks. Adaptive capacity also plays a role in determining agricultural productivity. For instance, communities in the Hindu Kush Himalayan region, spanning parts of Pakistan, India, and Nepal, are particularly vulnerable to climate change due to their heavy reliance on agriculture for livelihoods, geographical isolation, limited access to global markets, low productivity, and inadequate infrastructure.

Agriculture in South Asia is highly vulnerable to the impacts of climate change and its variability. According to projections by the IPCC in 2007, the region is expected to experience a temperature increase of 0.5–1.2 °C by 2020, 0.88–3.16 °C by 2050, and 1.56–5.44 °C by 2080, depending on various future development scenarios.

These long-term changes in temperature and precipitation patterns are likely to alter cropping seasons, suitability for crop cultivation, and contribute to an increase in the incidence of diseases and pests, all of which can have significant impacts on crop yields, production, and food markets. For instance, research by Abbas et al. (2017) indicates that between 1980 and 2014, the growing periods for spring maize in Pakistan have shifted earlier by an average of 4.6 days per decade, while the sowing of autumn maize has been delayed by 3 days per decade. These changes have had severe implications for yield. Ultimately, these shifts pose challenges to the livelihoods of millions of farmers in the region, particularly those with limited capacity to adapt to climate change.

Research indicates that the impact of temperature and precipitation changes on crop production and food security is expected to worsen in South Asia. These impacts have been studied by analyzing the relationship between crop yields and soil water availability during various stages of crop growth. Water and heat stresses during crucial periods such as crop establishment and key growth stages like flowering, pollination, and grain filling can severely affect many crops (Porter & Semenov, 2005). If left unaddressed, the changes in temperature and precipitation patterns in South Asia will significantly affect agriculture in the long term (Aggarwal & Sinha, 1993). However, the actual effects of climate change on agriculture vary depending on the type of crops, specific locations, and the ability of communities to adapt to climatic risks. Adaptive capacity also plays a role in determining agricultural productivity. For instance, communities in the Hindu Kush Himalayan region, spanning parts of Pakistan, India, and Nepal, are particularly vulnerable to climate change due to their heavy reliance on agriculture for livelihoods, geographical isolation, limited access to global markets, low productivity, and inadequate infrastructure (Rasul et al., 2019).

In recent years, numerous studies have delved into the effects of climate change on major food crops like rice, wheat, and maize in South Asia. Parry et al. (2004) highlighted potential reductions in crop yields, ranging from 2.5–10% in parts of Asia by the 2020s, escalating to 5–30% by the 2050s. Lal (2011) further examined the impacts of thermal and hydrological stresses on agricultural productivity in the region. He suggested that while the effects of global warming on food production may not be severe until the 2020s, maintaining irrigation water availability and effective pest control measures would be crucial. However, beyond 2050, productivity of summer crops is expected to decline due to heightened climate variability and increased pest pressure, particularly affecting winter crops with a projected 2 °C temperature rise. Looking ahead to the end of the century, a reduction of at least 4–10% in net cereal production in South Asia is anticipated if temperatures rise by 3 °C. Additionally, Knox et al. (2012) conducted an assessment of the expected impacts of climate change on the yield of eight major crops in the region. They forecasted an average yield reduction of 8% by the 2050s, with maize and sorghum yields expected to decline by 16% and 11% respectively, while rice yields are projected to remain relatively stable.

15.4 Gendering Impact of Climate Change on Agriculture

The effects of climate change have disparate impacts on individuals' economic security, health, and overall well-being, varying based on factors such as gender, race, ethnicity, class, caste, (dis)ability, sexual identity, and age. While climate disasters affect men, women, and people of all genders, those from marginalized communities, including Indigenous populations and individuals from low-income backgrounds, are particularly vulnerable (United Nations, 2019). However, due to entrenched gender inequalities within societal structures, these impacts are unevenly distributed, leaving women and marginalized genders at heightened risk. For example, men often encounter fewer barriers to migration for economic opportunities, leaving women to manage both paid and unpaid responsibilities while residing in areas disproportionately affected by climate disasters (United Nations, 2019). Additionally, women and girls' restricted access to social, political, and economic resources prior to disasters impedes their post-disaster recovery efforts (Global Facility for Disaster Reduction & Recovery, 2018). In instances such as the 2010 floods in Pakistan, mobility constraints and lack of identification cards hindered women's access to aid and services. Moreover, women, children, and the elderly are at increased risk of developing mental health issues, including depression, anxiety, and post-traumatic stress disorder, compounded by pre-existing vulnerabilities stemming from gender inequality, time and income poverty, and resulting physical and mental exhaustion, thereby exacerbating economic hardships due to climate change (American Psychiatry Association, 2023). Understanding how societal roles, gender norms, and the gendered structure of economies contribute to gender inequalities in access to income, land, livelihoods, and power is crucial in comprehending women's heightened vulnerability to climate change impacts. Notably, women typically shoulder approximately 76% of unpaid care responsibilities globally, translating to up to 65% of their total weekly work hours being unpaid (Addati et al., 2018; Butt et al., 2023) (Table 15.1).

Rural women are disproportionately affected by climate change due to their greater reliance on biomass, such as agricultural products, wastes, wood, and other forest resources, for energy and livelihood compared to males. They depend more on ecosystem services for food security compared to rural males, since they are often

Table 15.1 Gender-differentiated impacts of climate change on women

Climate change impacts	Impact exacerbate gender inequalities
Crop failure	Increases work load of women
Fuel shortage	Women have to consume more time in fuel collection
Water Scarcity	Women have to spend more time in water collection
Natural disasters	Greater incidence of mortality for women
Displacement	Forced migration make women more vulnerable
Conflict	Violence against Women

Source World Bank & FAO (2017)

engaged in agricultural production and natural resource management. Research done in Malawi by Asfaw and Maggio (2018) revealed that severe weather events had a more substantial impact on consumption and nutrition in places where women control a larger percentage of land. High climatic variability makes women in agriculture more sensitive and less resilient to shocks when compared to their counterparts. Gender disparities in reliance on natural resources and ecosystem services contribute to varying adaptive capabilities and levels of risk and susceptibility to biodiversity loss and changes in natural resource access and management. Women in many regions have restricted access to agricultural advice services and official rural organizations. This diminishes their chances to acquire knowledge on coping mechanisms and climate-resilient agriculture.

According to a study by World Bank (2009), restricted ownership and tenure stability significantly hinder women's ability to get financing, thereby undermining their ability to adapt to climate change. They are unable to get funding for climate-smart agricultural advancements due to the lack of legal ownership of the property. Women have limited access to services that could assist in acquiring new technologies, enhancing natural resource management practices, and implementing more efficient cropping and livestock management. These improvements could help women tackle natural resource degradation and enhance their ability to withstand climate change. Women participate in decision-making processes to a very little extent, have restricted access to and control over environmental goods and services, and are not involved in allocating the advantages of environment management. As such, women's ability to address climate change gets lessened. Women typically put in greater hours of work to support the household during extreme weather events like floods and droughts. This means they also have less opportunities for education and training, skill development, and income generation. Gender has a significant role in many situations when determining a person's sensitivity to potential danger. For example, research shows that during natural disasters linked to climate change, women and girls are more likely than males to go hungry (FAO, 2017; Oxfam, 2019).

For instance, in India, twice as many women as males claimed to have cut back on their food intake due to the drought (Lambrou & Nelson, 2010). Also a lower socioeconomic status, restricted access to warnings and shelter, and limited agency in making decisions about potential hazards are some of the factors contributing to women's higher susceptibility to harm during natural disasters (Table 15.2).

Discriminatory societal structures impede women's equal access to resources. Examples of these include formal inheritance laws and unwritten customs that deny women the ability to own land. These organizations additionally limit women's agency and their capacity to make decisions at the home and community levels, along with discriminatory social conventions and gender roles. When women do have access to resources, it's possible that they lack the agency to use and manage those resources in order to prepare for the effects of climate change. For instance, women in Bangladesh are less likely to diversify into sustainable livelihood options in the face of climate shocks. This is because of social norms that restrict their freedom, their lack of knowledge, and their restricted access to productive resources, women in Bangladesh (Rabbani et al., 2015). Severe weather conditions have a detrimental

Table 15.2 Gendered impact of climate change

Dimension	Indicators	Impact
Exposure & Sensitivity	Frequency and intensity of extreme events	Direct relationship
	Productivity of food grains	Inverse relationship
	Combined soil and vegetation degradation	Direct relationship
	Instability in area of major crops	Direct relationship
	Deficit /erratic rainfall	Direct relationship
Adaptive capacity	Literacy rate	Inverse relationship
	Degree of gender equity	Inverse relationship
	Population Density	Direct relationship
	Village with paved roads	Inverse relationship
	Women's participation in workforce	Inverse relationship
	Percentage household with electricity	Inverse relationship
	Percentage households with improved drinking water source	Inverse relationship
	Health infrastructure per 1000 population	Inverse relationship
	Proportion of groundwater availability	Inverse relationship
	Access to credit	Inverse relationship

Source Authors construction

effect on people's life and means of subsistence, which leads to food insecurity. Sea level rise, cyclones, tidal bore, water logging, and saline intrusion are the main effects of climate change on Bangladesh's coastal areas. Bangladesh is the sixth most susceptible nation in the world in terms of the total number of people impacted by sea level rise. The participation of women is often overlooked in the discussion of climate change adaptation, though it is one of the most crucial part in the process of inclusion. Livestock suffers of lack of fodder during and after disasters. Women grow Napier grass for cattle and gather water hyacinth. To lower the chance of disaster, women grow date palms and fan palms. The sap from this tree is a well-liked and pricey beverage. Since rivers and canals round Bangladesh's coastal region, a lot of people go fishing. Women process and dry the fish that males capture. In an effort to lower the chance of tragedy, women plant plants, including fruit trees. Fishing nets are used or sold by their male members. During and after a disaster, women prepare meals for the family. To get water, they employ tube wells or filters (Alam & Rahman, 2014). Climate change risk perception and management may be directly impacted by women's gendered roles (Rakib et al., 2017). Women reduce the family's risk by being prepared for calamities in advance. Dry firewood is stored for emergencies and mud stoves are made. People may survive a food shortage with the aid of dry foods. Women may put up greater effort when faced with a calamity, but for a variety of reasons, they are less adaptive than men due to less control over assets and resources.

In Nepal out of twenty policy documents on climate change only two integrate gender considerations with the effects of climate change (Paudyal et al., 2019).

According to Nepal Labour Force Survey 2017/18, 61% women are involved in subsistence farming, whereas only 47% of men are. Women also have limited access to timely weather forecast information which affects their options for farming, they lack independent sources of income, and may be unable to access credit. In addition, it has been noted that agricultural patterns and livestock output altered by climate change have an impact on the gendered division of labour, meaning that women are responsible for greater work in the home, on the farm, and providing care for others (Eastin, 2018; Rao et al., 2019). When there are risks of climate change, male out migration increases and women are left behind with greater responsibilities of households and care workload of women. Climate change has been observed to cause greater job pressures and loss of social and economic support for women in India who are impoverished, from lower castes or ethnic minorities (Rao & Mitra, 2013). One of the most crucial things for advancing gender inclusion in climate policies in Nepal is gathering gender-disaggregated data from all sectors involved in climate change. The absence of such information results in gender-blind climate change strategies, which may unintentionally increase gender-based vulnerability (Mainlay & Tan, 2012). Women also have limited access to timely weather forecast information which affects their options for farming, they lack independent sources of income, and may be unable to access credit.

Impact of climate change are not gender neutral. According to Germanwatch (2021), Pakistan is one of the top 15 nations in the world that is most susceptible to the effects of climate change. Over the past ten years, there has been an increase in the frequency of extreme weather events, which has resulted in several fatalities and immense misery for people. Natural disasters caused by climate change, like heat waves, droughts, storms, and floods, interrupt lives, destroy livelihoods, and damage property and land.

15.5 Conclusion and Way Forward

Despite their vulnerability, women are viewed as both active and effective agents and promoters of adaptation and mitigation of climate change, in addition to being victims of it. Women have long since acquired the knowledge and abilities necessary for gathering and storing water, preserving and rationing food, and managing natural resources. In Bangladesh, administered by the Local Government Division under the Ministry of Local Government Rural Development and Cooperatives and UNDP, the LoGIC project implements the Community Resilience Fund (CRF), designed to assist women facing the highest climate-related risks in enhancing their resilience to climate change through the adoption of climate-adaptive livelihood strategies. Through the CRF assistance, these women utilize community-driven methodologies to invest in climate-resilient livelihoods such as sunflower cultivation, the cultivation of climate-resistant varieties of rice, pulses, watermelon, and other similar initiatives. Collaboratively, women residing along the coast formulated business proposals and subsequently secured leases for arable lands from local landowners in their vicinity.

Subsequently, they diligently prepared the land for the cultivation of crop varieties resilient to climate change (ReliefWeb, 2024).

Expanding the Research on Gender Equality in Climate-Resilient Agriculture: Enhancing women's understanding of climate-resilient techniques leads to empowerment: Female participants in Vietnam who engaged in training programs focusing on pest management and livestock rearing reported notable boosts in confidence, elevated status within their households, and greater contributions to household income (Chi et al., 2015).

Aligning new technologies with the resources and needs of women farmers: While innovative agricultural practices can aid farmers in addressing climate change, access to assets, time, and resources may not be equally available to both genders. Factors such as secure land tenure, water access, information, and financial resources may vary, impacting individuals' ability to leverage new technologies effectively. Additionally, differing priorities in adopting these technologies may exist between women and men. Climate-smart agricultural practices that demand increased labour inputs initially may hinder their adoption by women farmers, irrespective of the long-term benefits they offer. More studies are necessary to explore the advantages and drawbacks of climate-smart agriculture for women. Additionally, it is crucial to investigate the effects of men's migration for employment on women, including shifts in power dynamics, decision-making, and agricultural productivity.

Inclusive Financial Strategies for Enhancing Resilience: Innovative financing approaches involve collaborating with local women's groups to facilitate access to credit and production assistance. An example of this is the International Fund for Agricultural Development (IFAD) method, which leverages District Value Chain Committees (DVCCs) built upon local women's groups to access financial resources and production support. In Northern Ghana, DVCCs oversee the planning, execution, coordination, and oversight of activities within maize, soya, and sorghum value chains. These committees comprise various stakeholders such as buyers, input suppliers (seeds and fertilizers), service providers (extension and tractor services), financial institutions including rural banks, and farmer-based organizations (FBOs). The empowerment outcomes for women farmers encompassed acquiring additional parcels of land, participating in management training sessions, gaining decision-making authority over their income, venturing into non-traditional activities such as purchasing and operating trucks, and creating employment opportunities for others through small and medium-sized enterprises (Huyer et al., 2021).

Agricultural extension services and climate information need to cater to the needs of both women and men: These services play a vital role in promoting agricultural development and ensuring food security, especially in the face of climate change. Weather forecasts, provided across various timeframes along with a range of new farm management options, can assist farmers in adapting to changing climatic conditions and optimizing input usage. Utilizing multiple communication channels may enhance the effectiveness of reaching women. This can entail enhancing women's access to various platforms such as radio broadcasts, extension events, SMS and voice messages, as well as engagement through community groups, health clinics, and schools. Additionally, considering that women may prioritize and oversee different facets of agriculture compared to men, tailored information suited to their specific needs and responsibilities is essential.

Institutionally addressing adaptation and mitigation for women: This entails a comprehensive approach that extends beyond a sole focus on agricultural productivity. Activities should encompass various aspects of achieving food security in the face of climate change, including income generation, access to savings and loans, and other support services aimed at enhancing women's food access. While national and international organizations often prioritize increasing agricultural production and enhancing natural resource management, there is a need for greater emphasis on strengthening communities' overall adaptive capacity to effectively confront the challenges posed by climate change. Institutions tasked with assisting farmers in enhancing their food security amidst climate change must address broader priorities that extend beyond agriculture and are relevant to the concerns of both women and men. A crucial role of these institutions is to focus on building women's adaptive capacity alongside that of men.

Diverse stakeholders such as policy-makers, governments, and CSOs must include the gender viewpoint when evaluating possible trade-offs. Progress seen from one viewpoint may have adverse consequences for women. Advancements in one field might potentially heighten women's financial reliance while reducing their wage prospects and social standing. Gender analysis is essential in all climate-smart agricultural strategies. There is strong evidence that climate change may worsen or intensify disparities. Resolving gender inequality is not only about correcting an injustice. It is also a significant chance to use skills, expertise, and capabilities that were previously overlooked and underutilized. Granting equitable access to land and other resources for both female and male farmers may result in a 'triple dividend' of increased gender equality, higher food security, and improved climate change adaptation and mitigation. It enables a cost-efficient and revolutionary method for climate-resilient agricultural progress. To achieve this goal, a thorough reassessment of existing agriculture methods and potential climate-smart advancements is required.

References

Abbas Ghulam, et al. (2017). Quantification the impacts of climate change and crop management on phenology of maize-based cropping system in Punjab, Pakistan. *Agricultural and Forest Meteorology, 24*, 42–55. https://doi.org/10.1016/j.agrformet.2017.07.012. https://www.sciencedirect.com/science/article/pii/S0168192317302289

Addati, L., Cattaneo, U., Esquivel, V., & Valarino, I. (2018). *Care work and care jobs for the future of decent work*. International Labour Organization (ILO). https://www.ilo.org/global/publications/books/WCMS_633135/lang--en/index.htm

Aggarwal, P. K., & Sinha, S. K. (1993). Effect of probable increase in carbon dioxide and temperature on wheat yields in India. *Journal of Agricultural Meteorology, 48*, 811–814.

Alam, K., & Rahman, M. H. (2014). Women in natural disasters: A case study from southern coastal region of Bangladesh. *International Journal of Disaster Risk Reduction, 8*, 68–82. https://doi.org/10.1016/j.ijdrr.2014.01.003

American Psychiatric Association. (2023, May). *Climate change and mental health connections.* https://www.psychiatry.org/patients-families/climate-change-and-mental-health-connections

Asfaw, S., & Maggio, G. (2018). Gender, weather shocks and welfare: Evidence from Malawi. *The Journal of Development Studies*. https://doi.org/10.1080/00220388.2017.1283016

Butt, A. P., Berkhout, E., Zaghbour, C. M., Bush, A., Verma, R., & Pheko, L. L. (2023). *Radical pathways beyond GDP: Why and how we need to pursue feminist and decolonial alternatives urgently.* Oxfam. https://policy-practice.oxfam.org/resources/radical-pathways-beyond-gdp-621532

Chi, T. T. N., Paris, T., Le Duy, T. T. T. A., & Loan, D. T. (2015). *Enhancing the roles of women in rice farming as an adaptation strategy to climate change risks: A case study in submergence villages in Hau Giang province, South Vietnam.* Cuu Long Rice Institute.

Eastin, J. (2018). Climate change and gender equality in developing states. *World Development, 107*, 289–305. https://doi.org/10.1016/j.worlddev.2018.02.021

Economic and Social commission for Asia and the Pacific. (2022). Asia-Pacific riskscape @ 1.5°C: Subregional pathways for adaptation and resilience. *Asia-Pacific Disaster Report 2022 for ESCAP Subregions*. https://www.unescap.org/kp/2022/asia-pacific-disaster-report-2022-escap-subregions-summary-policymakers

FAO. (2017). *Country fact sheet on food and agriculture policy trends: Mali. Fact sheet*. FAO Rome.

Germanwatch. (2021). *Global climate risk index 2021*. https://www.germanwatch.org/en/cri

Global Facility for Disaster Reduction and Recovery. (2018). Gender equality and women's empowerment in disaster recovery. *UN Women*. https://www.gfdrr.org/sites/default/files/publication/gender-equality-disaster-recovery.PDF

Huyer, S., Simelton, E., Chanana, N., Mulema, A. A., & Marty, E. (2021). Expanding opportunities: A framework for gender and Socially-inclusive climate resilient agriculture. *Frontiers in Climate, 3*, 718240. https://doi.org/10.3389/fclim.2021.718240

Knox, J., Hess, T., Daccache, A., & Wheeler, T. (2012). Climate change impacts on crop productivity in Africa and South Asia. *Environmental Research Letters, 7*, 034032.

Lal, M. (2011). Implications of climate change in sustained agricultural productivity in South Asia. *Regional Environmental Change, 11*, 79–94.

Lambrou, Y., & Nelson, S. (2010). *Farmers in a changing climate: Does gender matter? Food security in Andhra Pradesh, India*. FAO.

Lopes, A. P., & Jomalinis, E. (2011). Agroecology: Exploring opportunities for women's empowerment based on experiences from Brazil. *Feminist perspectives towards transforming economic power*. Association for Women's Rights in Development (AWID).

Mainlay, J., & Tan, S. F. (2012). *Mainstreaming gender and climate change in Nepal*. International Institute for Environment and Development. http://www.jstor.org/stable/resrep01234.1

Oxfam. (2019). Stories of climate change: Learn and think about the impacts of the climate crisis on people's lives. *Education Resource Brief*. Oxfam.

Parry, M. L., Rosenzweig, C., Iglesias, A., Livermore, M., & Fischer, G. (2004). Effects of climate change on global food production under SRES emissions and socio-economic scenarios. *Global Environmental Change, 14*, 53–67.

Paudyal, B. R., Chanana, N., Khatri-Chhetri, A., Sherpa, L., Kadariya, I., & Aggarwal, P. (2019). Gender integration in climate change and agricultural policies: The case of Nepal. *Frontiers in Sustainable Food Systems, 3*, 66. https://doi.org/10.3389/fsufs.2019.00066

Porter, J. R., & Semenov, M. A. (2005). Crop responses to climatic variation. *Philosophical Transactions of the Royal Society of London b: Biological Sciences, 360*, 2021–2035.

Rabbani, M. G., Khan, Z. M., Tuhin, M. H., Naznin, Z., Emran, D. A., & Karim, S. (2015). *Climate change and migration in Bangladesh: A gender perspective*. Bangladesh Centre for Advanced Studies for UN Women. https://asiapacific.unwomen.org/en/digital-library/publications/2016/01/climate-change-and-migration-in-bangladesh

Rakib, M. A., Islam, S., Nikolaos, I., Bodrud-Doza, M., & Bhuiyan, M. A. (2017). Flood vulnerability, local perception and gender role judgment using multivariate analysis: A problem-based "participatory action to Future Skill Management" to cope with flood impacts. *Weather and Climate Extremes, 18*, 29–43. https://doi.org/10.1016/j.wace.2017.10.002

Rao, N., & Mitra, A. (2013). Migration, representations, and social relations: Experiences of Jharkhand labour to Western Uttar Pradesh. *Journal of Development Studies, 49*, 846–860. https://doi.org/10.1080/00220388.2013.778399

Rao, N., Lawson, E. T., Elaine, T., Raditloaneng, W. N., Solomon, D., & Angula, M. N. (2019). Gendered vulnerabilities to climate change: Insights from the semi-arid regions of Africa and Asia. *Climate and Development, 11*, 14–26. https://doi.org/10.1080/17565529.2017.1372266

Rasul, G., Saboor, A., Tiwari, P. C., Hussain, A., Ghosh, N., & Chettri, G. B. (2019). Food and nutrition security in the Hindu Kush Himalaya: Unique challenges and niche opportunities. In P. Wester, A. Mishra, A. Mukherji & A. B. Shrestha (Eds.), *The Hindu Kush Himalaya assessment: Mountains, climate change, sustainability and people* (pp. 301–338). Springer.

ReliefWeb. (2024). *Cooperative farming makes Bangladesh's coastal women farmers climate-resilient*. https://reliefweb.int/report/bangladesh/cooperative-farming-makes-bangladeshs-coastal-women-farmers-climate-resilient

United Nations. (2019, June 12). *Differentiated impacts of climate change on women and men; the integration of gender considerations in climate policies, plans and actions; and progress in enhancing gender balance in national climate delegations*. United Nations FCCC/SBI/2019/INF.8. https://unfccc.int/sites/default/files/resource/sbi2019_inf8.pdf

World Bank. (2009). *Gender in agriculture sourcebook*. World Bank. https://documents1.worldbank.org/curated/en/799571468340869508/pdf/461620PUB0Box3101OFFICIAL0USE0ONLY1.pdf

World Bank & FAO. (2017). *Training module. How to integrate gender issues in climate-smart agriculture projects*. https://www.fao.org/documents/card/en/c/45d93533-c024-4a11-90ec-feea329b2e39/

Chapter 16
Emerging Role of Microbes in Bioconversion of Biomass to Value Added Products: Current Insight and Future Status

Chetan Kumar Joshi, Susheel Kumar Singh, Mansi Rai, Nidhi Singh, Rukhsar Khan, and Garima Awasthi

16.1 Introduction

Sustainable environment has appeared to be the crucial theme for researchers due to unnecessarygeneration of waste and its harmful effect on the living organisms. A recent report mentioned the agriculture sector has contributed the hugeaggregate production of agricultural waste/residues every year (Rai et al., 2021). In this regards, Belewu and Babalola (2009) were mentioned the production of agrowaste, like rice and wheat straw approximately 673.3 and 709.2 million metric tons (MMT), respectively. The estimated generation of agricultural residues are approximately 147.2 MMT in all over the world, includes processed agro-industrial wastes and crop

C. K. Joshi
Department of Zoology, Government Science College, Sabalpura, Sikar, Rajasthan, India

S. K. Singh (✉)
Department of Physics, Dr. Shakuntala Misra National Rehabilitation University, Lucknow 226013, Uttar Pradesh, India
e-mail: susheelsingh489@gmail.com

M. Rai
Department of Microbiology, Central University of Rajasthan, NH-8, Bandar Sindri, Ajmer 305817, Rajasthan, India

N. Singh
Faculty of Agriculture, Medi-Caps University, Indore 453331, India

R. Khan
Department of Life Sciences, I.E.S University, Bhopal 462044, Madhya Pradesh, India

G. Awasthi
Department of Biotechnology, Amity University, Lucknow, Uttar Pradesh, India

P. K. Rai and S. Rai (eds.), *Climate, Environment and Agricultural Development*, Advances in Geographical and Environmental Sciences,
https://doi.org/10.1007/978-981-97-8363-2_16

residues (Sadh et al., 2018), rather than it forest residues, food industry waste, household wet waste, poultry and dairy waste are also contributed significant portion for valuable product and bioenergy production (Xu et al., 2019). Thus, the whole world is facing problem of the buildup wastes and its disposal causes a severe problem in the ecological balance as well as burning concern of sustainable environment (Leow et al., 2018). Improper disposal of these wastes has raised numerous complications that alarmed the fluctuation in environmental pollution, soil physiochemical properties, soil fertility, agro-ecology, and deleterious effect on animal and plant vigor (Erna et al., 2019; Rai et al., 2021). Bhuvaneshwari et al. (2019) reported the different disposal strategies of untreated biomass, either through unplanned landfilling or burning (burning of parali), dumping into pits. This has raised a severe harmful impact on useful microbes, fertility of soil, reshaping and shifting of microbial communities, soil physiochemical properties, greenhouse gases emission, soil erosion, air pollution, and climate change (Bos & Hamelinck, 2014; Singh & Nain, 2014).

Agricultural activities are dominating in all over the world, that's why the global production of agriculture waste in a year is approximately 998 million tons, while 500 million tons, alone reported in India (Bhuvaneshwari et al., 2019; Loow et al., 2015). The basic composition of biomass residues and agro-industrialwastes comprises two major components, soluble components (e.g., organic acids, amino acids, and sugar) and insoluble chemical components for e.g., cellulose (40–50%), hemicelluloses (25–30%), and lignin (15–20%) (Shamsul et al., 2018; Rai et al., 2020). Recent studies have validated other essentialcomponents are resins, oil waxes, fats, pigment, mineral and protein. Though, due to high nutritional components, these plant and animal biomass waste are recognized as the chief source for other value-added product and bioenergy production (Rai et al., 2020; Xuet al., 2019). Therefore, plant, animal and agro-industrial biomass wastes are used as the cheapest source for microbial growth through utilizing nutritional components ofbiomass wastes and further production of value-added products through different biotechnological bioconversion technique(fermentation and bioconversion/biodegradationprocess). Thus, scientistsare exploringemerging capabilities of microbes for environmental concern and biomass wastemanagement as well as promoted natural microbial conversion of biomass into valuable products. Hence the reprocessing of these biomass wastes is burning concern of sustainable environment andrestoration of ecological balance (Bhunashwari et al., 2019).

Microbes are extensively distributed in the extreme environmental conditiondue to that they have diversenutritional versatility, and metabolic ability to grow and utilize different biomass waste as substrate.In this perspective, researchers explored diverse group of microbial communities including fungi,archaea and bacteria as leading natural bioconvertor isolated from diverse microbiome (Anal et al., 2020; Rai & Solanki, 2014). Therefore, the application of biomass residues bioconversion via biotechnological process displays immense consideration towards involvement of microbes for resolving the deleterious effect of biomass build-up and production of valuable products (antioxidants, enzymes, amino acids, organic acids, single-cell protein (SCP), edible mushroom, biofuel, animal feed or fodder, bioactive metabolites, nanomaterial and biofertilizers) (Nguyen et al., 2010). With the advent of novel biotechnological approaches, numerousinnovative prospects have unlocked for application of biomass waste with reducing the threats of environmental pollution.

In this chapter we discussed the significance of microbe-mediated biomass waste and residues bioconversioninto bioenergy and valuable product as well as recycled the generated plant and animal biomass for environmental/ecological balance. The ongoing scientific studies are demonstrated the emerging role of microbes in biomass conversion and recently biotechnological approaches/or bioconversiontechnologies into valuable products in order to minimize production costs and environmental concern. The aim of this chapter to reveal the current trend of bioconversion technologies for transforming biomass waste to valuable products, specifically the microbes involved in transformation, and different prospects for minimizing the threat of environmental imbalance.

16.2 Biomass Generation as Waste

Biomass wasteis a considerable term basically includes plant and animal origin residues as waste. To be more specific, plant biomass waste comprises agro-industrial waste (for eg.leaves,stalks, straws and stems of cereal grains (barley, corn, rice, andwheat), hulls, cobs, legume waste, fruit peels, bagasse, husks, food processing wastes (oil cakes)and any part of a processed plant source) (Yazid et al., 2017), forestry waste (hard and soft wood) and wood industry waste (wood chips and saw dust, timber slash, mill scrap) (Rojas-Higuera et al., 2017). Agriculture biomass waste is mostly classified into two types of wastes, i.e., agriculture residues and processed agriculture residues.Agricultural biomass residues are outcome of various post-harvest agriculture processes (Lim & Matu, 2015). The agriculturalbiomass residues chemical composition comprises lignocellulosic materials and polyphenolic compounds that undergo through complex processes for complete bioconversion. In contrast to that animal biomass waste includes dairy waste, poultry waste, cattle manure, fish/shrimp waste, pig bristles, etc.

16.2.1 Agricultural Biomass Waste

Agriculture biomass residues/wasteisgenrated through post-crop harvestingand farming process (Rajkhowa et al., 2019). A huge amount of agriculture biomass residues can be exploited as animal fodder, fertilizers, soil improvement, setup for small industries like mushroom beds, manufacturing of human useful things, and various other processes. These biomass residues containhulls, leaves, seed pods, stems, plant stalks, pulp, vegetable matter, molasses, mushroom bedding, seeds residues, husks, bagasse, straw, shell, roots, stubble, peel, etc. that is often of no use and will be thrown away without proper discarding (Lim & Matu, 2015). The plant biomass waste is abundant in lignocellulose, composed of polysaccharides like lignin, hemicellulose and cellulose (Singh & Nain, 2014). Numerousplant residues/waste that comprise up to 20–30% lignin–hemicellulose–have potential biotechnological values and their bioconversion and/or fermentation yield industrially important components including enzymes,antioxidants, organic acids, biofertilizers, antibiotics, biofuels, and biogas (Rai et al., 2020).

16.2.2 Processed Agro-Industrial Biomass Waste

Processed agricultural biomass residues/waste can be described as the generation of residues/waste after processing of crop into a valuable alternate resource. In India,roughly 20% biomass residues/wasteis generated through cereals, vegetables and fruits producers every year (Rai & Solanki, 2014; Sadh et al., 2017). A massive amount of processed agro-industrial residues/waste produced through the processing industries like fruit, beverages, juice, confectionery, snacks/chips, and oil industries anditscompositions(hemicellulose, cellulose, lignin, carbon, nitrogen, moisture, ash,)allow for production of different generation energy sources, like bio-ethanol, biodiesels, biogas, biofertilizers and other commercially products. Oilindustries are other potential sources of processed residues/wastes that comprise diverse substrate based oil cakes like groundnut oil cake, canola oil cake, cottonseed cake, olive oil cake, palm kernel cake,soybean cake, coconut oil cake, mustard oil cake, sesame oil cake, sunflower oil cake, and rapeseed cake (Ramachandran et al., 2007). These processed agro-industrial residues/waste are relatively cheap, useful, rich chemical constituents that can be utilized as alternative substrates for fermentation.

16.2.3 Animal Biomass Waste

Animal biomass waste is rich inorganic content, nitrogen sources,potassium and high amount ofrecyclable phosphorous. It includes diverse type of waste such as,household organic waste, poultry waste, dairy waste, fish/shrimp waste, cattle manure, and pig bristles (Pampuro et al., 2017). Organic animal biomass waste/residues are rapidly degraded in few days according to environmental conditions, while exoskeletons/endoskeleton of animalsis stable and recalcitrantto enzymatic or chemical breakdown, so it takes time to decompose (Pampuro et al., 2018).These approaches reduce the cost of biomass degradation rather innovative and potentiallyeconomical approach and generatenumerous value added products such as, biogas, organic compost,biofertilizersand bio-nanomaterials (Pampuro et al., 2018; Xu et al., 2019).

16.3 Emerging Role of Microbes and Bioconversion of Biomass

Microorganisms play indigenousrole in the recycling of plant and animal biomass wastes (Ferdes et al., 2020). They have inherent metabolic capability and biosynthetic ability to transform, modify, and utilize biomass waste to obtain value-added product and production of bioenergy as well as drag new insight towards microbe mediated natural bioconversion (Rai et al., 2020). Several recent studiesdemonstrated multifarious functions of microbes in concern of decomposition/biodegradation ofbiomass residues into valuable products and bioenergy. The unique metabolic machinery of microbesrepresent them as potential decomposerand its enzymatic activity enhanced bioconversion/biodegradationrateof biomass residues through natural decomposers.There is a necessity of evolutionary research on emerging role of microbes that transformbiomassresidues into high-quality, stable, and valuable products/bioenergy in a comparatively shorter period (Sharma et al., 2023). Recently developed biotechnological methods has made substitutions for large-scale production of valuable product/bioenergyto deliver an effective way to advance food processing and quality, human health, environmental protection, food safety, crop production and production of value-added products.

The researchers demonstrated the synergistic action microbes such as bacteria, actinomycetes, yeast, fungi and mycorrhiza, are massively involved in transformation/bioconversion of complex lingo-cellulosic biomass residues into simple molecules through the catalytic action of microbial enzymes such as glucanases, cellulases, pectin methyl esterase, glycosidase hydrolases, xylanase, hemicellulases, carbohydrate esterases and polysaccharide lyases (Gayen & Ghosh, 2011; Mahalakshmi et al., 2016). These degraded simplest biomass residuesfurther subjected for production of value-added products includesantioxidants, organic acids,paper, enzymes, fuel, chemicals, biofuels, biogas,textile, and agricultural inputs (Ferdes et al., 2020; Ingale et al., 2014; Rajkhowa et al., 2019). Several recent reports documented the isolation, characterization and application of microbes (bacteria, actinomycetes, yeast, fungi and mycorrhizal fungi) for significant conversion ofplant lingo-cellulosic (20–30%) abundant biomass waste (Vishan et al., 2017; Yildirim et al., 2015).These processes includes different groups of microbes thatperform better either singly or in combination with other organisms for the production of valuable products using newly developed biotechnological methods (Rajkhowa et al., 2019; Ribeiro et al., 2017). Microbes also play an indispensible role in decomposition of discarded biomass waste/residues and sustained the nutrient pool in the soils, which mobilized into the plants and microbial biomass and maintain ecosystem balance (Chen et al., 2016; Pampuro et al., 2018; Rai et al., 2020). Compostingis best way to conserve the nutrient pool in the soiland microbe mediated degradation of organic biomass residues providesa number of benefits like as enhanced soil fertility, soil conditioning and maintained biogeochemical cycles which can lead to improved agricultural yield, enhanced soil biodiversity, reduced ecological risks and a healthier environment (Rai et al., 2020). These benefits mark composting as aperfect option for processing of the huge quantities of biomass residues through a natural succession of microflora (Rai et al., 2020). Several microbes like bacteria, and fungi (*Trichoderma harzianum, Polyporus ostriformis, Pleurotus ostreatus,* and *Phanerochaete chrysosporium*) are known to degrade lignocellulosic biomass into compost (Rubio et al., 2013; Sarkar & Chourasia, 2017). A series of emerging microbes and their metabolic ability that help in fast biodegradation, decomposition, and bioconversion of biomass into valuable products are recorded in Table 16.1.

16.3.1 Bacteria Mediated Bioconversion

The bacterial communities are ubiquitously present in extremeenvironments, therefore their prolific growth, physiological adaptation and metabolic enzymes are capable to convert lingo-cellulosic biomass wastes into organic simplest molecules. Several researchers documented bacteria mediated bioconversion of cellulosic material (leaves, rice bran, wheat bran, paddy straw, maize straw, black gram husk, bagasse, apple, vegetable waste, forest litter, plum wastes, cereal wastes, straw, saw dust,trash etc.,) into soluble and nonhazardous products in concern of valuable items production, sustainable environment and human welfare (Kaur et al., 2015; Oliveira et al., 2017; Singh et al., 2019). Effective bioconversion of organic biomassthrough action of bacterial enzymes, such as xylanase, cellulase, manganese peroxidases, laccase, and lipid peroxidase, secretedby cellulose degrading bacteria include the *Bacillus megaterium, B. cereus, Amycolatopsis mediterranean, Xanthomonas campestries, Pseudomonas* spp. and *Serratia marcescens* had been reported by several researchers (Faisal et al., 2014; Sadh et al., 2018; Singh et al., 2019).

16.3.2 Fungimediated Bioconversion

Fungi are eukaryotic, aerobic,saprophytic microorganisms which comprises unicellular (yeasts) to filamentous hyphal/mycelial (moulds) (Rai et al., 2020). Fungal communities are also contributing significant role in bioconversion/decomposition of biomass residues and restore the nutrient pool in the soils,that's why considered as biodegraders, decomposers, and bioconverters of non-useful products (Gautam et al., 2012). They efficientlyconvert the natural polymers of biomass waste with the help of extracellular multifarious enzymaticmachineryand metabolic pathways,saprophytic nature induces growth over biomass residues, bioaccumulate the degraded simplest compoundfor fungal species growth and remove the lethal wastes from the ecosystem (Rai et al., 2020). Their mycelia provides anchor tocolonize, establish, and breach substrates rapidly that assist in transportation and redistribution of nutrients within their biomass (Rai et al., 2020). Numerous fungi such as, *Aspergillus awamori, A. niger, Trichoderma reesei, T. harzianum, Penicilliumbrasilianum, Pleurotuseryngii,*

Table 16.1 Illustration of recent studies of bioconversion of biomass waste of plant and animal origin using biotechnological approaches and emerging role of microbes in bioconversion/degradation of biomass residues into valuable products

Biomass Bioconversion	Microorganisms	Valuable products	Biotechnological approaches	Mechanism involved in bioconversionand Relevant Remark	Reference
Enzymes					
Potato peel	Xanthomonas campestries	Xanthan	Solid state fermentation	Bioconversion of biomass into its monomer with action of microbial enzyme	Vidhyalakshmi et al. (2012)
Rice husk, rice bran, Wheat bran, Wheat straw, Black gram husk, Corn flour, Sugarcane bagasse, Maize straw, and Paddy straw	Bacillus amyloliquefaciens, B.licheniformis, B. tequilensis	Amylase	Solid state fermentation	Degradation of biomass residues into organic material	Rai and Solanki, (2014), Kaur et al. (2015), Paul et al. (2020)
Soybean,rice bran, black, Wheat bran, Gram bran,Sugarcane bagasse, Groundnut oil cake, and Coconut oil cake	Achromobacter xylosoxidans, Aspergillus niger, Rhizopus microspores var. oligosporus	Amylase, cellulase and xylanase	Solid state fermentation	Degradation of cellulose and lignocellulosic biomass and enzyme production	Mahalakshmi and Jayalakshmi (2016), Fernández Núñez et al., (2017)
Groundnut oil cake, Coconut oil cake, and Soyabean oil cake	Pseudomonas spp., Aspergillusmellus, A. niger	Lipase	Solid state fermentation	Bioconversion of lipid-containing biomass and enzyme production	Faisal et al. (2014),

(continued)

Table 16.1 (continued)

Biomass Bioconversion	Microorganisms	Valuable products	Biotechnological approaches	Mechanism involved in bioconversionand Relevant Remark	Reference
Soybean meal waste, wheat bran, caorncob cassava peel, and citrus pulp	Rhizopusarrhizus,Mucorsubtillissimus and A. oryzae	Protease	Solid state fermentation	Bioconversion of protein biomass into monomers and enzyme produced	Thakur et al. (2015), Nascimento et al. (2015)
Mango and Banana peel, Vegetable waste and Agro-industrial waste	Penicilliumspp., Schizophyllum commune, Melanocarpus sp., Aspergillus sp., Trichoderma reesei	Lignocellulases, Cellulase	Solid state fermentation	Bioconversion of cellulose and hemicellulose-containing biomass and enzyme production	Saravanan et al. (2012), Ferreira et al. (2014), Obeng et al. (2017)
Alcohols					
Banana stem	A. ellipticus and A. fumigatus	Bioethanol	Fermentation technique	Bioconversion of lignocellulosic biofuels	Ingale et al. (2014)
Wheat straw and agrowate	Penicillium janthinellum	Bioethanol	Fermentation technique	Employing on cellulaseswithout carbohydrate-binding modulesand bioethanol production from processed agrowasteviamicrobial enzymatic action	Pakarinen et al. (2014), Singhania et al. (2014)
Starch containing agriculture waste	Clostridium beijerinckii	Butanol	Fermentation technique		Maiti et al. (2016)
Vegetable's waste-potato peel, carrot peel, and onion peel	Saccharomyces cerevisiae	Bioethanol	Fermentation technique		Mushimiyimana and Tallapragada (2016)

(continued)

Table 16.1 (continued)

Biomass Bioconversion	Microorganisms	Valuable products	Biotechnological approaches	Mechanism involved in bioconversionand Relevant Remark	Reference
Antioxidants					
Peanut oil cakes, Rice starch	Bacillus spp., B. subtilis,	Antioxidant peptides	Submerged Fermentation	Production of antioxidant peptide through microbial bioconversion	Zhang et al. (2014)
Tomato wastes, Beetroot pomace, Carrot pomace	Bacteria and Fungi	Lycopene,Betalains, Carotenoids	Fermentation technique	Bioconversion of vegetable waste product for antioxidant production through microbes and apply in different biotechnologicalpurposes and health-related problems	Martins and Ferreira (2017),Kushwaha et al. (2018)
Waste products ofgrapefruit	Aspergillus niger	Antioxidant	Solid State Fermentation	Bioconversion of grapefruit by-products through action of microbial enzymes and produced antioxidants	Larios-Cruz et al. (2019)
Grape pomace	Actinomucor elegans and Umbelopsis isabellina	γ-linolenic acid, carotenoids, phenolic compounds and flavonoids	Solid State Fermentation	Bioprocessing of fruit waste for antioxidant production through Zygomycetes fungi	Dulf et al. (2020)
Antibiotics					

(continued)

Table 16.1 (continued)

Biomass Bioconversion	Microorganisms	Valuable products	Biotechnological approaches	Mechanism involved in bioconversionand Relevant Remark	Reference
Coconut oil cake and ground nut shell	Amycolatopsis mediterranean	Antibiotics	Solid state fermentation	Bioprocessing of oil cakes for production of antibiotics	Vastrad and Neelagund (2011)
Corn cobs, sawdust, rice hulls, and groundnut shell	Bacteria, Fungi and Actinomycetes	neomycin, oxytetracycline, and rifamycin	Solid state fermentation	Bioprocessing of Agrowaste	Kashif and Shahid (2022)

P. ostreatus, Polyporus ostriformis and *Phanerochaete chrysosporium* are well-known to bioconvert/biodegradade lignocellulosic biomass through enzymatic action viz., xylanases, cellulases, pectinase, glyoxal oxidase, lignin peroxydases, laccase, manganese peroxydases, esterase, and glucosidase (Mahalakshmi & Jayalakshmi, 2016; Singh et al., 2019; Yildirim et al., 2015). Several researchers reported decomposition and degradation of diverse biomass residues such assisal leaves, rice straw, sugarcane bagasse, wood chips and woody shavings through various fungal genera, namely, *Phanerochaete chrysosporium, Trametes versicolor* and *Pleurotus fabellatus,* were showed enhanced transformation when applied in combination with plant biomass residues (Huang et al., 2009).

16.3.3 Factor Influencing Microbial Bioconversion

Biomass degradation is a sequentialand continuous process which immobilize transformed simplest compound for growth, and/or enhanced decontamination of numerous biomass waste encompassing high lignocellulosic material from the surroundings via action of bacteria, actinomycetes, fungi, nematodes, invertebrates and plants (Rai et al., 2020). The bioconversion rate of biomass depends on various elements; including, the organic component and biochemical nature of biomass waste, physio-chemical characteristics of the environment (soil physio-chemical properties, type of soil, pH, temperature, water content, the presence of oxygen or other electron acceptors, humidity, and nutrients), indigenous microbes and biomass residueaccessibility to microbes (Schröderet al., 2018; Singh et al., 2019). The physicochemical nature and organic content of the diverse biomass waste and metabolic specification of the microbes are governed microbial interaction with substrate and the rate of biodegradation/bioconversion process. However, microbial growth, biomass utilization capacity, catalytic activity and kinetics of biomass degradation are affected by soil features such as,soil texture, availability of nutrients,solubility in water, temperature, pH, redox potential, moisture, oxygen content and soil indigenous microbial community (Rai et al., 2020). These indigenous microbes grow, reproduce and survive through utilizing nutrients such as nitrogen, carbon, potassium and phosphorous,available in soil and maintained through biodegradation of biomass residue.

Another important physical factor is pH. pH regulates the growth, metabolic activity and survival of microbes as well asproperties of the biomass residue (basic, acidic, neutral and alkaline) that would be transformed into valuable product through microbial action. The fluctuation in pH values (higher or lower) revealed substandard results while good result obtained at optimum pH condition for proper degradation of biomass through metabolic activity of microbes. In contrast, there are some other important influential physical factorslike temperature that regulate the survival of microorganisms, physiological activity of microbes and decomposition process of biomass waste. Therefore,fluctuation in temperature regulates the metabolic or physiological process of microbes, so the metabolic turnover of microbes is not maximum at every temperature, as result of that the bioconversion processes either speed up or slow down. Basically, extracellular enzymes of microbes are participated in the degradation process of biomass is maximum at optimum temperature, whilethe specific composition of biomass residue requires specific temperature for degradation. The demand of oxygen is another conclusive factor for microbial growth, physiological process and development as well as bioconversion of biomass residues (Rai et al., 2020). The concentration of oxygen is altered for diverse organisms (anaerobic, aerobic, obligate and facultative aerobic or anaerobic) which expedite the biotransformation rate in a better way. Oxygenis a gaseous requirement for most aerobic organisms.Microbial bioconversion is processed (degradation, bioconversion and fermentation process) throughanaerobic and aerobic condition. Microbial growth is dependent on content of water to accomplish their metabolic activity, so moisture content has substantial effect on bioconversion of biomass waste (Rai et al., 2020).

16.4 Microbial Bioconversion of Biomass into Valuable Product

A series of valuable products, such as antioxidants, biofertilizers, enzymes, anino acids, biopesticides, bioethanol, organic acids, aroma compounds, biosurfactants, single cell proteins, antibiotics, animal feed, biogas, pigments, and vitamins has been produced by bioconversion of biomass residues through different biotechnological methods reported by numerous researchers (Rai & Solanki, 2014; Sadh et al., 2018; Tsouko et al., 2017). A broad range of emerging microbes are applied for the production of these bioproductsat the lab and industrial scale through bioconversion/fermentation of biomass residues through the microbial enzymes or active metabolites. For that reason, biotechnological methods and their developing technologies are efficiently used for the production of value-added products through bioconversion/fermentation of biomass residue as well asopened a new idea for recycle of biomass wasteand sustainable environment.

16.4.1 Enzymes

The basic component of biomass waste influences the production of diverse valuable bioproducts. In this order, broad range of plant origin biomass waste, including fruit peel waste, rice/wheat bran, coconut husks, sugarcane bagasse, corn cobs, pomegranate peel waste, oil cakes, candelilla stalks, soybean waste, black gram bran, and flour mill waste were used by numerous researchers for the production of enzymes (Rai & Solanki, 2014; Saharan et al., 2017). Recently, more than one combination and composition ofbiomass residues/waste were used for the growth of microbes and fermentation of numerous enzymes like as glucoamylase (Suganthi et al., 2011),amylase (Rai & Solanki, 2014), xylanase (Azzouz et al., 2021), invertase (Mehta & Duhan, 2014), lipase (Oliveira et al., 2017), cellulase (Hernandez et al., 2018; Timo et al., 2017), Pectin methyl esterase (Gayen & Ghosh, 2011), and β-glucosidase (Sadh et al., 2017). Several recent studies documented the enormous bacteria viz., *Bacillus* sp. (Sodhi et al., 2005), *Paenibacillus xylanilyticus, P. terrae, Pseudomonas aeruginosa* (Dharmendra, 2012), *Pediococcus acidilactici*, actinomycetes viz., *Streptomyces lividans, Kitasatospora* sp., *Cellulomonas* sp., *Nonomuraea* sp., *Micrococcus* sp., and fungi viz., *Aspergillus niger, A. terreus,* (Sethi et al., 2016), *Fusarium* sp., *Penicillium notatum* (Gayen & Ghosh, 2011), *Trichoderma reesei, Candida rugosa* (Rekha et al., 2012), *Thielavia* sp., (Garcia-Huante et al., 2017), for different enzymes production from biomass waste (Table 16.1).

16.4.2 Antioxidants

The natural antioxidants serve numerous useful properties like as anti-infammatory, antimicrobial, antiviral, hepato-protective, anti-cancer, and anti-tumor activity, have a tendency to be detoxify chemical substance of human body and stabilize natural biomolecules of human cells (Fierascu et al., 2019; Martí-Quijal et al., 2021). Likewise, antioxidants (β-Sitoterol, α-tocopherol, flavonoids, carotene, caffeic acid, terpinene, cymene, lycopene, betalains) are also produced through different biomass reidues/waste (apple seed, orange peel, coconut shell, pineapple waste, strawberry, lemon peel, citrus pulp, carrot pomace, lettuce, tomato peel, grape skin, black current and pomegranate) with the use of microbes, *Bacillus* spp., *B. subtilis, Aspergillusawamori, A. niger, A. oryzae, Rhizopus* sp., *Actinomucor elegans, Umbelopsis isabellina, Mucor circinelloides* (Dulf et al., 2020; Hameed et al., 2017; Rashad et al., 2015; Sadh et al., 2017) (Table 16.1).

16.4.3 Antibiotics

The microorganisms comprise inherent ability to produce diverse group of antibiotics (substances that effectively prevent the growth or destroy/kill pathogenic microorganisms at low concentrations) (Tripathi, 2008). Recent studies have revealed thepproduction of antibiotics (L-Asparagine, oxytetracycline, penicillin, rifamycin B, and other important antibiotics) from a number of waste biomasscomprisingcorn cobs, cassava waste, ground nut shell, coconut oil cake, sawdust, and rice hulls, are another promising valuable bioproduct (Vastrad & Neelagund, 2011) (Table 16.1). At present, diverse group of microbes are described to the production of antibiotics such as *Penicillin chrysogenum, Streptomyces rimosus, Amycolatopsis mediterranean*, and *Pseudomonas plecoglossicida* (Onyegeme-Okerenta et al., 2013). In recent time, there is an increasing interest in the production of antibiotics from biomass residues with the help of antibiotic producing microbes/or genetically modified organism and newly developed biotechnological approaches support for the screening of antibiotic producing microbes,yield, purification and quantification of products.

16.4.4 Biosurfactant

Biosurfactant is also useful product for mankind that can be produced by using biomass residudes/waste such as peanut cake, sunflower oil, castor oil, mustard oil, barley bran, and rice bran, through the action of microbes (*Pseudomonas aeruginosa*) (Saravanan & Vijayakumar, 2014).

16.4.5 Biofuel and Biogas

The rising demand of fuels is due to exponential growth of population and rapid industrialization that reasons the excessive exploitation of natural resources, hence the search of alternative low-priced energy source is on demand. Therefore, the production of cost effective energy source as biogas and biofuel from biomass waste/residues of animal and plant origin becomes thought-provoking substitute of natural fuels. Recent reports exposed the production of second or third generation bioethanol as biofuels from different plant origin biomass residues containing high lignocellulosic material like potato waste, wheat, cotton stalks, corn stalks, cassava waste, sugarcane bagasse, beans, sawdust, rice straw, sweet potato waste, sugar beet and vegetables/ fruit waste like carrot peel, potato peel, banana waste and onion peel (Cai et al., 2016; Saini et al., 2014). Basically microbes (bacteria, yeast, fungi) mediated biofuel production encompasses biochemical conversion of biomass residues/waste through different biotechnological techniques, like alcoholic fermentation, anaerobic digestion, and photobiological process. Several studies documented the promising role of microbes for second or third generation ethanol production, including as, *Saccharomyces cerevisiae* (Mushimiyimana & Tallapragada, 2016), *Aspergillus ellipticus* and *A. fumigatus* (Ingale et al., 2014), *Clostridium beijerinckii* (Maiti et al., 2016).

Biogas is considered as energy-rich gas and its production employed anaerobic digestiontechnique fortreatment ofbiomass residues/waste of plant andanimal origin.Anaerobic digestionemploy conversion of biomass waste into methane and carbon dioxide (collectively called biogas) and biofertilizer through the action of multifaceted microbial community, including *Bacteroides, Chlorofexi, Firmicutes, Thermotogae, Clostridium* sp., *Thermovirga* sp., *Soehngenia* sp., *Actinomyces* sp., *Synergistetes*, Methanobacteria-*Methanosaeta, Methanolinea, Methanothermobacter, Methanospirillum* and *Methanoculleus* (Wang et al., 2017; Zamanzadeh et al., 2017). In a recent report,biogas production has been documented from differentbiomass residues as well as two weeds, i.e., *Typhaan gustifolia* L. and *Eichornia crassipessolms* Paepatung et al. (2009). Similarly, several researchers documented production of biogas by sequential bioconversion of various agriculture biomass residues and slurries of animal biomass residues with thehelp of diverse group of microbes (Azman et al., 2015; Hernández-Beltránet al., 2019; Otto Wagner et al., 2018; Sun et al., 2015; Wang et al., 2016). Presently,several researches focused on the animal and plant biomass derived biofuels production through anaerobic digestion(biotechnological methods) is cost effective as well as ecofriendly and alternative source of energy for upcoming future (Ferdes et al., 2020).

16.4.6 Single Cell Protein and Mushroom Production

Single cell protein (SCP) are nutritionally rich with high content of protein,produced bywide range of microbes, is recently developedchoice with potential use andeconomical perspective. Therefore, several recent studies has focused on the production of side streams, asSCPby using microbes to improve theirprotein content, further it can be used as animal feed and food source also (Ritala et al., 2017). Presently, single cell protein obtained from the bioconversion of plant biomass wastes (cucumber and orange peels) using microbes viz., *S. cerevisiae*. In addition to that some filamentous fungi, likeand filamentousalgae like *Chlorella pyrenoidosa* (45%), *C. sorokiana* (46–65%),*C. vulgaris* (42–55%) and *Scenesdesmus obliquus* (30–50%) (Waghmare et al., 2018) are comprises high protein content, also involved in production of SCP.

Similary, mushroom production is now operated as a noticeable method for the economical points of view through the bioconversion of agro-based residues/ waste into protein rich food using numerous edible fungi (Randive, 2012). Mushroom used for different purposes, like as a protein-rich food, bioremediation tool and nutrient cycle management. It degraded lignocellulosic material of agriculture residue/waste (rice bran, wheat bran, paddy straw, banana stalks, vegetable waste and bahia grass) through action of dominating fruiting bodies as *Pleurotus tuber-regium, Pleurotussajor-caju, Pleurotuseous* and *Pleurotus platypus* (Lakshmi & Sornaraj, 2014).

16.4.7 Bioconversion of Biomass Residue/Waste in Biocompost/Biofertilizer

Bioconversion of biomass residue/waste is best approach to recycle this waste into high economic valuable biocompost/organic fertilizer and efficient compost (Monroyet al., 2009) for sustainable agriculture. Its effective utilization minimizes the application of chemical fertilizers (Fitriatin et al., 2016; Rajkhowa et al., 2019). Composting is a successive bioconversion of biomass residue/waste into a useful and naturally degradable resource that can achieved through degradation of lingocellulolytic microorganisms such as actinomycetes, bacteria, and several fungi (Bohacz, 2017). The application of compost in soil act as conditioner and perform different function such as improves physical, chemical, and biological properties of soil, enhance fertility of soil, recycle and restore nutrients, and improve plant health (Clara et al., 2017; Han et al., 2017). Several studies reported substantial degradation/bioconversion of biomass waste through actinomycetes and bacterialactions, such as, *Bacillus subtilis, B. licheniformis, B. polymyxa, B. cereus, B. pumilus, B. firmus, B. brevis, B. circulans, B. megaterium, Cellulomonas,Clostridium spp., Pseudomonas spp., Sporocytophaga, Micromonospora, Streptomyces,* and *Thermoactinomyces* (Bohacz, 2017) and fungi like *Trichoderma harzianum, T. viride, Pleurotus ostreatus, Polyporus ostriformis* and *Phanerochaete chrysosporium* are known to play important role in composting of lignocellulosic materials through production of cell wall degrading enzymes (Rai et al., 2016). The co-inoculation practices (organic manure and microbes) improve crop yield through different mechanisms like mineralization, carbon addition, nutrient acquisition, and phytohormone production (Meena et al., 2017). In this order, numerous reported beneficial fungal and bacterial species are applied for fortification of organic compost, includes *Azotobacter, Rhizobium, Azospirillum, Pseudomonas, Bacillus, Candida oleophila, Burkholderia cepacia, Coniothyrium minitans, C. sclerotiorum, Fusarium oxysporum (nonpathogenic), Aspergillus niger, Gliocladium spp., Phlebia gigantean, Pythiumoligandrum, Streptomyces griseoviridis* and *Trichoderma* spp. that significantly improve soil health, and crop yield (Rai et al., 2016).

16.5 Conclusion and Future Perspectives

Biomass residues/wastecomprises rich nutrient constituents and bioactive compounds such as carbohydrates, proteins and minerals that's why it worked as "raw material" for numerousmicrobial bioconversions. Basically, thisbiomass residues/wastenutrient composition facilitates suitable growth conditions for the microorganisms and through successive bioconversion/fermentation processes diverse group of value added products can be obtained. Due to this, emerging role of microbes is become noteworthy for recycling the biomass waste and minimizing the environmental risk, as well as production of valuable product through biotechnological approaches. The increasing interest in microbial based bioconversion of biomass waste and production of value added products, dragging the curiosity for development of genetically modified microbes (insertion of desired microbial genes with specific functions) having fastest bioconversion capacity into value-added products. Therefore, an integrated approach of microbe-mediated biomass waste bioconversion improves ecological services and creates awareness about the biomass waste recycling.

References

Anal, A. K. D., Rai, S., Singh, M., & Solanki, M. K. (2020). Plant mycobiome: Current research and applications. In M. K. Solanki, P. L. Kashyap & B. Kumari (Eds.), *Phytobiomes: Current insights and future vistas* (pp. 81–104).

Azman, S., Khadem, A. F., Lier, J. V., Zeeman, G., & Plugge, C. M. (2015). Presence and role of anaerobic hydrolytic microbes in conversion of lignocellulosic biomass for biogas production. *Critical Reviews in Environmental Science and Technology, 45*, 2523–2564. https://doi.org/10.1080/10643389.2015.1053727

Azzouz, Z., Bettache, A., Boucherba, N., Prieto, A., Martinez, M. J., Benallaoua, S., & de Eugenio, L. I. (2021). Optimization of-1,4-Endoxylanase production by an Aspergillusniger strain growing on wheat straw and application in Xylooligosaccharides production. *Molecules, 26*, 2527.

Belewu, M. A., & Babalola, F. T. (2009). Nutrient enrichment of some waste agricultural residues after solid state fermentation using Rhizopus oligosporus. *Journal of Applied Biosciences, 13*, 695–699.

Bhuvaneshwari, S., Hettiarachchi, H., & Meggoda, J. N. (2019). Crop residues burning in India: Policy challenges and potential solutions. *International Journal of Environmental Research and Public Health, 16*(5), 832.

Bohacz, J. (2017). Lignocellulose-degrading enzymes, free-radical transformations during composting of lignocellulosic waste and biothermal phases in small-scale reactors. *Science of the Total Environment, 580*, 744–754.

Bos, A., & Hamelinck, C. (2014). Greenhouse gas impact of marginal fossil fuel use. *Project number: BIENL14773.*

Cai, D., Li, P., Luo, Z., Qin, P., Chen, C., & Wang, Y., et al. (2016). Effect of dilute alkaline pretreatment on the conversion of different parts of corn stalk to fermentable sugars and its application in acetone–butanol–ethanol fermentation. *Bioresource Technology, 211*, 117–24.

Chen, Y., Guo, R., Li, Y. C., Liu, H., & Zhan, T. L. (2016). A degradation model for high kitchen waste content municipal solid waste. *Waste Management, 58*, 376–385.

Clara, L., Fatma, R., Viridiana, A., & Liesl, W. (2017). Soil organic carbon: The hidden potential. FAO. http://www.fao.org/3/a-i6937e.pdf. Accessed 8 Nov 2018.

Dharmendra, K. P. (2012). Production of lipase utilizing linseed oilcake as fermentation substrate. *Intern J Sci Environ Technol, 1*, 135–143.

Dulf, F. V., Vodnar, D. C., & Toşa, M. I., & Dulf, E. H. (2020). Simultaneous enrichment of grape pomace with γ-linolenic acid and carotenoids by solid-state fermentation with Zygomycetes fungi and antioxidant potential of the bioprocessed substrates. *Food Chemistry, 2020*(310), 125927.

Erna, F. G., Tenkegna, T. A., & Asfaw S. L. (2019). Effect of agro industrial wastes compost on soil health and onion yields improvements: Study at field condition. *International Journal of Recycling of Organic Waste in Agriculture, 8*, 166–171

Faisal, P. A., Hareesh, E. S., Priji, P., Unni, K. N., Sajith, S., Sreedevi, S., Josh, M. S., & Benjamin, S. (2014). Optimization of parameters for the production of lipase from Pseudomonas ssp. BUP6 by solid state fermentation. *Advances in Enzyme Research, 2*, 125–133.

Ferdes, M., Dincă, M. N., Georgiana, M., & G., Bianca, S., Zăbavă, T., & Paraschiv, G. (2020). Microorganisms and enzymes used in the biological pretreatment of the substrate to enhance biogas production: A review. *Sustainability, 12*, 7205.

FernándezNúñez, E. G., Barchi, A. C., Ito, S., Escaramboni, B., Herculano, R. D., Mayer, C. R. M., & de OlivaNeto, P. (2017). Artificial intelligence approach for high level production of amylase using Rhizopusmicrosporus var.oligosporus and different agro-industrial wastes. *Journal of Chemical Technology and Biotechnology, 92*, 684–692.

Ferreira, N. L., Margeot, A., Blanquet, S., & Berrin, J. G. (2014). Use of cellulases from trichoderma reesei in the twenty-first century—Part I: Current industrial uses and future applications in the production of second ethanol generation. *Biotechnology and Biology of Trichoderma, 2014*, 245–261.

Fierascu, R. C., Fierascu, I., Avramescu, S. M., & Sieniawska, E. (2019). Recovery of natural antioxidants from agro-industrial side streams through advanced extraction techniques. *Molecules, 2019*(24), 4212.

Fitriatin, B. N., Simarmata, T., & Hersanti, T. (2016). Straw composting with biological agent inoculation and application biofertilizer to increase rice production. *International Journal of Sustainable Agricultural Research, 3*, 49–53.

Garcia-Huante, Y., Cayetano-Cruz, M., Santiago-Hernández, A. et al. (2017). The thermophilic biomass-degrading fungus *Thielavia terrestris* Co3Bag1 produces a hyperthermophilic and thermostable β-1, 4-xylanase with exo-and endo-activity. *Extremophiles, 21*, 175–186.

Gautam, S. P., Bundela, P. S., Pandey, A. K., Jamaluddin Awasthi, M. K., & Sarsaiya, S. (2012). Diversity of cellulolytic microbes and the biodegradation of municipal solid waste by a potential strain. *International Journal of Microbiology*. https://doi.org/10.1155/2012/325907

Gayen, S., & Ghosh, U. (2011). Pectinmethylesterase production from mixed agro-wastes by Penicillium notatum NCIM 923 in solid state fermentation. *Journal of Bioremediation & Biodegradation, 2*, 119.

Hameed, A., Hussain, S. A., Yang, J., Ijaz, M. U., Liu, Q., Suleria, H. A. R., & Song, Y. (2017). Antioxidants potential of the filamentous fungi(mucorcircinelloides). *Nutrients, 9*, 1101.

Han, X., Xu, C., Dungait, J. A. J., Bol, R., Wang, X., Wu, W., & Meng, F. (2017) Straw incorporation increases crop yield and soil organic carbon sequestration but varies under different natural conditions and farming practices in China: A system analysis. Biogeoscience Discuss. https://www.biogeosciencesdiscuss.net/bg-2017-493/bg-2017-493.pdf. Accessed 21 Dec 2018.

Hernández-Beltrán, J. U., Hernández-De Lira, I. O., Cruz-Santos, M. M., Saucedo-Luevanos, A., Hernández-Terán, F., & Balagurusamy, N. (2019). Insight into pretreatment methods of lignocellulosic biomassto increase biogas yield: Current state, challenges, and opportunities. *Applied Sciences, 2019*(9), 3721.

Hernandez, C., Milagres, A. M. F., Vazquez-Marrufo, G., Munoz-Paez, K. M., Garcia-Perez, J. A., Alarcon, E. (2018). An ascomycota coculture in batch bioreactor is better than polycultures for cellulase production. *Folia Microbiology (Praha)*. (Accepted). https://doi.org/10.1007/s12223-018-0588-1

Huang, H. L., Zeng, G. M., Jiang, R. Q., Yuan, X. Z., & Yu, M. (2009). Fluorescence spectroscopy characteristics of humic acid by inoculating white-rot fungus during different phases of agricultural waste composting. *Journal of Central South University of Technology, 16*, 440–443.

Ingale, S., Joshi, S. J., & Gupte, A. (2014). Production of bioethanol using agricultural waste: Banana pseudo stem. *Brazilian Journal of Microbiology, 45*(3), 885–892.

Kaur, P. S., Kaur, S., Kaur, H., Sharma, A, Raj, P., & Panwar, S. (2015). Solid substrate fermentation using agro industrial waste: New approach for amylase production by Bacillus licheniformis. *Int J Curr Microbiol App Sci, 4*, 712–717.

Kashif, A., & Shahid, M. K. (2022). Agricultural wastes as an alternative source for the production of antibiotics: Recent developments and future perspectives. In S. K. Nayak, B. Baliyarsingh, I. Mannazzu, A. Singh & B. B. Mishra (Eds.), *Advances in agricultural and industrial microbiology*. Springer. https://doi.org/10.1007/978-981-16-8918-5_7

Kushwaha, R., Kumar, V., Vyas, G., & Kaur, J. (2018). Optimization of different variable for eco-friendly extraction of betalains and phytochemicals from beetroot pomace. *Waste and Biomass Valorization, 2018*(9), 1485–1494.

Lakshmi, S., & Sornaraj, R. (2014). Utilization of see food processing waste for cultivation of the edible mushroom Pleurotus fabellatus. *African Journal of Biotechnology, 17*, 1779–1785.

Larios-Cruz, R., Buenrostro-Figueroa, J., Prado-Barragán, A., Rodríguez-Jasso, R. M., Rodríguez-Herrera, R., Montañez, J. C., & Aguilar, C. N. (2019). Valorization of grapefruit by-products as solid support for solid-state fermentation to produce antioxidant bioactive extracts. Waste and Biomass Valorization, 10, 763-769.

Leow, C. W., Fan, Y. V., Chua, L. S., Muhamad, I. I., Klemeš, J. J., & Lee, C. T. (2018). A review on application of micro orgarnisms for organic waste management. *Chemical Engineering Transactions, 63*, 85–90.

Lim, S. F., & Matu, S. U. (2015). Utilization of agro-wastes to produce biofertilizer. *International Journal of Energy and Environmental Engineering, 6*, 31–35.

Loow, Y. L., Wu, T. Y., Tan, K. A., Lim, Y. S., Siow, L. F., MdJahim, J., Mohammad, A. W., & Teoh, W. H. (2015). Recent advances in the application of inorganic salt pretreatment for transforming lignocellulosic biomass into reducing sugars. *Journal of Agriculture and Food Chemistry, 63*, 8349–8363.

Mahalakshmi, N., & Jayalakshmi, S. (2016). Amylase, cellulase and xylanase production from a novel bacterial isolate Achromobacter xylosoxidans isolated from marine environment. *International Journal of Advanced Research in Biological Sciences, 3*, 230–233.

Maiti, S., Sarma, S. J., Brar, S. K., Le Bihan, Y., Drogui, P., Buelna, G., & Verma, M. (2016). Agro-industrial wastes as feed stock for sustainable bio-production of butanol by Clostridium beijerinckii. *Food Bioprod Process, 98*, 217–226.

Martins, N., & Ferreira, I. C. F. R. (2017). Wastes and by-products: Upcoming sources of carotenoids for biotechnological purposes and health-related applications. *Trends in Food Science & Technology, 62*, 33–48.

Martí-Quijal, F. J., Khubber, S., Remize, F., Tomasevic, I., Roselló-Soto, E., & Barba, F. J. (2021). Obtaining antioxidants and natural preservatives from food by-products through fermentation: A review. *Fermentation, 7*, 106.

Meena, K. K., Sorty, A. M., Bitla, U. M., Choudhary, K., Gupta, P., Pareek, A., Singh, D. P., ... & Minhas, P. S. (2017). Abiotic stress responses and microbemediated mitigation in plants: The omics strategies. *Frontiers in Plant Science, 8*(172). https://doi.org/10.3389/fpls.2017.00172

Mehta, K., Duhan, J. S. (2014). Production of invertase from *Aspergillus niger* using fruit peel waste as a substrate. *Intern J Pharm Biol Sci, 5*(2), B353–B360.

Monroy, F., Aira, M., & Dominguez, J. (2009). Reduction of total coliform numbers during vermicomposting is caused by short term direct efects of earthworms on microorganisms and depends on the dose of application of pig slurry. *Science of the Total Environment, 407*, 5411–5416.

Mushimiyimana, I., & Tallapragada, P. (2016). Bioethanol production from agro wastes by acid hydrolysis and fermentation process. *Journal of Science Industrial Research, 75*, 383–388.

Nascimento, T. P., Sales, A. E., Camila, C. S., Romero, R. M. P., Takaki, G. M. C., Teixeira, J. A. C., Porto, T. S., & Porto, A. L. F. (2015). Production and characterization of new fibrinolytic protease from Mucor subtillissimus UCP 1262 in solid-state fermentation. *Advances Enzyme Research, 3*, 81–91.

Nguyen, T. A. D., Kim, K. R., Han, S. J., Cho, H. Y., Kim, J. W., Park, S. M., Sim., & S. J. (2010). Pretreatment of rice straw with ammonia and ionic liquid for lignocelluloses conversion to fermentable sugars. *Bioresour Technol, 101*, 7432–7438.

Obeng, E. M., Adam, S. N. N., Budiman, C., Ongkudon, C. M., Maas, R., & Jose, J. (2017). Lignocellulases: A review of emerging and developing enzymes, systems, and practices. *Bioresources and Bioprocessing, 4*, 16.

Oliveira, F., Carlos, E. S., Peclat Veronica, R. O. L., Salgado, J. M., Bernardo, D. R., Maria, A. Z. C., Armando, V., & Isabel, B. (2017). Optimization of lipase production by Aspergillus ibericus from oil cakes and its application in esterification reactions. *Food Bioprod Process, 102*, 268–277.

Onyegeme-Okerenta, B. M., Okochi, V. I., & Chinedu, S. N. (2013). Penicillin production by penicillium chrysogenum PCL 501: Effect of UV induced mutation. *The Internet Journal of Microbiology, 12*(1).

Otto Wagner, A., Lackner, N., Mutschlechner, M., Prem, E. M., Markt, R., & Illmer, P. (2018). Biological pretreatment strategies for second-generation lignocellulosic resources to enhance biogas production. *Energies, 11*, 1797.

Paepatung, N., Nopharatana, A., Songkasiri, W. (2009). Bio-methane potential of biological solid materials and agricultural wastes. *Asian Journal on Energy and Environment, 10*, 19–27.

Pakarinen, A., Haven, M., Djajadi, D., Várnai, A., Puranen, T., & Viikari, L. (2014). Cellulases without carbohydrate-binding modules in high consistency ethanol production process. *Biotechnology for Biofuels, 7*, 27.

Pampuro, N., Bertora, C., Sacco, D., Dinuccio, E., Grignani, C., Balsari, P., Cavallo, E., Bernal M. P. (2017). Fertilizer value and GHG emissions of pellets from the solid fraction of pig slurry compost. *The Journal of Agricultural Science, 155*, 1646–1658.

Pampuro, N., Cafaro, F., & Cavallo, E. (2018). Reuse of animal manure: A case study on stakeholders' perceptions about pelletized compost in Northwestern Italy. *Sustainability, 10*, 2028.

Paul, J. S., Beliya, E., Tiwari, S., Patel, K., Gupta, N., & Jadhav, S. K. (2020). Production of biocatalyst α-amylase from agro-waste 'rice bran' by using Bacillus tequilensis TB5 and standardizing its production process. *Biocatalysis and Agricultural Biotechnology, 26*, 101648.

Rai, S., Kashyap, P. L., Kumar, S., Ramteke, P. W. (2016). Identification, characterization and phylogenetic analysis of antifungal Trichoderma from tomato rhizosphere. *Springer Plus, 5*(1939). https://doi.org/10.1186/s40064-016-3657-4

Rai, S., Solanki, A. K., & Anal, A. K. D. (2020). Modern biotechnological tools: An opportunity to discover complex phytobiomes of horticulture crops. In B. K. Kashyap, M. K. Solanki, D. V. Kamboj, & A. K. Pandey (Eds.), *Waste to energy: Prospects and applications* (pp. 3–28). Springer.

Rai, S., & Solanki, M. K. (2014). Optimization of thermostable alpha-amylase production via mix agricultural-residues and bacillus amyloliquefaciens. *Notulae Scientia Biologicae, 6*(1), 105–111.

Rai, S., Solanki, M. K., Anal, A. K. D., Sagar, A., Solanki, A. C., Kashyap, B. K., & Pandey, A. K. (2021). Emerging frontiers of microbes as agro-waste recycler. In B. K. Kashyap, M. K. Solanki, D. V. Kamboj, & A. K. Pandey (Eds.), *Waste to energy: Prospects and applications* (pp. 3–27). Springer.

Rajkhowa, D. J., Sarma, A. K., Bhattacharyya, P. N., & Mahanta, K. (2019). Bioconversion of agricultural waste and its efficient utilization in the hilly ecosystem of Northeast India. *International Journal of Recycling of Organic Waste in Agriculture, 8*(Suppl 1), S11–S20.

Ramachandran, S., Singh, S. K., Larroche, C., Soccol, C. R., & Pandey, A. (2007). A oil cakes and their biotechnological applications—a review. *Bioresource Technology, 98*, 2000–2009.

Rashad, M. M., Mahmoud, A. E., Ali, M. M., Nooman, M. U., & Al-Kashef, A. S. (2015). Antioxidant and anticancer agents produced from pineapple waste by solid state fermentation. *Intern J Toxicol Pharmacol Res, 7*(6), 287–296.

Rekha, K. S. S., Lakshmi, C., Devi, S. V., & Kumar, M. S. (2012). Production and optimization of lipase from Candida rugosa using groundnut oilcake under solid state fermentation. *Intern J Res Eng Technol, 1*, 571–577.

Ribeiro, N. D. Q., Souza, T. P., Costa, L. M. A. S., Castro, C. P. D., & Dias, E. S. (2017). Microbial additives in the composting process. *Ciência e Agrotecnologia, 41*, 159–168.

Ritala, A., Häkkinen, S. T., Toivari, M., & Wiebe, M. G. (2017). Single cell protein—state-of-the-art, industrial landscape and patents 2001–2016. *Frontiers in Microbiology, 8*, 2009.

Rojas-Higuera, N. S., Pava-Sánchez, A. M., Pinzón Rangel, D. L., Díaz-Ariza, L. A., Quevedo-Hidalgo, B., & Pedroza-Rodríguez, A. M. (2017). Bio-transformed sawdust by white rot fungi used as a carrier for plant growth-promoting bacteria. European Journal of Wood and Wood Products, 75, 263-273.

Rubio, R., Perez-Murcia, M. D., Agullo, E., Bustamante, M. A., Sanchez, C., Paredes, C., & Moral, R. (2013). Recycling of agro-food wastes into vineyards by composting: Agronomic validation in field conditions. *Communications in Soil Science and Plant Analysis, 44*, 502–516.

Sadh, P. K., Chawla, P., Bhandari, L., & Duhan, J. S. (2017). Bio-enrichment of functional properties of peanut oil cakes by solid state fermentation using Aspergillus oryzae. *Journal of Food Measurement and Characterization, 12*, 622–633.

Sadh, P. K., Duhan, S., & Duhan, J. S. (2018). Agro-industrial wastes and their utilization using solid state fermentation: A review. *Bioresources and Bioprocessing, 5*, 1.

Saini, J. K., Saini, R., & Tewari, L. (2014). Lignocellulosic agriculture wastes as biomass feedstocks for second-generation bioethanol production: Concepts and recent developments. *3 Biotech, 5*, 337–353. https://doi.org/10.1007/s13205-014-0246-5

Saharan, P., Sadh, P. K., & Duhan, J. S. (2017). Comparative assessment of effect of fermentation on phenolics, flavanoids and free radical scavenging activity of commonly used cereals. *Biocatalysis and Agricultural Biotechnology, 12*, 236–240.

Saravanan, P., Muthuvelayudham, R., & Viruthagiri, T. (2012). Application of statistical design for the production of cellulase by trichoderma reesei using mango peel. *Enzyme Research, 2012*, 157643.

Saravanan, V., & Vijayakumar, S. (2014). Production of biosurfactant by Pseudomonas aeruginosa PB3A using agro-industrial wastes as a carbon source. *Malays J Microbiol, 10*(1), 57–62.

Sarkar, P., & Chourasia, R. (2017). Bioconversion of organic solid wastes into biofortifed compost using a microbial consortium. *International Journal of Recycling of Organic Waste in Agriculture, 6*, 321–334.

Schröder, P., Beckers, B., Daniels, S., Gnädinger, F., Maestri, E., Marmiroli, N., Mench, M., Millan, R., Obermeier, M. M., Oustriere, N., Persson, T., Poschenrieder, C., Rineau, F., Rutkowska, B., Schmid, T., Szulc, W., Witters, N., & Sæbø, A. (2018). Intensify production, transform biomass to energy and novel goods and protect soils in Europe—a vision how to mobilize marginal lands. *Science of the Total Environment, 616–617*, 1101–1123.

Sethi, B. K., Jana, A., Nanda, P. K., Das, Mohapatra, P. K., Sahoo, S. L., & Patra, J. K. (2016). Production of α-amylase by aspergillus terreus NCFT 4269.10 Using pearl millet and its structural characterization. *Frontiers in Plant Science, 7*, 639.

Shamsul, N. S., Kamarudin, S. K., & Rahman, N. A. (2018). Study on the physical and chemical composition of agro wastes for the production of 5-hydroxymethylfurfural. *Bioresource Technology, 247*, 821–828.

Sharma, S., Sagar, A., Yadav, S. S., & Rai, S. (2023). Production of value added products from fruit juice residues using enzyme technology. In M. Kuddus & P. W. Ramteke (Eds.), *Value-addition in agri-food industry waste through enzyme technology* (pp. 221–236). Academic Press.

Singh, D. P., Prabha, R., Renu, S., Sahu, P. K., & Singh, V. (2019). Agro-waste bioconversion and microbial fortification have prospects for soil health, crop productivity, and eco-enterprising. *International Journal of Recycling Organic Waste in Agriculture, 8*(1), 457–472.

Singh, S., & Nain, L. (2014). Microorganisms in the conversion of agricultural wastes to compost. *Proceeding of the Indian National Science Academy, 80*(2), 473–481.

Singhania, R. R., Saini, J. K., Saini, R., Adsul, M., Mathur, A., Gupta, R., & Tuli, D. K. (2014). Bioethanol production from wheat straw via enzymatic route employing Penicillium janthinellum cellulases. *Bioresource Technology, 169*, 490–495.

Sodhi, H. K., Sharma, K., Gupta, J. K., & Soni, S. K. (2005). Production of a thermostable a-amylase from Bacillus sp. PS-7 by solid-state fermentation and its synergistic use in the hydrolysis of malt starch for alcohol production. *Process Biochemistry, 40*, 525–534.

Suganthi, R., Benazir, J. F., Santhi, R., Kumar, R. V., Hari, A., Meenakshi, N., Nidhiya, K. A., Kavitha, G., & Lakshmi, R. (2011). Amylase production by *Aspergillus niger* under solid state fermentation using agro-industrial wastes. *International Journal of Engineering Science and Technology, 3*, 1756–1763.

Sun, L., Pope, P. B., Eijsink, V. G. & Schnürer, A. (2015). Characterization of microbial community structure during continuous anaerobic digestion of straw and cow manure. *Microbial Biotechnology, 8*, 815–827, https://doi.org/10.1111/mbt2.2015.8.issue-5

Thakur, S. A., Nemade, S. N., & Sharanappa, A. (2015). Solid state fermentation of overheated soybean meal (waste) for production of protease using Aspergillus oryzae. *International Journal of Innovative Research in Science, Engineering and Technology, 4*, 18456–18461.

Timo, S., Raphael, G., Nora, B., Scott, E. B., & Steven, W. S. (2017). *Thermoascus aurantiacus* is an intriguing host for the industrial production of cellulases. *Current Biotechnology, 6*, 89–97. https://doi.org/10.2174/2211550105666160520123504

Tripathi, K. D. (2008). Antimicrobial drugs. *Essentials of medical pharmacology* (6th edn., p. 710). Jaycee Brothers Medical Publishers Ltd.

Tsouko, E., Kachrimanidou, V., Dos Santos, A. F., do Nascimento, V. L. M., Papanikolaou, S., de Castro, A. M., Freire, D. M., & Koutinas, A. A. (2017). Valorization of by-products from palm oil mills for the production of generic fermentation media for microbial oil synthesis. *Applied Biochemistry and Biotechnology, 181*(4), 1241–1256.

Vastrad, B. M., & Neelagund, S. E. (2011). Optimization and production of neomycin from different agroindustrial wastes in solid state fermentation. *International Journal of Pharmaceutical Sciences and Drug Research, 3*, 104–111.

Vidhyalakshmi, R., Vallinachiyar, C., & Radhika, R. (2012). Production of xanthan from agro-industrial waste. *Journal of Advanced Scientific Research, 3*, 56–59.

Vishan, I., Sivaprakasam, S., & Kalamdhad, A. (2017). Isolation and identification of bacteria from rotary drum compost of water hyacinth. *International Journal of Recycling of Organic Waste in Agriculture, 6*, 245–253.

Waghmare, P. R., Patil, S. M., Jadhav, S. L., & Jeon, B-H. (2018). Utilization of agriculture waste biomass by cellulolytic isolate. *Agriculture and Natural Resources, 52*, 399–406.

Wang, S., Hou, X., & Su, H. (2017). Exploration of the relationship between biogas production and microbial community under high salinity conditions. *Scientific Reports, 7*, 1149.

Wang, X., Li, Z., Zhou, X., Wang, Q., Wu, Y., Saino, M., & Bai, X. (2016). Study on the bio-methane yield and microbial community structure in enzyme enhanced anaerobic co-digestion of cow manure and corn straw. *Bioresource Technology, 219*, 150–157.

Xu, C., Nasrollahzadeh, M., Selva, M., Issaabadi, Z., & Luque, R. (2019). Waste-to-wealth: Biowaste valorization into valuable bio(nano)materials. *Chemical Society Reviews, 48*, 4791.

Yazid, N. A., Barrena, R., Komilis, D., & Sánchez, A. (2017). Solid-state fermentation as a novel paradigm for organic waste valorization: A review. *Sustainability, 9*(2), 1–28.

Yildirim, N., Yildirim, N. C., & Yildiz, A. (2015). Laccase enzyme activity during growth and fruiting of Pleurotus eryngii under solid state fermentation medium containing agricultural wastes. *International Journal of Pure and Applied Sciences, 1*, 64–71.

Zamanzadeh, M., Hagen, L. H., Svensson, K., Linjordet, R., Horn, S. J. (2017). Biogas production from food waste via co-digestion and digestion-effects on performance and microbial ecology. *Scientific Reports, 7*, 17664.

Zhang, Y., Liu, J., Lu, X., Zhang, H., Wang, L., Guo, X., Qi, X., & Qian, H. (2014). Isolation and identification of an antioxidant peptide prepared from fermented peanut meal using Bacillus subtilis fermentation. *International Journal of Food Properties, 17*, 1237–1253.

SPRINGER NATURE

GPSR Compliance

The European Union's (EU) General Product Safety Regulation (GPSR) is a set of rules that requires consumer products to be safe and our obligations to ensure this.

If you have any concerns about our products, you can contact us on ProductSafety@springernature.com

In case Publisher is established outside the EU, the EU authorized representative is:

Springer Nature Customer Service Center GmbH
Europaplatz 3
69115 Heidelberg, Germany

The manufacturer's authorised representative in the EU is Springer Nature Customer Service Centre GmbH, Europaplatz 3, 69115 Heidelberg, Germany. If you have any concerns regarding our products, please contact ProductSafety@springernature.com

Printed and bound by CPI Group (UK) Ltd, Croydon, CR0 4YY
01/10/2026
02225926-0008